Transportation of Dangerous Goods: Assessing the Risks

edited by

F. Frank Saccomanno
Department of Civil Engineering, and
Institute for Risk Research
University of Waterloo
Waterloo, Ontario, Canada

and

Keith Cassidy
Major Hazards Assessment Unit
Health and Safety Executive
United Kingdom

The figures and tables in the papers "A Novel Risk Assessment Methodology for the Transport of Explosives" by Nigel Riley on pages 47-74 and "Hazardous Goods Transportation. A Major UK Study" by Keith Cassidy on pages 483-509 have been previously published in the Health and Safety Commission publication "Major Hazard Aspects of the Transport of Dangerous Substances". The Institute for Risk Research acknowledges permission from the Controller of Her Majesty's Stationery Office to print these figures and tables.

ISBN 0-9696747-1-6

Institute for Risk Research
University of Waterloo
Waterloo, Ontario, Canada   N2L 3G1
Tel: (519) 885-1211, ext. 3355 Fax: (519) 888-6197

Managing Editor: Lorraine Craig

Canadian Cataloguing in Publication Data

Main entry under title:

Transportation of dangerous goods: assessing the risks

Proceedings of the First International Consensus Conference on the Risks of Transporting Dangerous Goods, held April 6-8, 1992, in Toronto.

Includes bibliographical references.
ISBN 0-9696747-1-6

1. Hazardous substances - Transportation - Congresses. 2. Hazardous substances - Risk assessment - Congresses. I. Saccomanno, Fedel Frank M., 1948- . II. Cassidy, Keith, 1939-   . III. University of Waterloo. Institute for Risk Research. IV. International Consensus Conference on the Risks of Transporting Dangerous Goods (1st : 1992 : Toronto, Ont.).

HE199.5.D3T73 1992   363.17'7   C93-094047-4

Other titles available from the Institute for Risk Research are listed at the end of this publication.

CONTENTS

Preface...vii
F.F. Saccomanno and K. Cassidy

List of Sponsors xv

**CHAPTER 1: APPLICATION OF QUANTITATIVE RISK
ASSESSMENT MODELS TO THE TRANSPORT
OF DANGEROUS GOODS**

Computer Support for Risk Assessment of Dangerous Goods
   Transportation 3
E. Weigkricht and K. Fedra

The Measurement of Risk from Transporting Dangerous
   Goods by Road and Rail 19
G. Purdy

A Novel Risk Assessment Methodology for the Transport of
   Explosives..47
N. Riley

Rail Transport Risk in the Greater Toronto Area...................75
E. Alp, R.V. Portelli and W.P. Crocker

Methodology for Assessing the Public Risk from Transporting
   Hydrogen   Sulphide..121
S.B. Russell and T.F. Kempe

Techniques for Risk Assessment of Ships Carrying
   Hazardous Cargo in Port Areas 153
J.R. Spouge

Risk Assessment of Transportation for the Canadian Nuclear
   Fuel Waste Management Program.............................183
T.F. Kempe and L. Grondin

CHAPTER 2: ANALYSIS OF DANGEROUS GOODS ACCIDENTS AND RELEASES

Historical Analysis of Dangerous Goods Spills in Alberta 213
S.P. Hammond and W.W. Smith

Characteristics of Motor Carriers of Hazardous Materials 229
L.N. Moses and I. Savage

Procedures for the Development of Estimated Truck Accident
  Rates and Release Probabilities for Use in Hazmat
  Routing Analysis 247
E.R. Russell, Sr. and D.W. Harwood

A Methodology for the Transfer of Probabilities Between
  Accident Severity Classification Schemes....................... 263
J.D. Whitlow and K.S. Neuhauser

A Regression Model for Estimating Probability of Vessel
  Casualties 279
T.K. Liu

Site Specific Transportation Risk Analysis for the Manitoba
  Hazardous Waste Management Corporation.................... 297
E.J. Yee

The Need and Value of Routing of Dangerous Goods
  Transports 315
G. Stjernman

Establishing Credible Risk Criteria for Transporting
  Extremely Dangerous Hazardous Materials.................... 335
R.A. Sivakumar, R. Batta and M.H. Karwan

CHAPTER 3: APPLICATION OF SIMPLE RISK ASSESSMENT METHODOLOGY

A Simple Consequence Model Based on a Fatality Index
Approach.. 345
L.H. Brockhoff and H.J. Styhr Petersen

A Simple, Usable, Empirical Risk Model for Small
Communities 365
E.R. Russell, Sr.

The Risks of Handling vs. Transporting Dangerous Goods 385
J.R. Kirchner and W.R. Rhyne

Transportation Also Begins and Ends with Risks................. 401
B.J. Griffin and R.G. Auld

CHAPTER 4: UNCERTAINTY IN RISK ESTIMATION

Analysis of Risk Uncertainty for the Transport of Hazardous
Materials... 423
F.F. Saccomanno and O. Bakir

International Discrepancies in the Estimation of the Expected
Accident and Tank Failure Rates in Transport Studies........ 439
P. Hubert and P. Pagès

Perspectives on a Transportation Corridor Risk Analysis 453
M. Abkowitz, K. Hancock and R. Waters

Hazardous Materials: A Comparison of the Severity of
Accidents from Transport and Fixed Installations.............. 469
P. Haastrup and H.J. Styhr Petersen

CHAPTER 5: RISK TOLERANCE, COMMUNICATION AND POLICY IMPLICATIONS

Hazardous Goods Transportation: A Major UK Study............. 483
K. Cassidy

A Comparison of Theoretical and Actual Consequences in
   Two Fatal Ammonia Incidents.................... 511
T.S. Glickman and P.K. Raj

Using a Delphi Technique to Develop High-Risk Scenarios
   and Countermeasures 531
E.R. Russell, Sr.

Risk Management for the Movement of Dangerous Goods........ 555
J.H. Shortreed, F.F. Saccomanno and K.W. Hipel

QRA-Aided Risk Management of Dutch Inland Waterway
   Transport... 581
H.L. Stipdonk and R.J. Houben

Transport Canada Dangerous Goods Accident Costing
   Study and Model...................................... 601
J. Wright

The Perceived Risks of Transporting Hazardous Materials
   and Nuclear Waste: A Case Study 617
A.H. Mushkatel and T.S. Glickman

CONCLUDING REMARKS

F.F. Saccomanno and K. Cassidy.................... 635

PREFACE

Background

A number of advances have been made in North America and Europe concerning the assessment of risks for the transport of dangerous goods. It was felt that the time was opportune to bring together experts in this field to assess the nature and extent of progress that has been made and to evaluate the reliability of the various risk models which have been developed and are currently in use.

The First International Consensus Conference on the Risks of Transporting Dangerous Goods was held in Toronto, Canada from April 6th to 8th, 1992. The central aim of this International Consensus Conference was to provide an open expert forum for the comparative assessment of quantitative risk assessment (QRA) models.

Over 30 state-of-the-art papers were presented dealing with a wide spectrum of topics, including risk estimation, data and accident analysis, risk perception, risk uncertainty, and issues of public policy. The results of a number of major quantitative risk assessment studies from Europe and North America were also presented for the first time to an audience that included risk analysts, decision-makers, carriers, representatives from the chemical industry and government officials.

This book is one of three reports produced as a result of this Consensus Conference. The two other reports include a documentation of the workshop consensus gathering exercise entitled "What is the Risk" (Institute for Risk Research, 1993) and an assessment of a corridor benchmark exercise involving the application of several risk models to a common dangerous goods transport problem entitled "Comparative Assessment of Risk Model Estimates for the Transport of Dangerous Goods by Road and Rail" (Saccomanno, Leeming and Stewart, 1993). Essentially, many of the papers in these Proceedings provide the necessary background information on the models that were applied in the corridor benchmark exercise, so as to facilitate a comparative technical assessment of the resultant risk estimates. A good technical appreciation of these models is also important in understanding the broader policy concerns raised by the consensus gathering exercise.

A number of these papers have been presented in modified form at other Conferences. However, in reviewing them for these Proceedings, we attempted to include original material that has not been published elsewhere. Many of the papers in this book are being published for the first time in a refereed document.

Organization and Overview of Papers

These Proceedings include a selection of twenty-nine papers presented at the Consensus Conference. These papers have been organised into five main chapters:

1. Application of QRA Models to the Transport of Dangerous Goods

2. Analysis of Dangerous Goods Accidents and Releases

3. Application of Simple Risk Assessment Methodology

4. Uncertainty in Risk Estimation

5. Risk Tolerance, Communication and Policy Implications

The papers in Chapter 1 describe several major studies dealing with the application of QRA to the problem of transporting dangerous goods. The first paper, by Weigkricht and Fedra, introduces a series of advanced QRA computer models developed by the International Institute for Applied Systems Analysis (IIASA) for application to extensive network problems. The results of three European studies are discussed: IRIMS carried out for the European Joint Research Centre, Haute Normandie carried out for INRETS in France, and XENVIS for the Dutch Ministry of Housing, Physical Planning and Environment. The second and third papers by Purdy and Riley present the results of two early versions of what is now becoming the Health and Safety Executive transportation RISKAT model. Purdy discusses the results of a corridor risk assessment involving the transport of flammable and toxic materials by road and rail in the United Kingdom. Riley applies RISKAT to the transport of explosives by road and rail along representative road and rail routes in the United Kingdom. Alp, Portelli and Crocker describe a comprehensive QRA application for the transport of dangerous goods by rail. This study was carried out as part of the 1987 Toronto Dangerous Goods Rail Task Force and makes use of the COBRA heavy gas dispersion model developed by Concord Environmental. In the next paper, Russell and Kempe apply QRA techniques to the transport of H_2S by road. This study is interesting in that it does not make use of an in-house QRA model, but draws extensively from results reported in the literature. The authors have demonstrated that the absence of an in-house QRA model should not preclude analysts from carrying out a comprehensive quantitative assessment of the risks of transporting dangerous goods. In the next paper, Spouge makes use of a specially modified version of SAFETI, a computer model developed by DNV Technica, to generate individual and societal risks isopleths for a port facility in the United Kingdom. As well as summarising risk criteria for LPG, this paper also presents a cost-benefit analysis of possible risk reduction measures. Kempe and Grondin describe a comprehensive QRA of the transport of nuclear fuel wastes to a designated disposal site. This study is limited to an assessment of immediate health risks, and is unique for North America in that societal and individual risks are expressed in terms of a "dangerous dose" criterion, rather than the more popular probability of death approach. This paper presents a condensed account of the larger nuclear waste transportation and disposal program of Ontario Hydro.

In Chapter 2, a less comprehensive view of QRA is taken, focusing essentially on accident-induced releases of dangerous goods during transportation and the transferability of the estimates from one jurisdiction to another. Hammond and Smith present an assessment of dangerous goods releases reported in the Alberta Public Safety Service data base. Despite adopting a simple analytical approach, this paper is important for describing a data base that is one of the most comprehensive "regional" information systems for road dangerous goods incidents in Canada. In the next paper, Moses and Savage report on a statistical comparison of road accident rates between general freight and dangerous goods carriers. The results of this study are based on 75,000 safety audits of major motor carriers in the United States. The authors' rather surprising conclusions suggest that some caution should be exercised in their interpretation and generalization, especially given the limited statistical evidence on which these results are based. Many QRA studies have in the past used accident rates derived from general freight and have applied these rates without modification to dangerous goods movements. This paper suggests possible inconsistencies with this approach, although for opposite reasons to those normally perceived for this type of comparison between carriers. Russell and Harwood present some useful statistics on road accident and release rates from the United States. The authors use general freight accident rates and apply these rates to dangerous goods shipments to obtain release rates on a per shipment basis. Rates from different states are compared with national rates. The release rates, in this study, have been averaged over all material types and accident situations. The extent to which this could introduce bias in the results in not discussed. Whitlow and Neuhauser report on an interesting technique for transferring estimates of accident severity from one jurisdiction to another. Accident severity affects the potential for release in an accident situation, and hence the risks involved. The transferability of estimates from one jurisdiction to another is recognised as a major source of uncertainty in risk estimation for the transport of dangerous goods. Liu reports on an accident rate and severity model for inland water transport, using national United States Coast Guard Casualty Maintenance data. This paper presents an interesting use of modifiers to transfer average national statistics to route specific conditions. Yee reports on a simple assessment of accident-induced releases for hazardous waste transport in the Canadian province of Manitoba. A number of interesting statistics are documented in this paper. Unfortunately, the author does not indicate how route control factors have been taken into account in his analysis, so that an objective technical assessment of this paper and its approach cannot be undertaken. Stjernman asserts, quite surprisingly given recent efforts in QRA, that since accident information for dangerous goods shipments is difficult to obtain from available data, the routing of dangerous goods should be carried out on the basis of protecting the most sensitive areas. This paper presents a rather interesting point of view on the justification (or lack of it) for QRA. Unfortunately, this assertion was not developed more fully in the paper, particularly with reference to stronger statistical evidence. Sivakumar, Batta and Karwan present the framework of a linear programming formulation for routing dangerous goods on a simple network. The objective function for this formulation is "conditional risk probability", defined as the product of route section risks and their probability of occurrence summed over the entire route, divided by the section probabilities of occurrence summed over the entire route. This model has not been calibrated or applied to an extensive network problem.

The third chapter in this book attempts to address the question: Can simple risk assessment models, many of them semi-quantitative, replace more comprehensive QRA computer models? In the first paper, Brockhoff and Styhr Peterson report on a simple risk assessment model for toxic and flammable materials, developed at the European Joint Research Centre. The authors apply their model to representative road and rail route sections. While a number of problems have been identified in this paper, the authors appear to be reluctant to suggest possible solutions to these problems. In the next paper, Russell presents a qualitative model of risk assessment that draws heavily on expert opinion. This model is designed for use in jurisdictions where more comprehensive QRA models are either unavailable or inappropriate given lack of user expertise and data. A very important question is left unanswered, however, whether simple risk models are adequate replacements for all problems currently using complex QRA formulations or whether this is a case of considering only those risks that can be simply modelled. In their paper, Kirchner and Rhyne demonstrate through simple examples what we all know (or feel) about the importance of loading/unloading and storage operations in determining transport risks. Simple QRA is applied to three representative transportation/handling cases: hydrogen fluoride gas, low level radioactive wastes, and flammable compressed gases in cylinders. Through these cases, the authors show that risks from storage and unloading/loading can exceed in-transit risks by a significant degree. Griffin and Auld provide a methodology for simple risk assessment of loading and unloading operations of high vapour pressure liquids. Several recommendations are made in this paper which could be useful to operators of shipping facilities. The scientific basis for much of the discussion, however, is notably absent from this paper, leaving many of these recommendations without a strong statistical support as to their effectiveness in reducing risks.

The fourth chapter in this book deals with the broader issues of uncertainty in risk estimation. Saccomanno and Bakir discuss two sources of uncertainty: uncertainty in risk estimation and uncertainty in the process. An approach is presented for considering both types of uncertainty, which is then applied to a simple risk formulation and a sample of estimates reported in the literature. Hubert and Pages address several inconsistencies in the estimation of accident and release rates as reported in a number of European and American studies. Possible reasons for these inconsistencies are discussed. Abkowitz, Hancock and Waters present the results of an application of QRA to a hypothetical road and rail corridor involving the shipment of chlorine, LPG and gasoline. The estimates of three "independent" sources are reported, one of which was used as the basis of the results attributed to VANA in the benchmark exercise associated with this Conference. Significant variability was observed much of which could be attributed to assumptions taken by the analysts. Haastrup and Styhr Petersen report on a comparison of risks between fixed installations and transport operations involving dangerous goods. This paper identifies a number of inconsistencies in the estimates of risk as reported by several sources. The statistical analysis is simple in that the authors do not attempt to place controls on the estimates to account for different types of materials, population distributions and weather conditions. The extent to which this could affect the results and conclusions reached by the authors in this paper has not been discussed.

The papers in Chapter 5 address several important policy issues regarding the use of QRA for the transport of dangerous goods. Four areas of concern include: risk tolerance, risk perception, risk communication and factors affecting the decision-making process. The first paper in this chapter, by Cassidy, provides an important discussion of the policy environment leading up to the 1991 Advisory Committee on Dangerous Substances report on the risks of transporting dangerous goods in "major hazard" quantities in the United Kingdom. A number of major conclusions and recommendations from the ACDS report are summarised and discussed. Glickman and Raj suggest a number of measures that can be adopted to make QRA more meaningful and realistic to decision-makers, most notably is the incorporation of sheltering considerations into the analysis. In the next paper, Russell attempts to identify certain high risk scenarios for the road transport of dangerous goods and to suggest practicable countermeasures, based on expert opinion. There is some concern that a mail out survey, such as is adopted here, is not the optimal way to arrive at consensus. To accomplish this, it is important that the experts "sit at the same table," and are allowed to express their views in a structured, unbiased and open format. The author argues that countermeasures suggested in the paper for each high risk scenario are effective, economic and practical. Unfortunately, the paper does not elaborate on how this has been achieved. Shortreed, Saccomanno and Hipel argue that QRA must be accompanied by effective support systems to aid decision-making and resolve conflicts. The arguments are enforced with reference to major stakeholders in the recently completed Toronto Dangerous Goods Rail Task Force study. In the next paper, Stipdonk and Houben present a model of the risks associated with inland water transport and use this model to predict the effect of alternative safety measures. This paper reports on some initial developments in an on-going study being carried out by the Dutch Ministry of Transportation and Public Works for the national inland water-way network. The authors view their approach as an input into a more comprehensive cost-effective evaluation of risks. In the next paper, Wright reports on a dangerous goods costing model developed by the Transportation of Dangerous Goods Directorate of Transport Canada. Seventy dangerous goods incidents in Canada reported over the last 10 years are examined and the monetary risks estimated. The author argues that expected monetary losses associated with dangerous goods incidents can and should be obtained as a critical input into the decision-making process. Mushkatel, Pijawka and Glickman deal with the broader issues of public perceptions as related to the risks of transporting hazardous nuclear wastes. These perceptions are assessed as a function of corridor location and distance from a potential incident. The authors point to the low level of credibility in the United States vis á vis what the public feels the government can or is willing to do to reduce these risks.

Chapter 6 includes a brief summary of the major conclusions and recommendations voiced by the participants at this conference regarding the QRA process and its application to the transport of dangerous goods. The central aim of this discussion is to make the process and its results more meaningful and accessible to users and decision-makers.

Acknowledgements

We take this opportunity to thank all those individuals who helped in the planning and administration of this Consensus Conference, especially those who served on the Conference Planning Committee, and as conference chairs and rapporteurs. We are grateful to Mrs. Adelind Mitchell of the Institute for Risk Research for her expert typing of the manuscript and to Lorraine Craig for her diligence in keeping the publication process on schedule. We reserve a special thank you for Professor John Shortreed of the Institute for Risk Research, who was instrumental in organising the Conference, and who was always available with helpful advice and suggestions.

We join the authors in the hope that this Conference will advance the state of knowledge of the risk analysis process and give rise to more informed decisions on how best to understand, control and where necessary reduce these risks. So far as we are aware, this was the first such Conference. We hope that when the lessons learned and reported here and elsewhere have been developed and applied, an opportunity will arise (and will be taken up) for a successor Conference. The problems addressed transcent national boundaries; and the continuation of this international effort is greatly to be encouraged.

References

Institute for Risk Research (1993). What is the Risk? Consensus Report. International Conference on the Risk of Transporting Dangerous Goods, April 6-8, 1992. Toronto, Canada. Prepared by Jeff Solway, Nashwaak Consulting. University of Waterloo, Waterloo, Canada.

Saccomanno, F.F., Leeming, D., and Stewart, A. (1993). Comparative Assessment of Risk Model Estimates for the Transport of Dangerous Goods by Road and Rail. Review of the benchmark corridor exercise carried out as part of the International Consensus Conference on the Risks of Transporting Dangerous Goods, April 6-8, 1992. Toronto, Canada. Institute for Risk Research, University of Waterloo, Waterloo, Canada.

F. Frank Saccomanno
Institute for Risk Research
University of Waterloo
Waterloo, Ontario, Canada

Keith Cassidy
Major Hazards Analysis Unit
Health and Safety Executive
United Kingdom

CONFERENCE SPONSORS

Institute for Risk Research,
University of Waterloo, Canada

Transportation of Dangerous Goods Directorate,
Transport Canada

U.K. Health and Safety Executive

Ontario Ministry of Transportation

Linde Division, Union Carbide

The Canadian Society for Civil Engineering

Chapter 1: Application of Quantitative Risk Assessment Models to the Transport of Dangerous Goods

Computer Support for Risk Assessment of Dangerous Goods Transportation

E. Weigkricht
K. Fedra
Advanced Computer Applications (ACA)
International Institute for Applied Systems Analysis
(IIASA) A-2361 Laxenburg, Austria

ABSTRACT

The increasing size and complexity of activities involving hazardous substances requires higher emphasis on the risk assessment in potentially harmful operations. Furthermore, the transportation of hazardous materials poses a considerable risk to public health and the environment. The analysis of alternative policies and operational choices for transportation by road, rail, or rivers, given a spatial pattern of production and storage facilities, consumption, environment and population, should be based on a detailed scientific assessment of numerous alternatives, involving technological, environmental, socio-economic and political elements in a comprehensive and directly usable form of risk analysis.

This paper describes three information and decision support systems which improve the availability and fast access to relevant data and information, and support the quick and effective generation, display, analysis, evaluation and comparison of different alternatives. The systems developed and implemented by IIASA's Advanced Computer Applications (ACA) project are:

- *IRIMS is a decision-oriented prototype system for the management of hazardous substances and industrial risk.*

- *For the region Haute Normandie a transportation risk assessment system has been developed.*

- *XENVIS, an interactive information and decision support system, is applied to problems of industrial risk and hazardous substances management (i.e., risk assessment for chlorine transportation) in the Netherlands.*

Common to all three systems is the integration of databases, modeling, and optimization techniques, the coupling with a geographical information system, as well as the use of symbolic user interfaces and computer graphics, to provide an efficient and convenient framework and information basis for decision making.

1. INTRODUCTION

The quantities of hazardous materials transported increased with economic development. There are different groups of people involved (e.g., ministries, industries, emergency services, transportation companies, general public), with different kinds of information requirements and interests.

To prevent accidents and to manage their consequences and impacts, it is necessary to investigate the risk associated with the transportation of hazardous materials. To minimize this risk, various types of information from different sources are needed quickly for further processing, inducting the use of advanced information technology. The use of interactive information and decision support systems, provided with an intelligent user interface is demonstrated with three application examples.

It will be shown that each of these systems uses a different approach and is tailored to user and application, but all illustrate three key concepts applied in the design of these systems, namely integration, interaction and visualization (see Fedra, 1990).

2. APPLICATION EXAMPLES

2.1 IRIMS

Under contract to the Commission of the European Communities' Joint Research Centre (JRC) in Ispra, Italy, ACA developed a demonstration prototype called IRIMS (Ispra RIsk Management Support) within the project *Decision-oriented Software for the*

Management of Hazardous Substances and Industrial Risk (see Fedra, 1986). The primary intended application of the decision support system was for regulatory purposes e.g., within the framework of the EEC post-seveso or similar legislation. The entire life-cycle of hazardous substances involves numerous aspects and levels of planning, policy and management decisions, as well as technological, economic, socio-political and environmental considerations. A modular design philosophy was adopted to allow the development of individual building blocks, which are valuable products in their own right.

IRIMS consists of different modules for information retrieval, simulation, optimization and multicriteria scenario evaluation. Included are a number of data bases (hazardous substances, industrial accidents reports, directives and regulations, industries, waste streams), a geographical information system, an environmental impact assessment module (for river water quality, long-range atmospheric transport, and groundwater quality management simulation) and an industrial structure optimization module.

Also part of the system is a transportation risk/cost analysis module (Figure 1) developed in collaboration with the Ludwig Boltzmann Institute (Kleindorfer and Vetschera, 1985).

Based on the European transportation network (major roads and railways) this module simulates the transportation of a hazardous substance between two cities. The user has to select a source and destination by picking them from the map. To define the substance to be transported, he can either load a substance directly from the data base, or define 'his' substance in terms of state (e.g., liquid), and flammability, water pollution, and reaction.

A first task can be interpreted as a shortest path problem (several algorithms can be found in literature to solve this problem). In a large network, however, the generation of all efficient alternatives would lead to response times too high for interaction with the system. That is why a heuristic approach has been implemented to define a restricted search area for the generation of several efficient route/vehicle combinations.

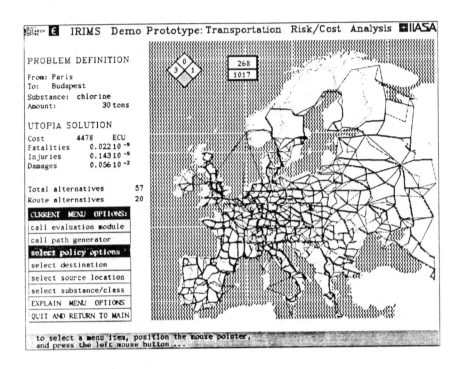

Figure 1: IRIMS: Risk/ cost analysis module - Generation of
candidate paths.

Second, for each of these alternatives a risk/cost evaluation is
performed. The cost evaluation is based on freight rates sampled
from commercial transport firms, assuming that firms involved
choose an alternative on the basis of costs, including insurance and
possible liabilities not covered by the insurance. The risk analysis
covers material damage as well as injuries and fatalities. Expected
values as well as variance of losses can be considered as decision
criteria.

There is no objective optimal solution as long as a trade-off between
criteria such as risks to health and safety of the general public as
well as costs to individual firms are involved. The selection of an
optimal alternative can now be performed by the user via an integrated multi-criteria evaluation package (DISCRET; Majchrzak,

1985, Zhao, Winkelbauer and Fedra, 1985; Figure 2), a powerful post-processor for discrete optimization of model-generated sets of alternatives.

The module allows the user to interactively determine his relevant criteria (minimize/ maximize/ ignore) out of the seven generated (cost, expected values and variance for injuries, damages and fatalities), to set constraints (minimum/ maximum values of criteria to be considered, e.g., 'the transportation costs should be less than 1500 ECU') and reference points. It also supports various display options (2D projections and frequency distributions), and the identification of non-dominated (pareto-optimal) alternatives. Finally, a satisfactory solution is identified, in accordance with the user's preferences.

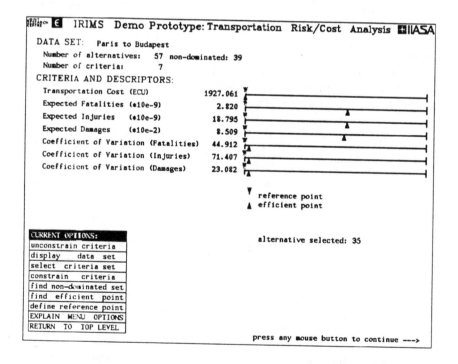

Figure 2:   IRIMS: Discrete multicriteria decision making - Find efficient point.

2.2 Case Studie Haute Normandie

In collaboration with INRETS (Institut National de Recherche sur les Transports et leur Sécurité) of the French Département d'Evaluation et de Recherche en Accidentoligie (DERA), an interactive prototype system has been developed (Fedra et al., 1989), integrating a path generator and risk assessment module with a set of data bases and a geographical information system.

The prototype is applied as a decision support for finding safe routes in the department Seine-Maritime, a rectangular area of 20 x 25 kilometers, covering 43 communal districts, where 16 chemical firms are located, generating an important flow of dangerous goods by road, railway and water.

The startup-screen of the system (Figure 3) displays the area of concern (most of the digitized cartographic data is drawn from the BDCarto, the national data base of digitized maps, developed by the Institut Geographique National (IGN)), showing land use (including high, medium and low density residential area, activities such as industries or sports grounds, agriculture, forest, heath, sea/estuary/port, rocks/sand, and water body), as well as the road network. Other cartographic data are only available under the geographical information system (GIS) component of the system.

Also used were additional data, mostly statistical (e.g., 1982 population census, SIRENE for data on industrial sites, 1988 inventory for 'communes' (INSEE, 1984a; INSEE, 1984b; INSEE, 1988)), including information on population, residences, schools and colleges, industries, hospitals, services and transportation infrastructure.

For the transportation system, each arc of the road and railway network has a set of properties e.g., function (e.g., regional), number of lanes, width, administrative category (such as railway SNCF or national road), name and/or road number, surface, node levels, ground position and miscellaneous (e.g., pay for use). This information is used for the risk calculations.

Auxiliary modules had to be implemented to link these additional data to the spatial data available (e.g., population data to areas) to integrate all data within the INGRES data base management system, accessible by the overall system.

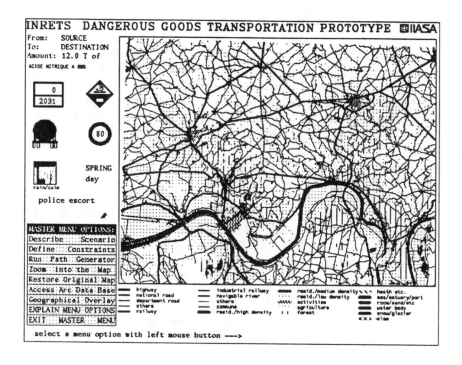

Figure 3: Startup screen of the transportation risk assessment system for the Haute Normandie.

In contrast to the previous system, we are now looking for the path with the lowest risk explicitly, within certain constraints that the user can set interactively during the operation of the system. He can pick nodes and/or arcs from the map and allow/ exclude its use by the path generator. It is also possible to exclude the crossing of certain types of land use, e.g., exclude arcs crossing residential area with high population density. Also the use of certain types of arcs can be excluded, e.g., allow only arcs with more than one lane.

Also offered are certain GIS functions such as the display of additional cartographic data, and zooming in and out of the map.

After a minimum scenario description of source and destination, the path generator can be activated, using Dijkstra's shortest (= minimal risk) path algorithm (see for example Boffey, 1982 or Aho, Hopcroft and Ullman, 1983).

The risk analysis is carried out separately for each arc considered. Depending on its properties, a certain accident rate is associated with each arc. The risk value used is a product of this rate and the population affected. The population affected depends on the impact zone (originally set to 1 kilometer left and right of the arc as a default value) and the population density within that area.

An auxiliary program calculates, for each arc, the distribution of the affected area within the different types of land use (e.g., 42.8 m^2 of agricultural land use are affected). To calculate the population affected by an accident on a specific arc, one just accumulates the area of land use types multiplied by the densities related to that land use. The system was set up to later allow the possibility to distribute land use per specific 'commune' , and as an overlay make use of the detailed statistical data already incorporated in the system.

Once the path with the minimal risk has been calculated, it is highlighted on the map. The number of arcs in use, the risk value, and a list of possible constraints set previously by the user are displayed. The user now has the choice of going back to the master menu or looking at the result in detail (see Figure 4), starting from the source: one after the other each arc used, as well as its description, is displayed. This includes its properties loaded from the data base, the calculated risk value, as well as size and types of land use affected by this arc.

For the first prototype a number of factors have been neglected or simplified considerably. In the meantime a number of improvements have been implemented already (see Lassarre et al., 1990) or are under development.

2.3 XENVIS

Under contract to the Dutch Ministry of Housing, Physical Planning, and the Environment (VROM) the X-based ENVironmental Information System XENVIS is being developed. Part of the system is an interactive and graphics-oriented framework and post-processor for the risk assessment package SAFETI (developed by

Technica Inc., Consulting Scientists and Engineers, London; Technica, 1984) to facilitate quick generation, display, evaluation and comparison of policy alternatives and individual scenarios.

Using the GIS part of XENVIS the user can select a number of overlay topics available to be displayed/ deleted on/ off the map. He can compose 'his' map, and all other modules of the system showing a map (e.g., the path generator) will display this selection on the screen (see Figure 5 displaying land use, meteorological data, chlorine industries and the railway network).

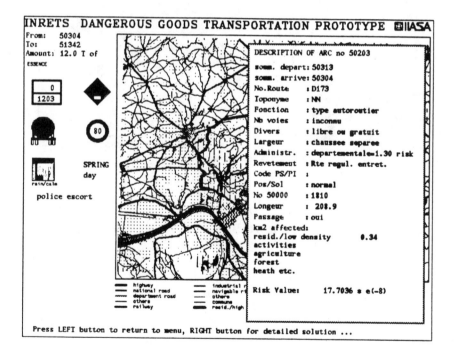

Figure 4:   Region Haute Normandie: Display properties of each arc in the generated path.

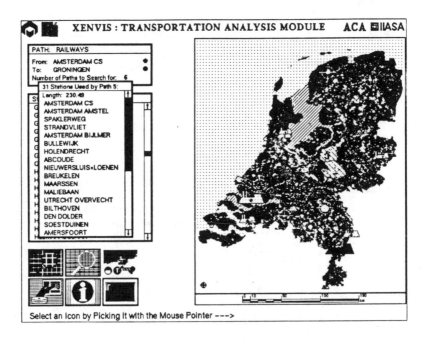

Figure 5: XENVIS: Transportation analysis module: Display a path including length and stations.

The XENVIS path generator offers the user to choose between railways and roads, to select source and destination of the transportation event by name or by picking them on the map, and to set the number of 'shortest' path alternatives to search for. Based on Dijkstra, the algorithm has been extended to find the n shortest paths from source to destination. For the railways, a penalty is increasing the length of a path that traces very sharp angles, i.e., would require the train to change direction.

After the calculations, the shortest path is displayed in red on the map. Also shown is the length of the path displayed, as well as a list of stations (for the railways) or cities (for the roads) used by that path. Upon request, the next shortest path (if any) will be displayed in red (see Figure 5), and the differences to the shortest path will be displayed in white. The length of the path and list of stations/ cities will be updated.

One of the options of the path generator module takes the current path and prepares it for direct input into the SAFETI package. If the path is longer than the maximum length SAFETI can take, it will generate several input files. Using SAFETI's original interface, the risk analysis can then be performed for each segment. SAFETI's output includes F-N curves and risk grids. The results are made available for graphical display and interpretation (see Figure 6). An auxiliary program will generate risk contours, that can be displayed as overlays on the map of the Netherlands.

3.   INTEGRATION, INTERACTION AND VISUALIZATION

Software for risk assessment, and transportation risk in particular, is usually complex, requires large amount of different input data, and a considerable amount of effort in both the problem formulation and the interpretation of results.

Many computer-based models and methods are potentially useful, and a large amount of formal, mathematical and computational methods have been developed in the area of transportation risk analysis. To turn a potentially useful method into one actually used however, requires a number of special features as well as an approach that takes psychological and institutional aspects as well as scientific and technical ones into account.

Tools that are easy to use, equipped with a friendly user interface, use problem-adequate representation formats and a high degree of visualization, customized for an institution and its specific view of problems, and that are developed in close collaboration with the end user, stand a better chance of being used than tools that are based on "only" good science. Good science is a necessary, but certainly not sufficient condition for a useful and usable information and decision support system; there are definite advantages to user participation, with consideration of questions of maintenance and the update of information requirements from the very beginning.

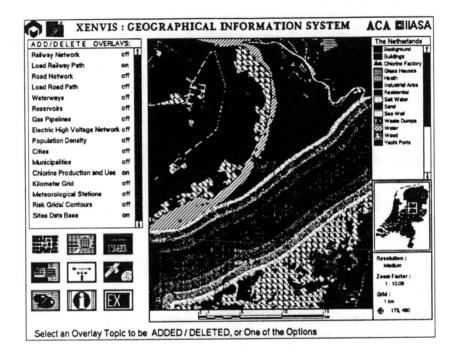

Figure 6: XENVIS: Display risk contours for a path previously
generated.

Advanced information technology provides the tools to design and
implement smart software, where in a broad sense, the emphasis is
on the man-machine interface. Integration, interaction, and
visualization, are key concepts that are common to the three
examples discussed above.

Integration implies that in any given software system for real-
world applications, more than one problem representation form or
model, several sources of information or data bases, and finally a
multi-faceted and problem-oriented user interface ought to be

combined in a common framework to provide a useful and realistic information base.

Interaction is a central feature of any effective man-machine system: a real-time dialogue allows the user to define and explore a problem incrementally in response to immediate answers from the system; fast and powerful systems with modern processor technology can offer the possibility to simulate dynamic processes with animated output, and they can provide a high degree of responsiveness that is essential to maintain a successful dialogue and direct control over the software.

Visualization provides the band-width necessary to understand large amounts of highly structured data and extract the information from these data, and permits the development of an intuitive understanding of processes and interdependencies, of spatial and temporal patterns, and complex systems in general. Many of the problem components in a real-world risk assessment or management situation, and certainly risk itself, are rather abstract: representing them in a symbolic, graphical format that allows visual inspection of systems behavior, and, in general, symbolic interaction with the machine and its software is an important element in friendly and easy-to-use computer-based systems. If risk can be seen as a colored cloud drifting over transportation networks, it becomes tangible.

4.   DISCUSSION

No matter how many numbers are used in a formal, mathematical analysis, risk assessment always contains a considerable soft, subjective component. By their very nature, basic data on usually -- and hopefully -- rare events are scarce, and thus, contain a considerable element of uncertainty. Frequencies of accidents, based on historical data must assume no structural change in the transportation system. It is, however, in many cases exactly these structural changes that need to be evaluated.

The estimation of consequences of accidents, on the other hand, not only involve considerable uncertainty, but also a considerable degree of subjective evaluation. And the trade-off between alternatives described with multiple criteria certainly draws on socio-political and ethical values more than on numerical values.

Another dimension is the inclusion of environmental
considerations. Risk assessment traditionally often concentrates
on human fatality risk, and sometimes property damage. There is,
however, increasing interest in including environmental
consequences into a truly multi-criteria risk assessment. The
consequences of, for example, spills of toxic material for surface
and groundwater bodies can induce some long-term damage, and
the clean-up costs can be staggering. The same holds true for oil
spills, as recent examples clearly show.

Evaluation in terms of the cost of repair is a well established concept
in environmental economics. It is however, the irreversible
damages to ecosystems that again introduce the multi-criteria
problem of evaluation.

While "smart" interactive software certainly cannot solve these
problems, it offers tools for a different paradigm for risk analysis.
Risk assessment, we argue, is by necessity a comparative, multi-
criteria analysis that aims at fostering social and political
consensus on acceptable risk as much as at any exact classification
or numerical description of risk. This requires the direct
participation and involvement of the decision maker, a term we use
to include all the participants or actors in the usually rather complex
and often messy decision making process.

Multiple actors, with often very different objectives, values, and
perceptions but also background and levels of formal education,
should at least share some basic level of information and have a
structured framework for debate on a technical and scientific basis.
Information and decision support systems for risk analysis can
provide this information basis and common framework and
language. They can, at best, introduce intellectual discipline and
rigor and help make the decision making process more rational --
as much as this is possible in the realm of risk assessment. The
integration of numerous sources of information and alternative
tools or models, direct access and the directly understandable
graphical formats of presentation can improve communication and
thus the information basis, and help structure and guide the debate
over often contentious issues.

Easy-to-use software stands a better chance to be used; and software that can, in principle at least, be used by all the actors in the decision making process, stands a better chance to be accepted.

User friendliness, however, is not only an important characteristic for access and communication in a diverse user group. For the technical professional, it offers the possibility of a more experimental use of software tools, the ability to work more efficiently and thus, effectively. Smart systems will take care of all the more mundane tasks of data handling, preparation and checking, and assist in presentation and analysis, freeing the analyst to concentrate on the real problems that require human creativity, which is somewhat more difficult to build into computers.

5. REFERENCES

AHO, A.V., HOPCROFT, J.E., & ULLMAN, J.D. (1983). Data Structures and Algorithms. Addison-Wesley Publishing Co.

BOFFEY, T.B. (1982). Graph Theory in Operations Research. The Macmillan Press Ltd.

FEDRA, K. (1990). Interactive Environmental Software: Integration, Simulation and Visualization. In Pillmann, W. and Jaeschke, A. [eds.] Informatik für den Umweltschutz.(Data Processing for Environmental Protection). Reprinted as IIASA RR-90-10. Proceedings of the 5th Symposium, 19-21 September 1990. Vienna. Springer-Verlag. pp.733-744.

FEDRA, K. (1986). Advanced Decision-oriented Software for the Management of Hazardous Substances. Part II: A Demonstration Prototype System. (Final Report to EURATOM CEC/JRC. Ispra. Italy). Reprinted as CP-86-10. International Institute for Applied Systems Analysis, A-2361 Laxenburg, Austria.

FEDRA, K., LASSARRE, S., & WEIGKRICHT, E. (1989). Computer-Assisted Risk Assessment of Dangerous Goods Transportation Case Study for Haute Normandie. Final Report. Submitted to the Institut National de Recherche sur les Transports et leur Sécurité International Institute for Applied Systems Analysis. A-2361 Laxenburg, Austria.

INSEE (1984a). Recensement de la population 1982--Description de la bande de données communales. INSEE. Paris.

INSEE (1984b). Recensement général de la population de 1982. Guide d'utilisation Tome 1. INSEE. Paris.

INSEE (1988) Recensement général de la population de 1982. General Census of the Population of 1982. Annuaire Statistique de la France 1988. Resultats de 1987. Vol.93. Institut national de la statistique et des études économiques. Paris, France.

KLEINDORFER, P.R., & VETSHERA, R. (1985). Risk--Cost Analysis Model for the Transportation of Hazardous Substances. Final Report Vol. I. October. Ludwig Boltzmann Institute, Vienna.

LASSARRE, S., FEDRA, K., & WEIGKRICHT, E. (1990). Computer-Assisted Routing of Dangerous Goods for Haute-Normandie. Presented at the specialty conference State and Local Issues in Transportation of Hazardous Materials: Towards a National Strategy. Held May 14-17, 1990 in St-Louis, Missouri.

MAJCHRZAK, J. (1985). DISCRET: An Interactive Package for Discrete Multicriteria Decision Making Problems. Report, June 1985. Systems Research Institute, Warsaw, Poland.

TECHNICA (1984). The SAFETI Package. Computer-based System for Risk Analysis of Process Plant. Vol.I-IV and Appendices I-IV. Technica Ltd., Tavistock Sq., London.

ZHAO, C., WINKELBAUER, L., & FEDRA, K. (1985). Advanced Decision-oriented Software for the Management of Hazardous Substances. Part VI: The Interactive Decision-Support Module. CP-85-50. International Institute for Applied Systems Analysis, A-2361 Laxenburg, Austria.

The Measurement of Risk from Transporting Dangerous Goods by Road and Rail

Grant Purdy[1]
DNV Technica Ltd
Highbank House
Exchange Street
Stockport SK3 0ET
Cheshire UK

ABSTRACT

In any debate about the transport of dangerous goods where the effectiveness of existing legislative controls is challenged, it is very important that there is a full understanding of the magnitude of the risks involved and the causes and major contributors so that properly informed decisions can be made. This paper gives details of the methodology developed for the analysis of the risks arising from the carriage, in bulk, of toxic and flammable substances by road and rail as part of a major study into the risks faced by the British population from the transport of dangerous substances.

This paper concentrates on the novel aspects of the study and in particular consequence and human impact modelling. In particular, models are given for the interaction of passenger and dangerous goods trains taking into account the ability of signals and other systems to detect and stop approaching trains. In the case of road transport, the models allow for the characteristics of different road types and the behaviour of motorists to be simulated.

The relative risks of transporting hazardous materials by road or rail are explored and it is shown that the inclusion of motorist and

---

[1]The views expressed in this paper are the author's and do not necessarily represent those of DNV Technica or of the Health and Safety Executive.

rail passenger populations significantly affects the calculated risk levels. It is concluded that the safe routing of materials with large hazard ranges may be more easily achieved by road. While, the natural separation afforded by the rail system may make this mode more suitable for lower hazard materials. However, it is concluded that in Britain, there is no evidence to support, on safety grounds, a general transfer of hazardous goods from road to rail or the reverse.

1. INTRODUCTION

Last year (1991) saw the publication by the UK's Health and Safety Commission (HSC, 1991) of the results of a 5 year study into the transport of dangerous goods in Britain. That study, by a sub-committee of the Commission's Advisory Committee on Dangerous Substances, considered the risks to the British population from the carriage of dangerous goods by rail, road and by sea in the light of the present regulatory and voluntary controls and the need for and possible nature of additional controls.

When the sub-committee first commenced its work it was reassured by the very good historical record in Britain. But it soon found that it had no way, other than by subjective appreciation, to gauge the effectiveness of the complex range of controls which exist and which, either by design or as a side benefit, act to protect members of the public. These controls appeared to be effective but the Committee members were concerned at the potential for large numbers of injuries and fatalities if a tanker of hazardous material were to become involved in an accident. Such an accident has rarely occurred in the UK and the number of casualties has, in all cases, been small. However, when the Committee considered the potential for disaster if, for example, a chlorine tanker ruptured in a built up area, they felt that they needed much more information on the likelihood such an event might occur before they could come to any decision on the tolerability of the risk. To satisfy these two needs - to gauge the effects of existing controls and safeguards and to understand more fully what was the chance of a disaster - the Committee decided that some form of quantified risk assessment should be carried out to provide them with tangible results and a means of making consistent and defendable decisions.

This was the first occasion when the risk to a nation from the transport of hazardous materials had been measured to such a degree and the study involved considerable research in order to develop suitable methods of analysis. Further research was also needed to understand the results which the analysis produced. While studies looking at the risks from transporting hazardous materials have been and are being carried out elsewhere (and all these were reviewed), none of these methodologies were found to be fully appropriate for the UK study. In general this was because:

- elements of the methodology could be considered 'obsolete';

- they had been developed to reflect a transport system or a system of regulatory control that was somewhat different to that in the UK;

- they had been developed specifically to investigate one aspect of transportation, for example, the safe routing through a city area, and did not have wider applicability.

For these reasons a 'new' approach was necessary: specific to the British situation, which sought to minimise uncertainty while providing 'transparency' of the risk calculation process so that the decision makers could understand and have confidence in the results. It also had to allow for assumptions to be easily changed so that the models could be used as 'testbeds' for the gauging the effectiveness of changes to the system of control. The approach was developed by two technical working parties (one for marine, the other for land based transport) on which sat members of the Health and Safety Executive, its contractors, industry, the emergency services and academia.

This paper is concerned with the work of the technical working party for land-based transport and the modelling associated with the transport of non-explosive substances in bulk (called 'the UK Study' throughout the rest of this paper). While the techniques of analysis were developed in the context of the British situation, many of the lessons learnt and insights gained have much wider application.

2.   OBJECTIVES OF THE RISK ANALYSIS

The choice of consequence and impact models and indeed the manner of conducting a risk analysis depends on the eventual use of the results; who will use them and for what purpose. In this case, it would not have been useful to expend effort developing complex and indepth analyses where, for example, there were great uncertainties in frequency data or the decision making process could not accommodate significant levels of precision. This is one of the most important principles which guided the development of models and techniques for this work, while we sought methods of analysis which optimised accuracy, this was often at the expense of unnecessary precision.

Similar considerations applied to the type of risks measured and the presentation of the results. Some transportation studies have concentrated on individual risk calculations, presenting the results as contours or risk transects diagrams showing individual risk against distance from the transport route. While such studies may be useful for routing exercises, where a new transport corridor is being selected, unless there is good evidence on the relative distribution of failure events along the route (i.e., 'high spots'), individual risk results can add little to the understanding of risk from a transport operation. The risk numbers produced are normally so small as to be beyond the normal range of human comprehension. Most importantly, this type of treatment fails to address the public's (and the politician's) major concerns; not the risk to individuals, but that to society at large: the risk of a disaster. This involves not only considerations about the potential for transported hazardous substances to cause multiple fatalities but also the likelihood these might occur because a loss of containment accident coincides in time and space with a human population. Societal risk is therefore not only a more appropriate measure but it also seems to yield more useful results. It leads naturally, via the generation of expectation values (average number of lives lost) to considerations about the need for, and cost benefit of risk reduction measures. Societal risk analysis does involve many generalising assumptions and averaging but these are not inconsistent with the 'smeared out' nature of the risk associated with transport along a route.

3.   FREQUENCY ANALYSIS

For those countries or regions with a history of hazardous goods accidents, consulting the historical record is normally the first step in any study of risk. Indeed, if enough incidents have (unfortunately) occurred, the modelling of the possible consequences and impact of such events may be of secondary importance. In Britain, however, we have suffered few such incidents. Those that have occurred have normally involved flammable liquids and no person has yet died as the consequences of a leak from a damaged tanker (road or rail) holding liquefied flammable or toxic gases such as LPG or chlorine. For this reason, the UK Study adopted a some-what different approach to obtaining the release frequencies for hazardous substances in transit.

One possible approach would have been to use an event tree such as that in Figure 1. This is similar to the developed by Hubert et al. (Hubert, 1985) from French data. This builds on data from accidents involving hazardous goods vehicles to synthesise a puncture rate for a hazardous goods tanker. However, there is no evidence to suggest that the drivers of hazardous goods vehicles will act in a similar manner to drivers of other vehicles nor that such vehicles will suffer equipment and other failures at the same rate as other similar vehicles containing other bulk materials. The value given to the critical probability associated with 'escalation' to puncture is critical yet very uncertain. Even for countries where good data exist there are always the uncertainties associated with under-reporting.

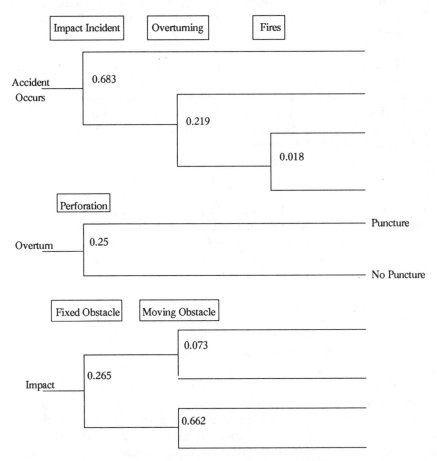

Figure 1: Incident data analysis after Hubert (1985).

An analysis of the available UK data on rail and road incidents involving tankers containing hazardous materials showed that releases could occur from two sources, firstly by puncture or rupture following collision, roll-over or derailment, or secondly, from failure or mal-operation of the tanker equipment. For the rail mode there was sufficient data on 'thin walled' (~6mm) wagon accidents to generate a frequency for punctures and equipment leaks directly. Over 6 years, 80 cases of spills due to 'equipment leaks' and 4 incidents involving substantial spillage following puncture were found. This data suggested a puncture frequency of 6.3×10^{-8} per tank wagon km.

For road transport, 25 incidents were found over a four year period. Analysis of these data yielded a spill frequency of 1.4×10^{-8} per loaded tanker km for large spills (>1500 kg) from collisions etc and 0.7×10^{-8} per loaded tanker km for large spills arising out of equipment failure.

While motor spirit spill frequencies could be obtained directly from this analysis, there are no incidents recorded in the UK where properly designed road or rail tankers for pressurised liquefied flammable or toxic gases have been punctured. For these it is therefore necessary to adopt a synthetic approach to deriving appropriate spill frequencies; a rate generated by statistical techniques from an 'accident free' history provides a useful 'upper bound' check. For transport by rail, the technical working group used an analysis by ICI Transport Engineering Division to give spill frequencies for ammonia, chlorine and LPG. This analysis considered the historical accounts of puncture of 'thin walled' wagons and estimated in each case the conditional chance of failure if the vessel concerned had been a 'thick walled' (typically 11-16mm) LPG/Ammonia or Chlorine containing vessel.

Although data on US rail incidents is easily available, it was felt the differences between the design standards and operating practices made this data inapplicable to the British situation. However, for road transport, the differences were less important and could be identified with some confidence. Because of this, US road data could be used and, by appropriate modification to exclude those events which could not or were unlikely to occur in Britain, spill frequencies were derived. This process is described in detail in the Study report (HSC, 1991). Fault tree analysis was used to develop the possible causes and events which could lead to equipment leaks. These were then used to derive appropriate equipment spill frequencies for both rail and road transport of LPG, ammonia and chlorine.

In summary, therefore, the following spill frequencies were derived for this work:

Table 1:   Frequency of spills against cause, substance and transport mode.

SUBSTANCE	ROAD TRANSPORT		RAIL TRANSPORT	
	Puncture/ Rupture (x 10^{-10} per wagon km)	Equipment Leak (x 10^{-10} per wagon journey)	Puncture/ Rupture (x 10^{-10} per wagon km)	Equipment Leak (x 10^{-10} per wagon journey)
Motor Spirit *	190	70	630	- **
Chlorine	0.8	36	9	310
Ammonia	4.8	70	25	130
LPG	4.8	52	25	83

*   for large spills only
**   not considered as such small spills are unlikely to affect members of the public

Table 1 gives base event frequencies. For flammable events, it is also necessary to consider the probability that a spill will then be ignited and whether this will take place initially or at some later time once a flammable cloud has developed. In some cases it was possible to estimate ignition probabilities from accident data, but the under-reporting of spills which have failed to ignite makes these figures unreliable. In most cases, ignition probabilities have to be estimated using synthetic techniques or by expert judgement. This is simplified in transport situations as often the spill causing event involves sufficient energy to cause ignition or other sources (for example other road vehicles) are nearby. For the UK study we used the following values:

Table 2: Ignition probabilities for flammable substances.

SUBSTANCE	TYPE OF IGNITION	RAIL		ROAD	
		Small Spill	Large Spill	Small Spill	Large Spill
Motor Spirit	Immediate	0.1*	0.2*	0.03	0.03
Motor Spirit	Delayed	0.0	0.1	0.03	0.03
Motor Spirit	None	0.9*	0.7	0.94	0.94
LPG	Immediate	0.1	0.2	0.1	0.2
LPG	Delayed	0.0	0.5	0.5	0.8
LPG	None	0.9	0.3	0.4	0.0

* Derived from historical data.

4.   CONSEQUENCE ANALYSIS

As with all forms of such quantified risk analysis, the selection of a representative set of failure cases and assignment of the corresponding spill sizes/rates are the most important steps to producing an accurate characterisation of risk. An optimum set of cases has to found which while minimising computational effort do not unduly compromise accuracy. Fortunately in the transport situation there are several constraints which act to limit the range of possible events:

- for multi-compartment tankers, the simultaneous loss of contents from more that one compartment is extremely unlikely;

- small releases of flammable material are unlikely to ignite or cause hazard as they are rapidly dispersed as the tanker moves and even when stationary, the normal 'open' aspect of a transport situation will aid dilution;

- above a certain hole size, either the release of pressurised, liquefied gas will be so rapid that it can be considered equivalent to an instantaneous release on vessel rupture or the hole will be sufficiently large to lead to the pressure vessel 'un-zipping';

- in the rail environment, ignited jets of LPG are unlikely to create significant hazard unless they impinge on other LPG tankers which then BLEVE. However, the BLEVE frequency used should normally include such a cause.

For the UK transport study, one hole size together with total vessel rupture and a nominal equipment leak rate were used. Taking three release sizes only is judged to be on the borderline of what is acceptable for this type of study but the selection was driven by the limited data available on how to partition the base puncture frequency between different hole sizes. In the absence of any corroborative data, it was assumed that 10% of the releases from pressure vessels were instantaneous and could be modelled as the entire loss of contents. In the case of toxic materials, the cloud contained 100% of the tanker contents, for LPG twice the adiabatic flash fraction was assumed to enter the vapour cloud. Sensitivity testing to a 99%/1% split or a 50%/50% split showed that this assumption was not critical.

Particular care is needed to take into account the physical aspects of the spill environment when the consequences are modelled. Factors such as the containment effect of roads, drains etc can significantly affect the shape and dimensions of the hazard zone. This is particularly true of spills of flammable liquids where the hazard zone is only slightly greater than the area of confinement provided by the road or rail corridor. Furthermore, on the road, surface water drains will limit the size of any liquid pool.

For Motor Spirit spills we therefore considered two cases: either the tanker remained on the highway or rail corridor in which case the spill was confined by kerbs, drains etc or the tanker left the road or rail line and was modelled as a circular pool. The pool will (in both cases), if ignited, reach a maximum size where the regression rate was equal to the spill rate. Spreading pool expressions such as those given by Shaw and Briscoe (1978) can be coupled with a 'drain model' and a fire model (Mitzner and Eyre, 1982) to estimate the maximum area of road affected. As the thermal hazard decays rapidly as a receptor moves away from a burning pool, most people who were exposed outside the pool could escape and the area of the pool can be taken as the hazard zone. For example, we calculated that a 25 kgs^{-1} continuous release from a leaking rail wagon (32 te) would produce a pool of radius 24 m. This would persist for about 20 minutes but if the vapours ignited during this time, the pool size would regress to 12 m. Table 3 shows similar results for road tanker spills.

Table 3: Motor spirit road tanker pool areas.

SPILL SIZE	POOL AREA (m²)	
	Immediate Ignition	Delayed Ignition
25 kg s⁻¹	314	908
4000 kg	707	1018
8000 kg	1385	1964
12000 kg	2124	3019

The possibility of a 'soft' BLEVE fireball due to heating of a Motor Spirit tanker in a fire has been considered but does not seem likely. Analysis of the Summit Tunnel Fire Incident (Jones, 1985) has shown that even under severe heating conditions, Motor Spirit tankers will not rupture if 3 out of 4 relief valves work or will take at least an hour of prolonged heating if only two operate.

For LPG the type and extent of the hazard depends on the mode of release and whether and when it is ignited. Figure 2 in an example, for continuous LPG releases, of the event trees which can be drawn to rationalise this potential for escalation. Similar trees exist for instantaneous releases of LPG and for spills of Motor Spirit.

For LPG, standard consequence modelling approaches can be adopted. However, when applied to transportation accidents, certain special considerations apply:

* for BLEVE's, the resulting fireball will contain a large propor-tion of the vessel content's as the vessels are always full and because British LPG road tankers do not normally have relief valves;

* BLEVE's are much more unlikely on the road as one of the primary causes, jet fire impingement for one tanker to another, is highly improbable;

* Vapour Cloud Explosions (VCE's) are very unlikely on both road or rail given the open aspect and limited amount of confinement available. Given the small contribution they will therefore make to the overall risk, a simple consequence model (such as TNT equivalence) is appropriate;

* outside the flammable cloud, the probability of death due to the effects of overpressure from a VCE is low and can be ignored.

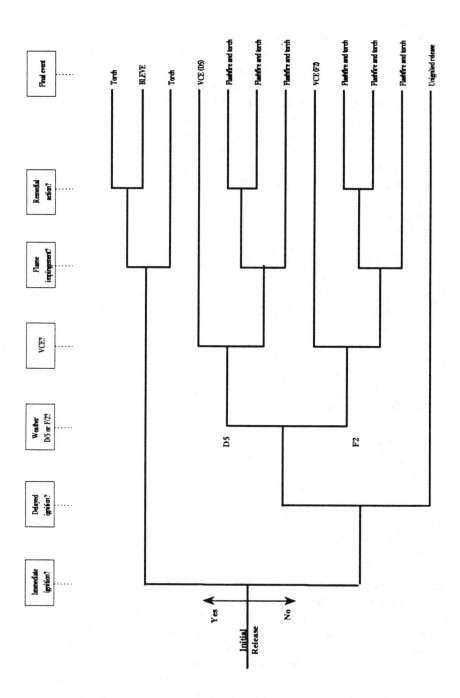

Figure 2: Event tree for continuous LPG releases.

The risk from released toxic gases such as ammonia and chlorine is very dependent on the accuracy of the dispersion modelling. As societal risk is to be calculated, the crosswind extent of the cloud is as important as the downwind hazard range. The societal risk estimation involves the calculation of the numbers of fatalities from the areas of land which experience more than a criteria toxic load. The use of simple gaussian models which do not allow for negative buoyancy effects such as cross and up wind spreading will therefore produce inaccurate (likely to be optimistic) results. The release orientation in relation to the wind can be an important consideration and the modelling of the initial momentum driven jet seem important pre-requisites to the use of an accurate dense gas dispersion code.

There is also a strong dependency between the crosswind and downwind extent of the cloud and the level of atmospheric turbulence. This is characterised by the use of appropriate parameters to represent different Pasquill Stability categories. For the UK study, only two categories; D with a windspeed of 5 ms^{-1} and F at 2 ms^{-1}, were used for reasons of computational efficiency. This choice may have had a significant effect on accuracy, and more categories - four or six are usual - are to be preferred for toxic gases. Given the relatively short range of flammable hazard zones, two categories are probably adequate in this case.

The above discussion only applies strictly to above ground, open air releases on a flat, unobstructed terrain. There has been considerable interest recently about the carriage of hazardous materials through tunnels and the assessment of the associated risks needs special consideration. In the confined space of a tunnel, the spread of the hazardous consequences is very much affected by the air flow and the channelling effect of the tunnel. For example, the blast wave from a vapour cloud explosion could be expected to be transmitted largely undiminished along a tunnel. One of the most serious hazards is the hot, often poisonous smoke and products of combustion which can travel significant distances along the roof of the tunnel away from a fire. The UK study did not pay particular attention to tunnels as they did not constitute a large proportion of any of the routes studied and, in general, the only members of the public who would be affected would be those using the same road or rail tunnel. Further work is required to refine the current analysis methods so that a more accurate estimate can be made of the contribution of tunnels to an overall route risk. At present, few decisions involving the control of dangerous goods through tunnels seems to be based on any form of risk assessment.

5. IMPACT ANALYSIS

While the modelling of consequences and the estimation of frequencies are important components of the risk analysis approach, of equal importance is the estimation of the number of people who will be killed or injured by a particular hazardous event; Societal risk places equal emphasis on both the frequency of occurrence and number of fatalities. However, we find that this aspect of analysis has been little developed elsewhere and it was given particular attention in the UK Study. In particular, it seemed important to us to include all the population who may be affected by a dangerous goods incident. This includes motorists on a road where an incident occurs or members of the public travelling as passengers on trains which become involved on the rail. If only those people who live near the transport route are considered in the analysis, a very incomplete picture may be presented of the risk and its major contributors. This could lead to erroneous conclusions about the nature of and benefits from risk reduction strategies.

5.1 Off Route Population Density Measurement

For long transport routes, the population distribution along the route has to be characterised by a limited number of population categories, each representing an average situation. For the UK study we chose the four categories shown in Table 4.

Table 4: Off-route population categorisation scheme.

POPULATION CATEGORY	AVERAGE DENSITY (km^{-2})
Urban	4210
Sub-urban	1310
Built-up Rural	210
Rural	20

The length of the transport route along side of which each category of population exists can be obtained using computerised techniques for handling census and other demographic information. Much use is now being made of Geographical Information Systems (GIS) to handle such data although we found that a manual technique, using maps, provided a level of accuracy that was very acceptable given the many other uncertainties in this work. One refinement of the approach was to note those lengths of track or road alongside which either population of the same class exists on both sides and those where, for instance, the rail line or road has formed a natural barrier and there is one side of urban development while the other is rural. To prevent 'double counting', in the 'one sided' case, for directional hazards - for example a torch flame or toxic cloud - the frequency of the event is halved and, for events with circular hazard ranges such as BLEVE fireballs, the number of fatalities is halved.

It is also important to take into account the natural separation that occurs between off-route populations (typically residential) and the road or rail line. In Britain, there are very few locations where population comes within 25 m of a rail line and so when the impact of an event is being calculated, this 25 m 'swathe' must be excluded. This approach also acts to 'screen out' small, low consequence events from the analysis.

5.2   Off-road and Motorist Population Modelling

In the road situation there is a smaller but nevertheless important separation between the road and the off-road population. The width of the separation depends essentially on the class of road. It may be only the width of a pavement on an urban, single carriageway road but it may be much larger for a motorway. Furthermore, there are large sections of some routes where 'ribbon development' in a narrow strip alongside the road produces a very high population density (for example shopping areas) with open, low population density land beyond. To accommodate all these situations and to encompass the variation in the on-road, road user population density, a zoning scheme was developed. This is shown in Figure 3 for a dual carriage way road. The zone structure is described in Table 5.

This scheme also allowed us to model the response and density variations in the motorist population following an accident involving the release of hazardous material. We find that even at night, on main roads and especially motorways and dual carriageways, traffic rapidly builds up behind an accident leading to a very high population density on that carriageway. On the opposite carriageway, the traffic slows down due to the 'ghoul' effect; again increasing the population density. By assuming that 10% of traffic comprises heavy goods vehicles that occupy 20 m of lane length while other vehicle are 4 m long, an average vehicle population of 1.5 gives a Zone d population density of 0.056 m^{-2} for motorways and 0.05 m^{-2} for other roads. Zone d ahead of the accident is essentially clear. For the other carriageway, we have assumed that the curiosity of the motorists produces a density of 0.5 that of Zone d, but in both directions.

Table 5: Population zoning structure for roads.

ZONE	NAME	DESCRIPTION
a	Off-route population	This is similar to that used in the rail study but may be `depleted' if there is ribbon development.
b	Dense population	This allows for a high population density immediately adjacent to the road.
c	Clear Zone	Motorways and Dual Carriageway roads are likely to have a significant gap between the road edge and the population.
d	Motorists, Accident Side	Road User population which `backs-up' behind the accident.
e	Motorists, Other Side	Road user population on other side of carriageway.
f	Clear Zone	Same as Zone c.
g	Dense population	Same as Zone b.
h	Off-route population	Same as Zone a.

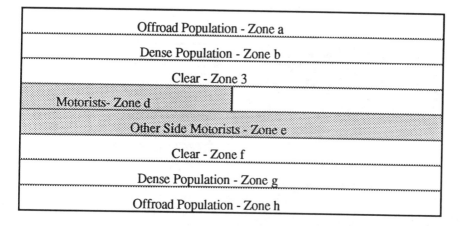

Figure 3: Population zoning scheme for dual carriage-way roads.

This scheme also allowed us to model those events which have directionality, for example a toxic gas release influenced by wind direction and its momentum driven phase. There are, of course, an infinite range of possible directions, but these can be reduced to the 4 cases shown in Figures 4a and 4b. The cloud is represented as either travelling perpendicular to or along the carriageway. In the along the carriageway case, the plume can either travel in the direction of the affected carriageway or opposite to it. For perpendicular case, the plume either travels off the road from accident or across the other carriageway. There is a further complication with instantaneous releases of dense gas where we would predict some gravity driven movement of the cloud, up-wind.

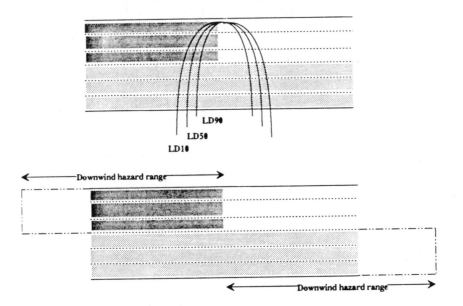

Figures 4 (a) and (b): Model for motorist fatalities, wind across and along carriageway.

5.3 Human Impact Measurement - Flammable Substances

For flammable and explosive events, we find that consequence models predict a fairly sharp cut-off between the point where people exposed will suffer very serious and likely to be fatal injuries. For flammable events we therefore adopted an impact model which had two 'steps':

- above the LD_{50} hazard range, all die;
- between LD_{50} and LD_{01} 25% of people die;
- beyond the LD_{01} all survive.

Where the LD_{50} and LD_{01} are very close together, this can be simplified to a single step where everyone inside the LD_{50} hazard range dies. This is particularly true for Motor Spirit where only those within the pool fire are assumed to die.

This approach is only true for overpressure events and thermal events to people out-of-doors. For non-continuous thermal events such as flash-fires, people indoors are assumed to survive; even if their homes catch on fire.

For motorists, it can be assumed that vehicles provide very little protection against fires and explosions. Those in cars are effectively trapped and escape from the road is not easy in congested traffic.

5.4 Human Impact Measurement - Toxic Gases

To allow for the accurate representation of the variation in human susceptibility and to enable the implementation of the zoning schemes for on and off-route populations, it was necessary to use a graduated approach to dose-effect modelling for toxic gases. The normal manner of doing this is to use 'probit' equations which seek to represent that variation in the percentage of a population that will die against a received 'toxic load' assuming a long-normal relationship. These have the general form:

$$Pr = a + b \ln (C^n.t) \qquad (1)$$

We used three levels of impact, LD_{90}, LD_{50} and LD_{10} for this study and assumed that the proportion of the population that will die in the area between LD_X and LD_Y will be $(X+Y)/2\%$.

It has been shown (Purdy and Davies, 1985) that going or being indoors provides considerable mitigation against the effects of toxic gases. The impact on people indoors can be calculated by using a simple gas infiltration model which allows for the exponential build up of concentration indoors ($C(O)$) while the gas cloud is present outside:

$$C(I) = C(O) . 1 - \exp(-\lambda t) \tag{2}$$

where $C(O)$ is the outside concentration, λ is the ventilation rate and t the duration of exposure, and a decay phase once the cloud has passed but people still remain indoors:

$$C(I) = C(M) . (-\lambda t) \tag{3}$$

where $C(M)$ is the maximum indoor concentration reached. The integration of these expressions with respect to time with the concentration raised to a power n (taken from the probit equation) yields a toxic load ($\int C^n . dt$). This can be compared with the probit relationship to give an expected % fatalities.

Figure 5 shows some of the potential options available to a person who is affected by a toxic gas. For people out-of-doors this can be rationalised into a simple model:

- at or above a concentration (C_1) a person will be unable to take any action and is likely to die;

- below this concentration, down to C_2, there is a chance that he or she can escape indoors. C_2 can be set so that chance is (say) 0.2;

- below that concentration there is a higher probability of escape but of those who remain outside the proportion who die is given by $(X + Y)/2$ where the area falls between the LD_X and LD_Y hazard ranges.

This model is shown in Figure 6. Therefore, for hazardous event E in weather j, the number of people out-of-doors likely to be killed is:

$$N_{O,Ej} = D_q . P_{Oj} [A_{C1} + A_1{'}(1 - P_{e1}) + (1 - P_{e2})(.95.A_2{'} + .7A_3{'} + .3A_4{'})] \tag{4}$$

where

$P_{O,j}$ is the proportion of the people who might be out of doors in weather j,

P_{e1} is the chance of escape within concentration C_2,

Pe_2 is the chance of escape within concentration C_3, and

area A_1' is $A_{C2} - A_{C1}$,

A_2' is $A_{C3,90} - A_{C2}$,

A_3' is $A_{C3,50} - A_{C3,90}$,

A_4' is $A_{C3,10} - A_{C3,50}$.

In reality, this expression is more complex as in for some release, $A_{C3,90} \leq A_{C2}$ or even $A_{C3,90} \leq A_{C1}$ and $A_{C3,50} \leq A_{C2}$.

Once people have escaped indoors, they may still be subjected to a fatal toxic load of gas. The number of fatalities indoors therefore comprises those who are already indoors and perish together with the proportion of the 'escapees' who also die. It is given by:

$$N_{I,E,j} = D_q.(1 - P_{O,j}) + D_q'.(.95A_{D,90} + .7A_5' + .3A_6') \tag{5}$$

where

$A_{D,90}$ is the area covered by the indoor LD_{90} isopleth,

A_5' is $A_{D,50} - A_{D,90}$

A_6' is $A_{D,10} - A_{D,50}$

The proportion of those who escape indoors who subsequently die will depend on whether they escape from $\geq C_2$ concentration, P_{e1} who go indoors, or $\geq C_3$ concentration, i.e P_{e2} go indoors. D_q' is the average population density of escapees.

For motorists, the protection afforded by their vehicles is very limited. Work by Cook (1988) shows that the 'Ram' effect of the car, even without a fan switched on provides a very high level of ventilation. Therefore we have assumed that these people are effectively out-of-doors and the expression given above, without the terms for escape can be used:

$$N_{M,E,j} = D_M [.95.A_{90} + .7(A_{90} - A_{90}) + .3(A_{10} - A_{50})] \tag{6}$$

where A_x is the area of carriageway (one side) which will experience toxic load LD_x or more. This area is given by:

$$\text{Area} = \text{hazard range to } LD_x \text{ x carriageway width} \tag{7}$$

D_M is the motorist population.

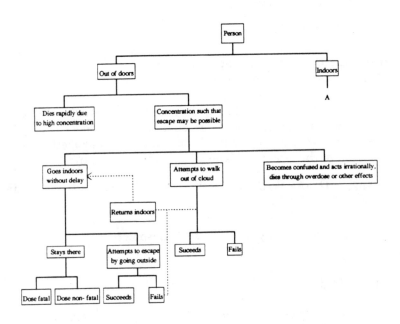

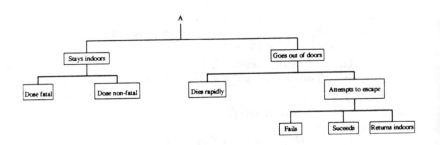

Figure 5: Range of options available to individual affected by toxic gas.

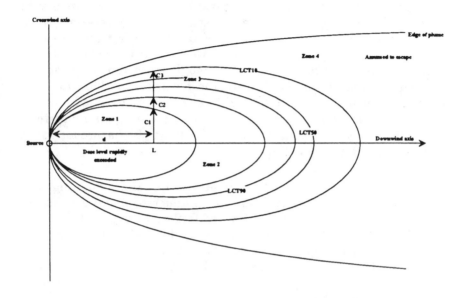

Figure 6: Model for toxic gas impact.

5.5 Rail Users (Passengers) Interactions.

In Britain, the rail network is used for both goods and passenger transport. This raises the possibility that one or more passenger trains may interact with a hazardous goods incident causing fatalities on the passenger train. Most other studies have failed to consider this 'extra' population but our work shows that they can make a significant contribution to the risk and that steps to prevent and minimise such interactions need to be considered.

On British Rail, the signalling system is principally concerned with preventing collisions by trains running on the same track. Signalling failure was the cause of one the UK's most serious transport incidents involving a hazardous substance. This occurred at Eccles, near Manchester in December 1984 when a passenger train ran into the back of a 14 wagon goods train hauling 'gas oil'. Three tanks ruptured forming pool fires and a 'fireball' which caused three fatalities and 76 injuries.

Events involving LPG and liquefied toxic gases have the potential to cause many more fatalities and our analysis has mainly considered the interaction of passenger trains with incidents involving rail wagons of these materials. These materials have long range effects which could affect a passenger train properly stopped by the signalling system, the so called 'obedient' train. Moreover, there is a possibility, although more remote, that the passenger train might collide with the hazardous goods train and cause the release, might collide with a previously derailed train or might, as this is specifically not prevented by the signalling system, be affected as it attempted to pass by the scene of a hazardous goods incident on an adjacent line.

This is a complex study which requires that the signalling and emergency systems on British Rail to be understood and adequately represented. Using a combination of fault and events trees, the PASSTRAM (Purdy, 1993) model was developed to allow the frequency and consequences of such interaction to be calculated for a route that involves sections along which different passenger train types, of different frequencies and passenger numbers travel at different times of the day.

6. CASE STUDY - A COMPARISON OF TRANSPORTING CHLORINE BY ROAD AND BY RAIL

To demonstrate the use of the models described in this paper and to bring out many of the points made above, we have carried out calculations of the societal risks associated with the transport of the same annual tonnage of chlorine between two locations by road or rail. At present, this trade is conducted by road between these two sites, approximately 100 km apart but a change of mode is a realistic possibility.

The route, in the north west of England, is at present served by road tankers several times a day. This one route constitutes a significant proportion of the national annual tonnage of chlorine transported by road in Britain. The journey is 103 km long, of which 80 km is motorway, and the rest is mostly single carriageway. The route travels past, but not through, three large towns and only about 1 km of the route has 'urban' population on at least one side with 19 km with suburban population on one or both sides. Most of the rest is rural. The exact breakdown is:

Table 6: Population distribution along study road route.

ROAD TYPE	POPULATION TYPE	DENSITY (km^{-2})	NO. OF SIDES	LENGTH (km)
Motorway	Urban	4210	2 1	0.0 1.0
	Sub-urban	1310	2 1	1.0 13.0
	Built-up Rural	210	2 1	2.0 17.5
	Rural	20	2 1	45.5 31.5
Single Carriageway	Urban	4210	2 1	0.0 0.0
	Sub-urban	1310	2 1	4.0 1.0
	Built-up Rural	210	2 1	4.0 0.5
	Rural	20	2 1	13.5 1.5

The road tankers which travel this route make 1,743 journeys a year carrying 17.5 te each time.

The alternative delivery by rail would require 1,052 29 te tankers a year. The rail route is about 97 km long but passes through 3 major towns with populations of 176,000, 81,674 and 126,000 respectively. The route includes 6 km of urban and 20 km of sub-urban population. Most of the route is also used extensively by passengers trains; part is the main West Coast main line between London and Scotland. The passenger train traffic is:

Table 7: Passenger train traffic on study route.

	INTERCITY		PROVINCIAL	
SECTION	Day	Night	Day	Night
Warrington	47	18	32	3
Kirkham	10	1	125	11

Using the techniques described above we have calculated the following levels of societal risk for the different modes:

Table 8: Societal risk results - Transport by rail.

	FREQUENCY OF N OR MORE FATALITIES $(X10^{-6} yr^{-1})$					
	1	10	30	100	300	1000
Passengers	39.5	39.5	39.5	10.6	0.0	0.0
Off-rail Population	105.0	47.8	27.8	26.8	11.9	5.2
Both Populations	107.3	68.0	56.8	41.5	13.6	5.7

Table 9: Societal risk results - Transport by road.

	FREQUENCY OF N OR MORE FATALITIES $(X10^{-6} yr^{-1})$					
	1	10	30	100	300	1000
Motorists	16.7	10.5	8.8	4.8	1.7	0.0
Off-road Population	15.5	5.9	2.9	1.4	0.9	0.0
Both Populations	19.0	13.6	10.3	6.2	2.6	0.1

These results are also shown in Figure 7 as FN curves.

It can be seen that:

- the risk by rail is approximately 5 times that by road;
- risks to rail users is about double that to motorists;
- risks to off-rail populations are approximately 8 times higher than those to off-road populations;
- the road risk is dominated by that due to motorist involvement.

These results are due to a common factor in British transport systems; most of our rail system was built over 100 years ago and was intended to go from town to town while most of our major roads have been built over the last 20 years and have been specifically routed to take traffic away from centres of population.

It would be possible to construct a route which would be more favourable to rail, but in reality the historical legacy of our transport systems will always tend to produce lower risks for the transport by road of materials with long hazard ranges. The risks from the transport of these substances will be lower if the route followed avoids centres of population and this is more easily achieved in Britain by road rather than rail. Substances with a shorter range effect such as motor spirit, should normally be more safely transported by rail since there is already a very worthwhile separation between the rail line and people who live nearby and passenger train involvement is likely to be restricted to direct collisions when, at worst, only a few passengers may be affected.

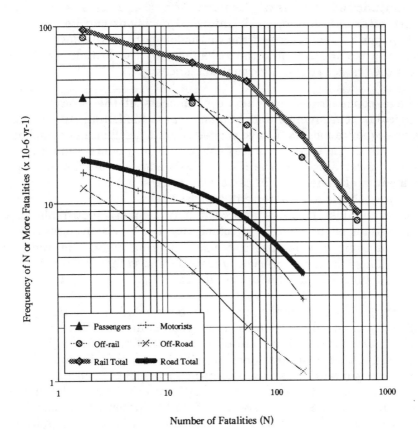

Number of Fatalities (N)

Figure 7: Societal risk result for road/rail comparison.

It is clear that in Britain it is not possible to say that transport of hazardous substances by rail is safer than by road or, indeed, vice versa. However, there seems to be no case on safety for the British Authorities to enforce modal transfer. This contrasts with the situation in other countries such as Germany where legislation now requires transfer to rail for longer journeys.

7. CONCLUSIONS

Throughout Europe, concern is being voiced about the transport of dangerous goods and the risks posed to members of the public. Legislators are shifting their attention from the problems of fixed major hazard installations to addressing what is the most appropriate means to control the risk from hazardous materials in transit.

It is very important that there is a full understanding of the magnitude of the risks involved and the causes and major contributors so that properly informed decisions can be made. In this paper I have described the methodology that was developed as part of a major study into the risks faced by the British population from the transport of dangerous substances.

I have concentrated on the novel aspects of the study and in particular consequence and human impact modelling. In the case of consequence models, I have suggested that the choice of model and the depth of the analysis must be driven by an understanding of the overall uncertainties of the risk analysis and the contribution each element makes to that uncertainty. Where it matters, the most accurate models are appropriate; for less sensitive elements, a more simple and less rigorous approach may be more justifiable. The final arbiter of the degrees of complexity and precision necessary is the end user; in this case a decision making body. The analysis methodology must be sufficiently transparent so that the results can be understood and used with confidence.

The modelling of human impact has been a feature of this paper reflecting the need perceived by those conducting the UK Study to be more rigorous in the treatment of this aspect of hazardous goods risk analysis. Other workers have not dealt with this in such detail before but our work has shown that the inclusion of motorist and rail passenger populations can significantly affect the calculated risk levels and can therefore have a profound effect on any conclusions which are drawn on the need for further legislative controls and the nature of those controls.

In support of these points and to demonstrate the use of the models that were built, the relative risks of transporting chlorine by road or rail has been explored in a realistic case study. From this it can be concluded that the safe routing of materials with large hazard ranges may be more easily achieved by road. For lower hazard materials, the natural separation afforded by the rail system may make this mode more suitable. However, in Britain, there appears to be no evidence to support, on safety grounds, a general transfer of hazardous goods from road to rail or the reverse.

Acknowledgements I would like to thank the Directors of DNV Technica for their support in preparing this paper. I should also like to acknowledge the very considerable contributions made to this work over many years by past colleagues in HSE, notably Geoff Grint, Nigel Riley, Linda Smith and Stewart Campbell and by the members of the technical working party.

9. REFERENCES

COOKE, M. (1988). Risk assessment safety: A case study of two chlorine plants. Thesis for MSc, UMIST, Manchester.

HUBERT, P. (1985). Les Risques d'accident majeurs dans les transports de maitre dangereuses. Centre d'Etude sur L'evaluation de la Protection Nucleaire, D 114, France.

HSC (1991). Major Hazard Aspects of the Transport of Dangerous Substances. Advisory Committee on Dangerous Substances, HMSO, ISBN 0 11 885676 6.

JONES, A. (1985). The Summit Tunnel Fire. IR/L/FR/85/86, HSE Report.

MITZNER, G.A., & EYRE, J.A. (1982). Large scale LNG and LPG pool fires. In The Assessment of Major Hazards (pp. 146-164). I Chem E, Rugby.

PURDY, G., & DAVIES, P.C. (1985). Toxic gas incidents - some important considerations for emergency planning. In Multistream, the Subject Group Symposium (pp. 257-268). I Chem E, Rugby.

PURDY, G. (1993). Risk analysis of the transportation of dangerous goods by road and rail. J. Haz. Mat., 33, 229-259. Elsevier, Amsterdam.

SHAW, P., & BRISCOE, F. (1978). Evaporation from spills of hazardous liquids onto land and water. UKAEA, Safety and Reliability Directorate, R210, Culcheth, UK.

A Novel Risk Assessment Methodology for the Transport of Explosives[1]

Nigel Riley
Health and Safety Executive
St. Annes House
Bootle, L20 3MF
Merseyside UK

ABSTRACT

A comprehensive quantified risk analysis of the national traffic in explosives by both road and rail was carried out in the United Kingdom. The methodology addressed the wide variety of explosives articles and substances transported and analysed their inadvertent initiation by fire, impact and indirect causes. Effects and consequence models were derived from wartime and other experience after discussion with representatives of the Ministry of Defence and other explosives experts. The national societal risks, derived for typical routes, provided valuable insights into the levels of risk, the main contributors and how the risk to the exposed populations varied. The risks derived for road transport are higher than for rail and are consistent with the historical record. This reflects both a greater tonnage by road and more explosives with a mass explosion hazard. An important finding of the study is that road users are exposed to the majority of the risk in the road mode whereas it is those living near the route in the case of rail. The most likely cause of an explosion during road transport is fire on the vehicle spreading to the cargo. Measures are proposed to reduce the risk from the transport of explosives in future.

---

[1] The tables and figures were previously published in the Health and Safety Commission publication <u>Major Hazard Aspects of the Transport of Dangerous Substances</u>, Appendix 10 ISBN 0118856766. This material is reproduced with the permission of the Controller of Her Majesty's Stationery Office.

1. INTRODUCTION

Accidental explosions are rare events. When a van carrying explosives caught fire and exploded on an industrial estate in Peterborough, UK (HSE, 1990) the then Transport Minister informed Parliament that a committee was studying the problem would present a report on all forms of dangerous goods in due course (Hansard, 1989). The committee had been charged by the government in 1985 with the task of pursuing the major hazard aspects of transport in the UK to establish whether further controls voluntary or mandatory, were necessary. The committee members were drawn from industry, academia, transport organisations, emergency services, independent consultants and the government authorities both local and national. The Committee's working parties studied chlorine, ammonia, LPG and motor spirit transported by road and rail. The marine working party assessed bulk dangerous chemical cargoes shipped into UK ports. A comprehensive assessment of the transport of explosives was planned for the second phase of the work, to be completed after a first report (HSC, 1991). The parliamentary statement was the spur that, ensured the first report included explosives.

2. THE NATURE OF EXPLOSIVES

The term 'explosives' encompasses a wide range of substances and articles which can produce a pyrotechnic or a practical effect by explosion. They include:

a) Explosive substances, solids or liquids (or mixtures), capable of producing gases by chemical reaction at such temperature, pressure and speed that damage may occur to their surroundings;

b) Pyrotechnic substances designed to produce heat, light, sound, gas or smoke as a result of non-detonative self sustaining exothermic chemical reactions.

c) Explosive articles, those devices where one or more explosives are confined within the same containment.

The wide variety of explosives substances and articles transported by road and rail in the UK and their very different properties required a more elaborate preliminary analysis than had been the case for unique substances in the earlier studies.

Our approach was to categorise explosives into groups with similar properties i.e., approximately the same vulnerability to stimuli and comparable effects on initiation. The categorisation was based upon the United Nations labelling scheme which allocates explosives to "Class 1" (UN, 1988). The Hazard Divisions (HD) within "Class 1" are subdivided as follows:

HD 1.1 Substances and articles which have a mass explosion hazard (i.e., affecting the entire load virtually instantaneously) e.g., dynamite cartridges packed in wooden boxes

HD 1.2 Articles which have a projection hazard but not a mass explosion hazard e.g., mortar bombs packed in metal boxes.

HD 1.3 Substances and articles where the main hazard is from fire with a minor blast or projection hazard or both, but no mass explosion hazard e.g., soft-packaged propellant.

HD 1.4 Substances or articles presenting no significant hazard e.g., boxes of small arms ammunition.

Neither HD 1.5 nor HD 1.6 were transported in the UK at the time of the study and therefore were not taken into account. For completeness, the divisions are included here:

HD 1.5 Substances that are very insensitive but have a mass explosion hazard.

HD 1.6 Articles that are extremely insensitive and do not have a mass explosion hazard.

After consulting experts within the Ministry of Defence (MoD) and the civil explosives industry we chose a seven group scheme. HD 1.1 was further divided into three groups, HD 1.3 into two groups and HD 1.2 and HD 1.4 remained as single groups. The sub division addressed the heat sensitivity of the different explosives. The categorisation scheme is summarised in Table 1.

Table 1: Summary of categorisation scheme.

HD 1.1			HD 1.2	HD 1.3		HD 1.4
M	N	P	Q	R	T	Q
Heat sensitive substances in flammable packaging	Heat sensitive articles - not readily ignitable	Heat insensitive substances	Heat sensitive articles	Heat sensitive substances in flammable packaging	Heat sensitive articles and substances in non-flammable packaging	Heat sensitive articles which present no great hazard

3. RAIL TRAFFIC DATA

Data on the rail movements of explosives was kindly provided by British Rail and show that 8132 wagons loaded with explosives each travelled an average distance of 320 km on the network in the year, April 1988 to 1989. The records were not sufficiently detailed to permit categorisation. Consequently, for one month a log of all explosives movements at the busiest marshalling yard was undertaken by British Rail. Details of the explosives transported were either corroborated from MoD sources or were kindly supplied by Royal Ordnance plc. The explosives movements were sorted into the appropriate groups and then assigned to the four hazard divisions, as shown in Table 2. 44% of the explosives wagons surveyed belonged to HD 1.4 i.e., articles which would not produce a significant hazard in the event of accidental initiation.

Table 2: Breakdown of rail cargoes by hazard division.

Hazard Division	Percentage of Wagons Containing Explosives Belonging to Specified Hazard Division
HD 1.1	21%
HD 1.2	12%
HD 1.3	10%
HD 1.4	44%
HD 1.1/1.3	1%
HD 1.1/1.4	3%
HD 1.1/1.2/1.4	1%
HD 1.1/1.2/1.3/1.4	1%
HD 1.2/1.3	1%
HD 1.2/1.4	4%
HD 1.3/1.4	2%

4.   ROAD TRAFFIC DATA

Many organisations transport explosives by road in the UK. Data was provided by MoD (Army, Navy and RAF) and the principal commercial explosives transport companies. The omissions from our data were:

- movements by small manufacturers. We compensated for this by scaling up the collated data in the light of known movements from manufacturing sites;

- movements where data could not be provided in the timescale. These included movements by visiting U.S. Forces (believed to be few but large) and foreign road haulage companies carrying imports (believed to be small);

- movements on private/light goods vehicles. Such vehicles may carry up to 50 kg (110 lb) of certain types of explosives including HD 1.1 under the terms of the legislation. Data on the annual loaded mileage is not readily available and had to be omitted from our study.

Road cargoes were analysed in a similar manner to rail movements. After grouping the different explosives, they were allocated to the appropriate hazard division as indicated in Table 3. The majority of the road mileage in the UK (94%) is by vehicles carrying some HD 1.1 material, and over half of this is due to loads in the most sensitive group in flammable packaging: Category M.

Table 3: Breakdown of road cargoes by hazard division.

Hazard Division	Proportion of Mileage Covered by Vehicles Carrying Loads Belonging to Specified Hazard Division
HD 1.1	94%
HD 1.2	2%
HD 1.3	4%

5.   INCIDENT FREQUENCIES

Details of explosives events and dangerous occurrences were obtained from a variety of sources, including the MoD, the Explosives Inspectorate and the Explosives Incident Database Service (EIDAS) operated by the Safety and Reliability Directorate (SRD) and the Health and Safety Executive (HSE). Although there have been at least seven incidents in the past 40 years in the UK involving the ignition of explosives on road and rail vehicles, only one caused a fatality (HSE, 1990). We chose the 40 year period to exclude the incidents that occurred in the late 1940's where the influence of less stringent controls on wartime manufacturing and transport procedures may have been a significant factor.

The historical data was used to determine rates for dangerous occurrences involving fire and impact. Conditional probabilities for a subsequent explosive reaction were derived from a combination of accident data, trials data and expert judgement. The initiation of unsafe explosives was also based upon the historical record.

5.1 Impact

Two levels of impact severity were considered in this study:

a) Crashes/collisions of rail and road vehicles and derailment of rail wagons;

b) Crushing of explosives spilled in a previous crash or derailment.

The capacity of correctly packaged explosives to withstand 12 metre drop tests onto a hard, unyielding surface means that for substances and articles the likelihood of impact induced initiation is low. The impact-induced initiation rate derived is:

$$1.10^{-11} \text{ per wagon-km.}$$

Crushing of explosives beneath the wheels of a rail vehicle is more likely to induce initiation. We analysed the signalling systems and emergency arrangements on British Rail and derived a rate for the crushing of explosives by passing trains of:

$$1.10^{-10} \text{ per wagon-km}$$

5.2 Heat

Heat insensitive explosives in a fire are more likely to burn than explode e.g., HD 1.1 substances belonging to Category P. Taking this into account we derived an overall fire-induced initiation rate of:

$$6. \ 10^{-10} \text{ per wagon-km.}$$

5.3 Initiation of Unsafe Explosives

Two events, one road and one rail in the past 40 years and estimated movements of 1.10^9 to 2.10^9 km enabled us to derive an initiation rate of 1.10^{-9} per vehicle-km/wagon-km. Clearly, a more detailed examination of the procedures to prevent unsafe explosives and articles of various kinds from being loaded onto transport vehicles would have been a major study in itself and could not be pursued as part of this work.

5.4 Partitioning Explosives Events

Event trees were used to partition the explosives events by type and quantity of explosives initiated. The overall scheme is complex and shown for rail in Figure 1. Each of the boxes in Figure 1 leads to its own event tree. An example is shown in Figure 2, illustrating the different explosives events that could occur as a result of a fire on an explosives rail wagon. The event trees were quantified from traffic data and analysis of the different types of explosives.

A similar approach was adopted for the road events, the main difference being that instantaneous communication was not considered as only single loads are involved.

6. EXPLOSIVES CONSEQUENCE MODELS

In order to predict the number of fatalities from each explosives event we used the best effects functions available to us. We recognise that they are limited by current knowledge and that research projects in progress at the time of this study may lead to more appropriate functions in the future.

6.1 Hazard Division 1.1 (Mass Explosion Hazard)

HD 1.1 material, when initiated, could result in a mass explosion causing damage and injury due to blast, fragments, ground shock and heat. Experience shows that blast and fragments are the dominant mechanisms of harm from this type of explosion. However, in built up areas it is not the direct effect of blast on people in the open that is the most significant cause of fatalities but the collapse of buildings on the occupants. Naturally, the buildings do provide a degree of protection against fragments for the people inside.

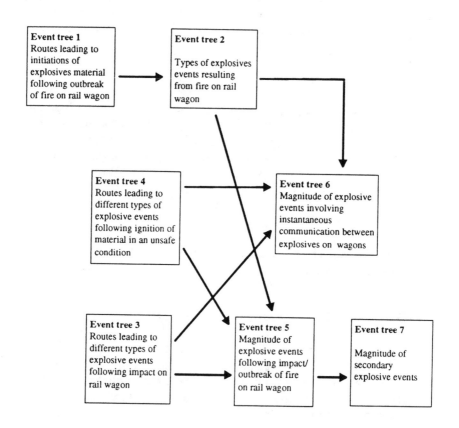

Figure 1: Plan of event trees.

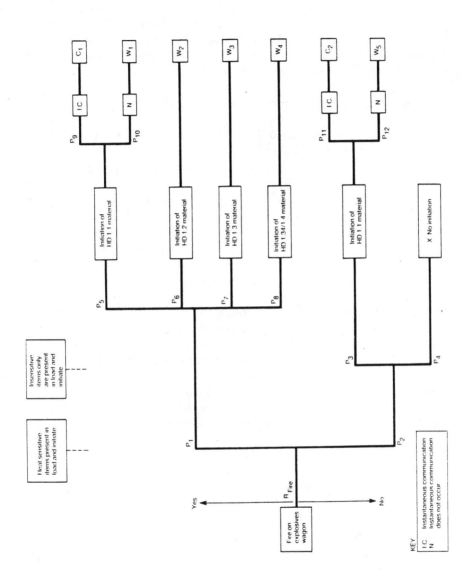

Figure 2: Event tree outlining possible types and sizes of explosives events on rail wagons.

The model we adopted was based on an analysis of 12 wartime V2 rocket attacks (Marshall, 1987). Such missiles hit the ground travelling at a velocity of approximately Mach 3 and thus there was no warning of the attacks. The analysis allowed estimates to be made of the overall percentage of fatalities among groups of people (both indoors and out) at various distances from an explosion of specified size. The results are presented in Figure 3.

6.2 Hazard Division 1.2 (Fragment Hazard)

A cargo consisting of HD 1.2 articles is unlikely to detonate en masse. In the event of initiation, a series of explosions involving a few articles at a time is more likely, creating a fragment rather than a blast hazard. The number of articles involved in each event will depend upon the type of article and the method of packing. For this study we assumed that no more than 50 kg NEQ (Nett Explosive Quantity) of articles would initiate simultaneously. We understand that this is generally conservative. Survivors of the initial explosion were assumed to take shelter before any further events occur, thus reaching a position of reasonable safety.

Fatality probabilities due to fragments for people indoors and outdoors were derived from data kindly supplied by MoD. A lethal fragment was judged to be one with a kinetic energy in excess of 80 J (Kelly, 1979). Two graphs were produced showing fatality as a function of distance. (As the data from which the graphs were produced is classified confidential they are not reproduced here). However, for people in the open at 200m, the average fatality probability is 0.02 and for those indoors an order of magnitude lower.

There is a considerable variation in fragment densities from HD 1.2 articles. Our analysis is based upon a more energetic type of munition and although the results have been "factored down", an element of conservatism in the analysis may remain.

6.3 Hazard Division 1.3 (Fire Hazard)

Ignition of HD 1.3 substances and articles could give rise to one of two types of fire, idealised and non-idealised. In an idealised fire, the whole mass of explosive burns simultaneously and it is over very quickly. A non-idealised fire which is hampered by the packaging and spacing between packages burns for much longer and in an extreme case could give rise to sequential fires of one article at a time.

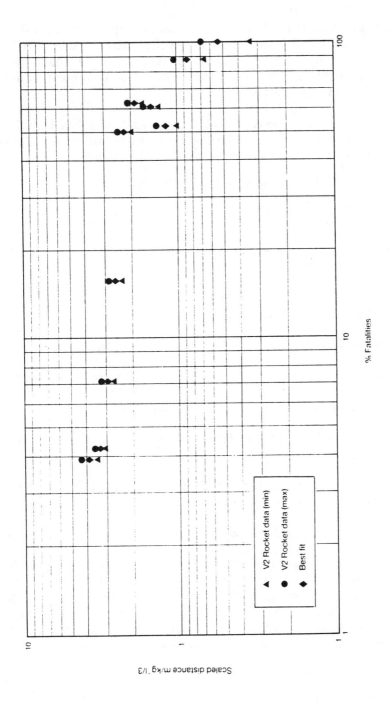

Figure 3: HD 1.1 mass explosion hazard to people.

We judged the probability of an idealised fire involving HD 1.3 substances packaged in soft materials such as cardboard to be unity. We assumed HD 1.3 substances packed in metal boxes and cartridges of HD 1.3 would invariably give rise to a non-idealised fire in the event of ignition.

Fatality probabilities were derived from models developed by SRD for the two types of fire. Radiation doses were related to fatality probability using the Eisenberg probit (Eisenberg, 1975).

$$Y = 2.56 \ln (\text{dose}) - 14.9$$

where the dose is expressed in s $(kWm^{-2})^{4/3}$

The graph for 99%, 50% and 1% lethality from idealised fires is shown in Figure 4.

7.   HAZARD ANALYSIS

The explosives consequence models were applied by determining:

a)   the various sizes of explosives loads carried by road and rail;

b)   the hazard ranges for the different load sizes;

c)   estimates of the population densities bounded by the hazard ranges;

d)   estimates of the numbers of fatalities within each hazard range in the event of an explosion.

7.1 Typical Load Size Selection

It was not a practical proposition to calculate the hazard ranges for all the explosives loads transported by road and rail. Therefore, we derived representative groups of loads for notional sizes of cargo. The technique chosen was the so called 'maximum entropy analysis' (Johnson, 1979) (where entropy is a quantitative measure of the uncertainty expressed by a probability distribution).

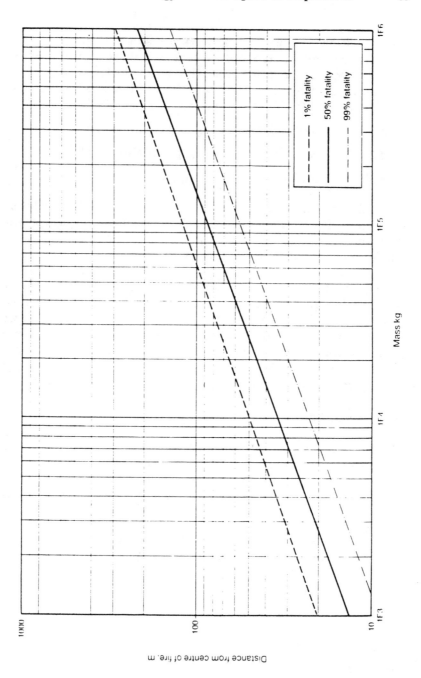

Figure 4: HD 1.3 thermal radiation hazard from an idealised fire to people outdoors.

The conditions of the analysis were set so as to split the data on various sizes of road and rail cargoes into three bands. Each band was assigned an average quantity of explosive (NEQ) and an associated probability. An example of this analysis for road vehicles transporting HD 1.1 explosives is given below:

Band 1   Mean NEQ = 316 kg probability = 0.18

Band 2   Mean NEQ = 1778 kg probability = 0.65

Band 3   Mean NEQ = 14125 kg probability = 0.14

Road vehicle loads do not exceed 16000 kg NEQ (72581b) those for rail vehicles do not exceed 20000 kg (9072 1b). However, the potential for instantaneous communication between explosives loads on rail vehicles means that larger events are possible.

In some cases, the data sets contained no more than two values of load size, in which case partitioning into three bands was clearly inappropriate. The results of the maximum entropy analysis were used to quantify the final nodal probabilities in the event trees. Figure 5 provides an example where the frequency estimates for various events on road vehicles are divided according to the quantity of explosives initiated.

7.2 Hazard Ranges

For each of the representative explosives loads moved by road and rail the hazard ranges were determined using the consequence models. For example, the hazard ranges from the blast effect model for the three representative load sizes to 90%, 50% and 10% fatalities are as follows:

	316 kg	1770 kg	14125 kg
L90	6 m	10 m	21 m
L50	14 m	26 m	51 m
L10	18 m	32 m	64 m

As with the other en-route consequence calculations, we considered that the majority of fatalities would occur within the 10% contour. We recognized fatalities could occur at greater ranges and the recommended evacuation distance for 14 tonnes of high explosives is 600 metres (MoD, 1989). Similar tables were constructed for the representative loads carried on all road and rail vehicles (except HD 1.2 where a limit of 50 kg NEQ was applied).

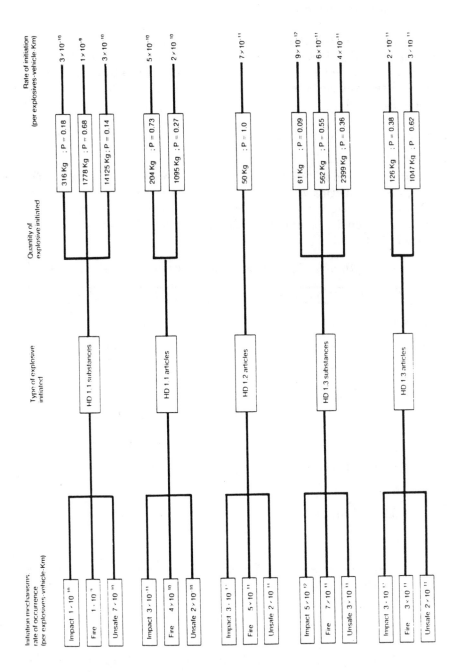

Figure 5:   Event tree outlining possible types of sizes of explosives
events on road vehicles.

7.3 Population Densities

We first considered the residential population living alongside the transport route. Few people live within 25 metres of a rail track, we therefore assumed a 'clear zone' on both sides of the track. There is a similar separation between roads and houses but the width of the zones is smaller.

In calculating the population densities we adopted the four categories:

Urban, suburban, built up rural and rural (HSC, 1991). The population densities were 4210, 1310, 210 and 20 people per square kilometre respectively. We also took account of the variation in population density on opposite sides of each route, as shown for a rail route in Table 4.

Table 4: Length of various sections of representative rail - route A.

Type of Off-Rail Population	Length of Section
URB 2 sides	9.13 km
URB 1 side/Sub 1 side	18.26 km
URB 1 side/BUR 1 side	6.09 km
URB 1 side/RUR 1 side	7.61 km
Sub 2 sides	9.64 km
Sub 1 side/BUR 1 side	22.21 km
Sub 1 side/RUR 1 side	21.50 km
BUR 2 sides	29.41 km
BUR 1 side/RURAL 1 side	71.00 km
RUR 2 sides	125.16 km
Total	310.00 km

The fatalities in the residential population (N) were calculated on the following basis:

$$N = D [0.95 A_{90} + 0.7 (A_{50} - A_{90}) + 0.3 (A_{10} - A_{50})]$$

Where D is the population density and Ax is the area bounded by the Lx hazard contour and the clear zone along the route. We assume that between X% and Y% lethality contours (X + Y)/2% fatalities will occur.

On the road route, the off-road population varied between the different sections, as shown in Table 5.

Table 5: Length of various sections of representative road route.

Road Class	Type of OFF-Road Population	Length of Section
Motorway	Sub 2 sides	0.28 km
	Sub OS/BUR NS	0.28 km
	Sub NS/BUR OS	0.84 km
	Sub OS/RUR NS	0.84 km
	Sub NS/RUR OS	0.56 km
	BUR 2 sides	0.14 km
	BUR OS/RUR NS	3.08 km
	BUR NS/RUR OS	2.58 km
	RUR 2 sides	13.16 km
Dual Carriageway	URB NS/SUB OS	0.28 km
	URN BS/BUR OS	0.56 km
	Sub 2 sides	0.84 km
	Sub OS/BUR NS	0.56 km
	Sub NS/BUR OS	0.84 km
	Sub OS/RUR NS	0.28 km
	Sub NS/RUR OS	0.67 km
	BUR 2 sides	0.84 km
	BUR OS/RUR NS	0.28 km
	BUR NS/RUR OS	1.12 km
	RUR 2 sides	2.18 km
Single Carriageway	URB NS/BUR OS	0.56 km
	Sub 2 sides	0.11 km
	Sub OS/BUR NS	1.12 km
	Sub NS/BUR OS	0.84 km
	BUR 2 sides	6.72 km
	BUR OS/RUR NS	4.82 km
	BUR NS/RUR OS	4.76 km
	RUR 2 sides	37.36 km
	Total	86.50 km

We also considered the extent to which rail passengers and road users could be involved in an explosives incident. Data supplied by British Rail enabled us to determine the population densities on trains travelling the routes used by explosives wagons. On the road, we assumed that traffic would be held up on the carriageway behind an incident. On single carriageway roads, incidents were assumed to block the road in both directions and the same population density was applied to the queue of traffic on both sides of the road. Generally, dual carriageways and motorways have a central barrier. Traffic was assumed to continue flowing on the other side of such roads but at a reduced density to allow for slowing of the moving traffic due to the "ghoul effect".

In calculating fatalities among train passengers we considered three scenarios:

i)   a train is stopped at a signal and is affected by an explosion near by;

ii)   a passenger train collides with an explosives train on the same track;

iii) a passenger train passes over explosives spilled in a previous collision or derailment.

For cases (ii) and (iii) only the passengers within the hazard range of the explosion are assumed to be affected. In case (i) passengers will only be affected if the hazard range exceeds the distance to the signal. If this occurs, fatalities are assumed to arise in the coaches within the hazard range. For example:-

$$N_{50} = D \, (INT \, Nc \, F_{50} + 0.5)$$

where N_{50} is the number of passengers within the L_{50} Contour

       D is the density of the passenger population
       Nc is the number of carriages on the train
and F_{50} is the proportion of the train within the L_{50} contour.

The total numbers of fatalities was given by:

$$N = 0.95 \, N_{90} + 0.7 \, (N_{50} - N_{90}) + 0.3 \, (N_{10} - N_{50})$$

This assumption is conservative for explosives such as HD 1.1 since it assumes a linear relationship with distance whereas blast overpressure decays exponentially.

8.   RISK ANALYSIS

The main aim of this analysis was to assess the national risk. Therefore, after consultation with British Rail we selected two routes A and B to reflect the differing patterns of rail traffic. The routes are not identified for security reasons but include both intercity and provincial traffic. On route A the former predominates and on route B the latter, plus some freight-only track.

Road routes tend to be of two types: trunk deliveries between manufacturing sites and central storage depots, and local deliveries between the depots and customers premises. The representative road route chosen is operated by Nobel's Explosives Company. Although a trunk route it has many of the geographical features of a local delivery journey. Again the route is not identified for security reasons.

8.1 National Societal Risk of Transporting Explosives by Road

The representative route was partitioned into different sections according to road type and off-road population density, as indicated in Table 5. For each of the 28 sections, 11 representative types and size of cargo were calculated giving a total of 308 frequencies for explosives events for which the corresponding fatality estimates were derived. The spreadsheet enabled the fatalities in the on-road and off-road populations to be addressed separately as well as the total in each event.

The national societal risk was derived in a two stage process:

i) Each section of the route was taken in turn and the fatalities summated for events producing numbers of fatalities exceeding the following numbers:

>1, >3, >10, >30, >100, >300, >1000

(An example of the computation is given in Figure 6).

ii) the frequency totals for each section are summated to give the overall frequencies for the events leading to the numbers of fatalities in the various bands considered. The spreadsheet also calculates the contributions to the overall risk from the on-road and off-road populations, as indicated in Table 6. This demonstrates clearly that road users experience a higher risk than the population living along the route as shown in Figure 7. This arises because routes taken by explosives vehicles are selected to avoid population centres as much as possible.

National risk		
Two sided population		
Dual Carriageway	All population	
Population Density:	Nearside: Offside:	4.2e-3/sqm 2.1e-4/sqm
Length of pop.den:		.56 km
Length of route:		86.5 km
No. of movements:		46243 yr-1

		Frequency	Fatalities
Init'n HD 1.1 Substances		8.890e-6 3.3583-5 6.914e-6	8 21 55
Init'n HD 1.1 Articles		1.299e-5 4.806e-6	6 16
Init'n HD 1.2 Articles		1.818e-6	9
Init'n HD 1.3 Substances	Idealised	2.349e-7	1
	Idealised	1.436e-6	5
	Idealised	9.398e-7	16
Init'n HD 1.3 Articles	Non-Idealised	5.366e-7	5
	Non-Idealised	8.754-7	15

Societal risk			Frequency of N or more fatalities				
N	> = 1	> = 3	> = 10	> = 30	> = 100	> = 300	> = 1000
	72.03	72.79	47.12	6.91	.00	.00	.00

Figure 6: Explosives road risk analysis.

8.2 UK National Societal Risk of Transporting Explosives by Rail

The representative rail routes were partitioned into different sections according to the passenger traffic density, i.e., intercity, provincial and freight only track which carried no passengers. For each section of the route, 8 representative types and size of cargo for single wagons were calculated, and a further 32 events which represented the incidents involving a whole train. Different spreadsheets were constructed for the rail passenger and off-rail populations. The rail passenger spreadsheet took account of passenger trains stopped at nearby red signals as well as those directly involved with the explosives train or spilled explosives in an incident. Thus the societal risks to the on-rail and off-rail populations were derived separately. A further programme was written to calculate the total societal risk, distinguishing those events where fatalities were predicted for both the on and off rail populations from events where no passenger train was present and only off-rail fatalities occurred.

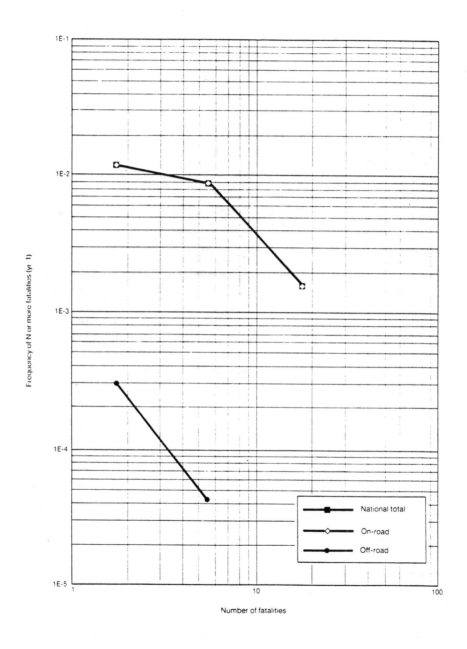

Figure 7: National societal risk - Transport of explosives by road.

Table 6: National societal risk for transport of explosives by road.

N	Frequency of N or more fatalities (10^{-6} yr^{-1})		
	Off-Road	On-Road	National Total
1	769	11279	11279
3	112	11243	11243
10	17	6683	6712
30	0	373	381
100	0	0	0

The national societal risk was derived by scaling up the rail traffic on the selected routes to the volume of national explosives traffic and the average length of route (320 km). The results for Routes A and B are shown in Tables 7 and 8 respectively. The results are remarkably similar, particularly for the risks to the off-rail population as Figures 8 and 9 indicate. The slightly higher passenger risks on Route A reflect the greater volume of passenger traffic along this route, but in both cases the risks to the off-rail population are the dominant contribution to the overall risks. This reflects two factors: firstly, the low probability that an accident involving an explosives train would also involve a passenger train and secondly the higher proportion of track in urban areas as compared with the road route.

Table 7: National societal risk for transport of explosives by rail (obtained by scaling up from route A).

N	Frequency of N or more fatalities (10^{-6} yr^{-1})		
	Off-Rail	Passengers	National Total
1	619	7	626
3	304	7	311
10	105	6	111
30	0	2	2
100	0	0	0

Table 8: National societal risk for transport of explosives by rail (obtained by scaling-up from route B).

N	Frequency of N or more fatalities (10^{-6} yr^{-1})		
	Off-Rail	Passengers	National Total
1	664	4	668
3	427	4	431
10	207	3	210
30	0	1	1
100	0	0	0

9. COMPARISON OF RAIL AND ROAD RISKS IN THE UK

The results show higher overall levels of risk from road transport; approximately an order of magnitude greater than the risks from rail transport. The annual probability of a fatal explosives incident involving rail transport is assessed at about 10^{-3}. Comparing this with actual experience, there has been only one fatal initiation of an explosives load on a road vehicle in the past 40 years in the UK and none on rail wagons.

One of the main conclusions of the study is that it is inappropriate to draw conclusions regarding the relative safety of the two modes for the following reasons:

• Different types and quantities of explosives are moved by the two modes of transport. Most of the explosives transported by rail are almost entirely military. In contrast, over half of the explosives loads in our road data were civil explosives. In practice, this means a much higher ratio of HD 1.1 to HD 1.2 and HD 1.3 are moved by road;

• Road explosives traffic exceeds rail traffic by a factor of two or three in terms of both mileage and tonnage;

• Differing proportions of risk apply to rail passengers and other road users. A more meaningful comparison of inherent risk is obtained by calculating the societal risk per tonne-km as shown in Table 9. Even then, the differences due to the different types of explosives remain.

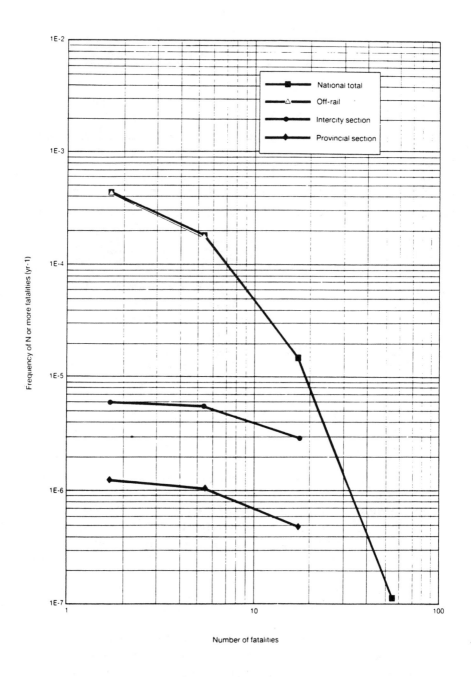

Figure 8: National societal risk - Transport of explosives by rail - Route A.

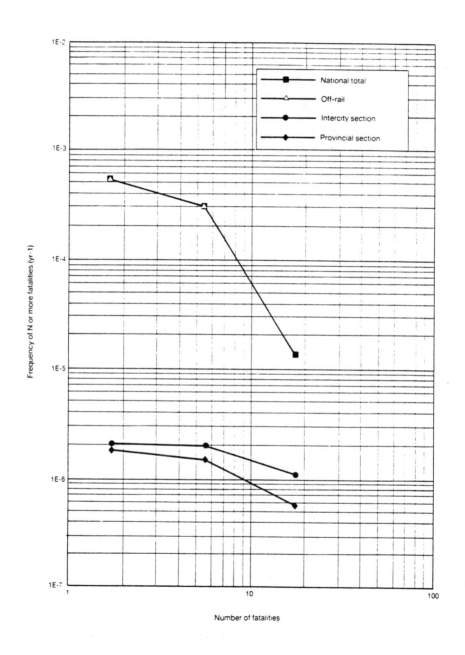

Figure 9: National societal risk - Transport of explosives by rail - Route B.

Table 9: Comparison of estimated societal risk for transport of explosives.

Frequency of N or more fatalities per tonne km year			
	N = 1	**N = 3**	**N = 10**	**N = 30**
Road	3×10^{-9}	3×10^{-9}	2×10^{-9}	1×10^{-10}
Rail Route A	1×10^{-10}	6×10^{-11}	2×10^{-11}	4×10^{-13}
Rail Route B	1×10^{-10}	8×10^{-11}	4×10^{-11}	2×10^{-13}

Higher levels of risk are obtained for road transport and this is due to two factors:

- the higher rate of events initiated by fire for road vehicles: 2×10^{-9} per vehicle-km compared with 7×10^{-10} per wagon-km for rail wagons. As this is the dominant mode of initiation, this is particularly important. Overheated tyres are considered to be the most likely cause of fires on explosives vehicles.

- An explosion on a road vehicle is highly likely to involve fatalities among other road users, particularly on a busy motorway, whereas an explosion on a rail line is much less likely to affect rail passengers. The risks to off road and off rail populations are broadly similar.

A further contrast between the two modes is the potential for communication between explosives loads on rail wagons. In the case of loads of HD 1.1 this communication may be instantaneous giving rise to much larger events following a rail accident.

10. UNCERTAINTY IN THE RISK ESTIMATES

The study has had to rationalise a very complex overall pattern of traffic. This has inevitably introduced an element of uncertainty into the analysis. The results are also subject to statistical uncertainty particularly as some of the accident rates are derived from very small numbers of events.

We have not been able to quantify confidence limits for our overall risk estimates but analogy with other studies leads us to believe the limits will be approximately plus or minus an order of magnitude.

The difficulty in defining such limits does not detract from the usefulness or validity of using the results to compare levels of risk within the analysis or to evaluate risk reduction due to remedial measures.

11. RISK REDUCTION MEASURES PROPOSED FOR FURTHER STUDY

As fire is the dominant initiation mechanism for road transport, this suggests that the following proposals be considered:

* improved fire protection of the load carrying compartment of explosives heavy goods vehicles to a standard which will withstand fire attack from fuel and tyre fires to allow time for emergency response;

* active and passive systems to prevent the spread of fire inside the load carrying compartment: e.g., automatic fire extinguishing systems and fire-resistant packaging;

* phasing out nitro glycerine - based blasting explosives and replacement with explosives that are less sensitive to fire.

En route risks on the rail network are low and therefore the scope for justifiable expenditure on further precautions is limited. The following proposals should be considered:

* Combustible materials such as wooden panels and dunnage should be removed from rail wagons and replaced with non-combustible items;

* Effective separation of wagons carrying HD 1.1 explosives to prevent instantaneous communication. Barrier wagons may be used to achieve this, carrying inert loads or less hazardous explosives e.g., HD 1.4.

Finally, we have already referred to the uncertainties to which our estimates are necessarily subject. In view of this we would emphasize that our risk estimates should be used with care and not taken out of context of this study.

Acknowledgements I would like to extend my thanks to my colleagues Dr. Peter Moreton of SRD, Mr. Grant Purdy of Technica for their support in this work. I would also like to thank the members of the ACDS Major Hazard Transport Sub-committee and its technical working party without whose cooperation this work would not have been possible.

12. REFERENCES

EISENBERG N.A. et al. (1975). Vulnerability Model. A Simulation System for Assessing Damage resulting from Marine Spills. US Coastguard, Nat. Tech. Inf. Service. (AD-A015, 245).

Hansard 23 March 1989 p126.

HEALTH AND SAFETY COMMISSION (1991). Major Hazard Aspects of the Transport of Dangerous Substances, HMSO.

HEALTH AND SAFETY EXECUTIVE (1990). The Peterborough Explosion, HMSO.

JOHNSON, R. W. (1979). Determining Probability Distributions by Maximum Entropy and Minimum Cross Entropy, APL79, Rochester, Proceedings.

KELLY, P.G. (1979). Fragment Injury Criteria. US(ST) - 1WP/11-79.

MARSHALL, V C (1987). Major Chemical Hazards, Ellis Norwood.

MoD (1989). Hazardous Load Warning Sheet for Road Movement of Explosives. F/Mov 774A Revised 3/89.

UNITED NATIONS (1988). Recommendations on the Transport of Dangerous Goods, 5th rev. ed., New York, (ST/SG/AC.10/1/rev 5).

Rail Transport Risk in the Greater Toronto Area

E. Alp
R.V. Portelli
W.P. Crocker
Concord Environmental Corporation
Downsview, Ontario
M3H 2V2

ABSTRACT

A major quantitative risk assessment was conducted for the Canadian Federal Government to assess the transportation of dangerous goods in the Greater Toronto Area. A detailed description is provided on the methodology used in the frequency and consequence analyses. The dangerous goods considered were gasoline, propane, chlorine, ammonia, oleum, explosives and phosphorus. Toxic, fire and explosion hazards were identified and the consequences were quantified using mathematical models. The effects analysis was performed using probit equations. Emphasis of the study was to evaluate the relative risk to the public from possible alternatives to the current system. Results of the study are presented in the form of societal and individual risk.

1. INTRODUCTION

Hazardous materials management in an industrial society is becoming increasingly important due to increased use of hazardous materials and the increased demand for disposing of hazardous waste products. Safe transportation of hazardous goods is an integral part of hazardous materials management, and achievement of higher safety levels requires systematic identification of the areas in the transportation system components which contribute most to risk and their subsequent modification to reduce risk for more effective management.

Risk is a measure that includes both the consequences of a hazardous release event and its frequency of occurrence. In simple terms it can be expressed as:

$$\text{Risk} = \text{Frequency} \times \text{Consequences} \qquad (1)$$

Previous studies have attempted to provide some preliminary estimates of risk due to dangerous goods (DG) transport in the Greater Toronto Area (GTA). These are the studies of Wade (1986) and Burton and Post (1983), which used qualitative methods for arriving at risk estimates. However, in view of the concerns for public safety, the present study was commissioned by the Canadian Federal Government for a quantitative risk assessment. Application of quantitative risk assessment techniques has become an internationally accepted and integral part of the determination of the safety of the public exposed to major hazards from industrial operations such as dangerous goods production, storage and transportation. The work presented herein constitutes a thorough quantitative risk assessment and as such is a major step forward from previous studies. The main objective of this quantitative assessment was to provide the "hard" information base to make risk management decisions regarding various alternatives such as new rail routes, operational changes and technology improvements.

For an overview of the method used in conducting this study, refer to Figure 1, where the risk assessment methodology is depicted in simplified form. The process begins with the identification of hazardous release events. Then, estimates of the consequences and frequencies for these events are made. These results are then used to calculate risk for each of the rail segments along the routes within a particular rail system alternative.

The first set of analyses is done for the existing rail system. Risk estimates for this case are considered to be the existing or baseline risk. The analysis is then repeated for the alternative routes and operating strategies and other risk mitigation measures. The alternatives are shown in Figure 2 and described in Table 1. Risks associated with the various alternatives can then be examined relative to the baseline risk.

There is a large variety of hazardous events with a continuum of consequences. For a risk assessment study of this type, where the emphasis is to provide estimates of relative risks of various alternatives for the purpose of making risk management decisions, it is not

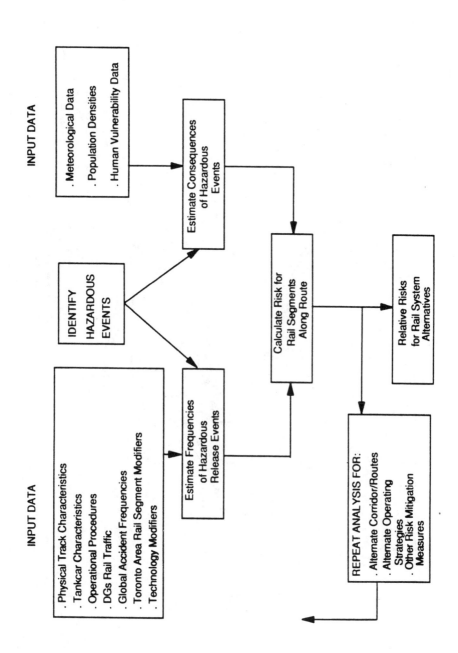

Figure 1: Risk assessment methodology.

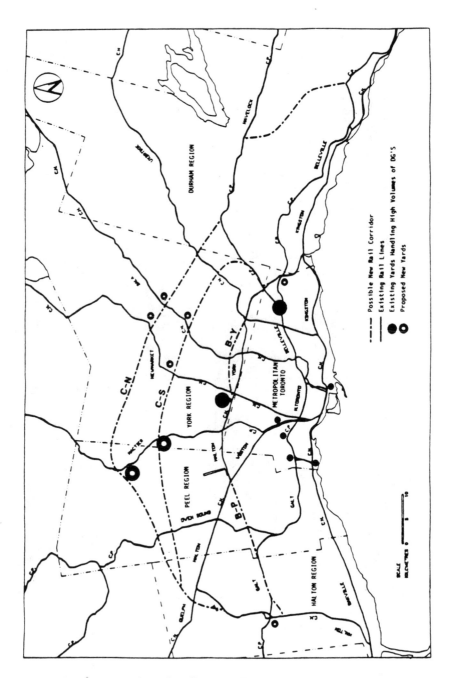

Figure 2: Region of study: Greater Toronto area.

Table 1: Rail system alternatives.

Alternative	Network	CN Traffic Routing	CP Traffic Routing
A1*	Existing	Existing	Existing
A2*	Existing with additional connections	Existing (A1)	Diversion of CP mainline DG trains to Halton/York Subs
B1A	Route B-P (Existing CN York Sub)	Milton cutoff for all through Halton Sub trains	Route B for mainline DG trains only, Agincourt yard
B1B	Route B-Y (with Thornhill diversion)		
B2A*	Route B-P (existing CN York Sub)	Same as B1	Route B for all mainline trains, Agincourt yard
B2B*	Route B-Y (with Thornhill diversion)		
B3A	Route B-P (existing CN York Sub)	Same as B1	Route B for all mainline trains, new Woodbridge yard
B3B	Route B-Y (with Thornhill diversion)		
C1*	Route C-N (north route)	Route C-N for mainline trains carrying DG's, setoff through DG cars at Vandorf & Cherry to avoid brining them into MacMillan yard	Route C-N for all mainline trains, new yard in Bolton area
C2	Route C-N (north route)	Route C-N for all mainline trains, new yard in Vandorf area	Same as C1
C3*	Route C-S (south route)	Route C-S for mainline trains, carrying DG's, setoff through DG cars at Elgin and North Maple to avoid bring them into MacMillan yard	Route C-S for all mainline trains, new yard in Bolton area
A1T*	Existing with new technology	Existing	Existing

* Indicates the alternatives for which detailed risks were assessed.

necessary to examine each and every possible hazardous event in detail. It is sufficient to group these events and identify a representative hazardous event for that category. To this end, release events that could lead to similar consequences were grouped together, and surrogate events, which would conservatively represent (i.e., not underestimate) the consequences of the events in their respective categories, were selected for each category. This grouping process is an essential part of performing risk assessments.

Selection of representative chemicals (see below) and meteorological parameters was carried out in a manner consistent with the approach of selecting representative hazardous events.

The above provides a brief overview of the risk assessment methodology. What follows is an explanation of each of the steps discussed above in more detail.

2.   IDENTIFICATION OF HAZARDOUS EVENTS

The potential hazards with which this study is concerned are acute hazards:

- acute toxicity (due to toxic clouds),

- flammability (due to flammable clouds),

- thermal radiation (due to jet and pool fires and fireballs),

- blast wave (from VCEs, BLEVEs, detonation, confined explosions), and

- missile damage.

In this study, a total of seven dangerous goods categories have been utilized. The seven chemicals representing these seven categories are listed below. All of these chemicals are in the TIPS list of priority chemicals (Environment Canada, 1985).

- Ammonia

- Chlorine

- Explosive (black powder)

- Gasoline

- Oleum (grade 30%, i.e., 30% excess free SO_3; this is the highest grade transported by rail (Environment Canada, 1984)).

- Phosphorus

- Propane

In grouping the Canadian TDGR classes into these categories, classes or divisions whose expected consequences are similar, were grouped together. In addition, some classes with low traffic volumes were grouped with classes with similar, but more severe potential for harm.

Within each category, chemical selection was based on the chemical (i) having a large traffic volume and (ii) posing a more severe potential hazard than the other chemicals in the category.

A breach of containment for rail tankcars carrying dangerous goods is initiated through three types of accident-induced loadings: crush load, impact load and puncture probe. In addition to these loadings, a breach can also take place as a result of direct fire-induced damage to the container and overpressure in the container caused by heat transfer. The causes of these types of releases can be determined by developing fault trees, with the loss of containment from a rail tankcar as the undesired event. Figure 3 shows a simplified form of the detailed fault tree developed for this study.

Table 2 shows how chemicals belonging to all TDGR (Transport of Dangerous Goods Regulations) classes/divisions were grouped into one of the seven chemical categories represented by each of the seven chemicals.

Releases were postulated to occur during normal transport or following transportation accidents. Given that a transportation accident occurs, it must be of sufficient severity to cause a release. A review of the detailed fault tree allows sequence of events to be identified, each of which would lead to a release. For each release sequence, the relative magnitude of the release rate was also deduced.

The release rate categories used in this risk assessment are:

- Small release rate - no external fire present (SRR/NEF): This category includes leaks from valves, fittings and the tank shell during normal transport and also releases from damaged valves due to transport accidents. It was assumed that a 0.02 m diameter hole at the bottom of the tank characterizes the representative release rate for this category. It was also assumed that the whole tank cargo is released from a stationary railcar.

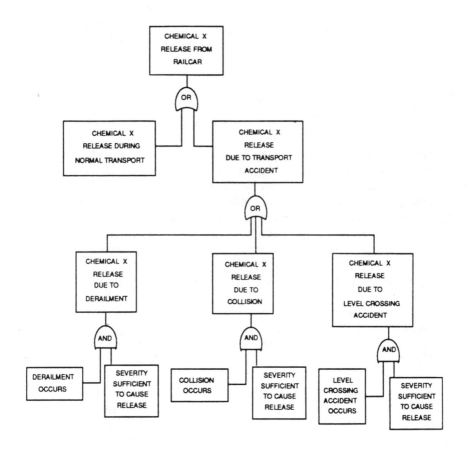

Figure 3: Top level of fault tree.

Table 2: Selected representative chemicals.

TDGR Class	Class Description	Chemical Selected	Comments
1	Explosives	Black Powder	Model blast wave.
2.1	Flammable Liquefied Gases	Propane	Model heavy gas/mist cloud dispersion, jetfire, fireball.
2.2	Liquefied Gases	-	Group with 2.1
2.3	Toxic Gases	-	Group with 2.4 as chlorine.
2.4	Corrosive Gases	Chlorine and Ammonia	Cl_2 Traffic known, remainder of class 2.4 taken as NH_3. Model heavy gas/mist cloud dispersion.
3.1	Flammable Liquids (F.P. $<-18°C$)	Gasoline	Model pool fire, jetfire.
3.2	Flammable Liquids ($-18 <$ F.P. $< 23°C$)	-	Group with 3.1
3.3	Flammable Liquids (F.P. $> 23°C$)	-	Group with 3.1
4.1	Flammable Solids	-	Group with 4.2
4.2	Spontaneously Flammable Solids	Phosphorus	Model P_4O_{10} cloud dispersion.
4.3	Substance that on contact with water emit flammable gases	-	Group with 2.1
5	Oxidizing substances and organic peroxides	-	Group with "Special Dangerous Commodity" Traffic (as H_2O_2) with 6.1; model H_2O_2 mist as SO_3.
6.1	Poisonous substances	oleum/SO_3	Model H_2SO_4 heavy mist cloud dispersion.
6.2	Infectious substances	-	Not addressed.
7	Radioactive Materials	-	Not addressed.
8	Corrosive substances	-	Not addressed.
9	Miscellaneous	-	Not addressed.

Notes: F.P.   = Flash Point
TDGR   = Transport of Dangerous Goods Regulations

- Large release rate - no external fire present (LRR/NEF): This category includes breaches of containment that result from transport accidents (derailments, collisions and level crossing accidents). It was assumed that a 0.30 m diameter hole at the bottom of the tank characterizes the representative release rate for this category. Catastrophic tank failure in non-fire situations is remote and as such was not selected to define the representative release rate. It is assumed that the whole tank cargo is released.

- Small release rate - with external fire present (SRR/WEF): This category covers releases from the pressure relief valves (PRV) in fire situations. PRV's are set to open at 75 percent of the tank test pressure as specified by Canadian Transport Commission (CTC) regulations. It was assumed that the release occurred from a fully open PRV, that remained fully open for the duration of the release. It is assumed that only half of the tank cargo is released. If the external fire continues after half of the tank cargo has been vented, at that point catastrophic failure of the tank is assumed.

- Large release rate - with external fire present (LRR/WEF): This category covers catastrophic tank ruptures induced by external fires. An example of this event is a BLEVE, which may involve any of the pressure-liquefied gases. Full-scale fire tests on rail tankcars reveal that the tanks fail from localized weakening of the shell when the tank is about half full of liquid (Manda, 1975). Thus, it is assumed that the tanks will rupture at half-full.

Releases from residue cars were felt to be significant only for ammonia, chlorine, propane and gasoline. The ammonia and chlorine releases were treated in a similar manner as the full cars (except the release quantities). For propane and gasoline residue cars, the release event is taken as a release of energy that could occur from a confined explosion caused by a reversing flame.

Some information has already been presented about the various types of hazards that could occur from various types of releases. In this risk assessment, event trees were used to identify the hazards that could result from each representative release rate.

Figures 4 and 5 show event trees for propane SRR/NEF and LRR/NEF events. It can be seen from these figures that not only are hazards chemical-dependent, they are also release rate-dependent.

Thus a hazardous event, H, includes both hazardous releases from containers resulting from accidents, and external factors which affect the eventual outcome.

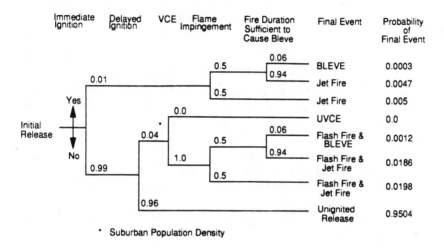

Figure 4: Event tree for propane - SRR/NEF.

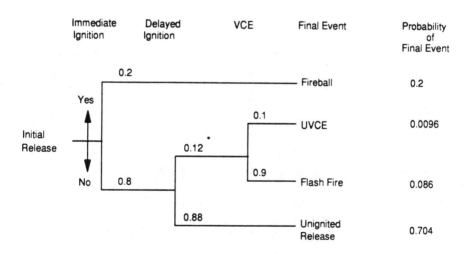

Figure 5: Event tree for propane - LRR/NEF

3. FREQUENCIES OF HAZARDOUS EVENTS

The frequency of a hazardous event ($F_H{}^*$) is dependent on the occurrence of an accident (F_A) of sufficient severity (P_S) to cause a hazardous release that is influenced by external factors (P_E). In mathematical terms it can be expressed as:

$$F_H{}^* = F_A\, P_S\, P_E \qquad (2)$$

The frequencies of occurrence of accidents were determined from a statistical analysis of accidents from Canada-wide data that were modified to reflect Greater Toronto Area track conditions. These frequencies were then input to the fault tree along with post-accident severity probabilities to estimate the frequency of occurrence of each type of release rate. These release rate frequencies were then input to event trees, which incorporate the influence of external factors to estimate the frequencies of occurrence of hazardous events.

Some hazardous events, namely toxicity hazards and the flammability hazard of propane are strongly influenced by the prevailing meteorology, thereby affecting the strength of the hazard (e.g., toxic load) at a point (x,y) away from the affected tankcar. The frequency of a hazardous event at that point, F_H, for these hazards is dependent on the joint frequency of occurrence of the hazardous event and weather condition, P_W:

$$F_H = F_H{}^* P_W \qquad (3)$$

The weather condition, in turn, is the joint probability of occurrence of the representative meteorological categories.

For hazardous events not dependent on meteorology:

$$F_H = F_H{}^* \qquad (4)$$

Figure 6 shows the risk assessment methodology used in this study and includes the frequency assessment methodology.

The starting point of the frequency assessment is the global accident frequencies. These include transport accidents and leaks during normal transport (any release of a dangerous good is reportable under CTC regulations).

The types of transport accidents considered (as opposed to leaks during normal transport) are derailments, collisions and level crossing accidents for mainline operations (which includes both mainline and branchline), and derailments and collisions for yard operations.

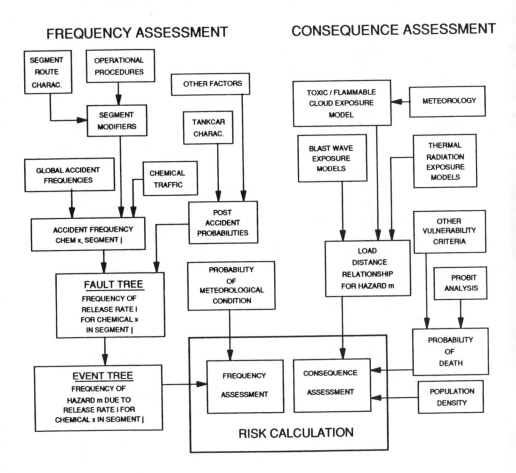

Figure 6: Risk assessment components.

Transport accident data for the Canadian railroad network (CN and CP networks only) were obtained from an analysis of the CTC accident data by grouping them into several dozen accident cause classes developed for the present study for the years 1980-1985.

The global accident frequency data prepared for this study represent average conditions for the Canadian rail network. In order that the transport accident frequencies reflect the local track infrastructure and operating strategies within the GTA, segment modifiers were developed for track characteristics, speed and signal system after examining the rail system and dividing it into about fifty segments of approximately uniform characteristics from the point of view of accident potential (Uniform Accident Environment or UAE segments).

The next step in the analysis is to combine the global accident frequencies and the segment-specific modifiers together with the chemical traffic on that segment and estimate the chemical-specific and segment-specific accident frequencies. For this purpose, an accident frequency matrix was developed on a spreadsheet to organize the frequency data and compute these accident frequencies. The computing steps for derailments and collisions and for level crossing accidents are described below.

Derailments and Collisions: Global accident frequencies that are expressed in various divisor units (train-miles, car-miles and locomotive-miles) depending on the accident cause were all converted to car-miles using appropriate multipliers estimated for each UAE segment. These accident frequencies ($F_{i,j}$) were then modified by segment-specific frequency modifiers to get modified accident frequencies and then multiplied by the modified chemical traffic as shown below:

$$F_{i,j,x} = [F_{i,j} \bullet MTS \bullet MSP \bullet MSG] \bullet [T_\chi \bullet MTR]/L_S \qquad (5)$$

where,

$F_{i,j,x}$ = Frequency of a derailment or collision due to accident cause i, for chemical χ in segment j (#/year-km)

$F_{i,j}$ = Pre-modified frequency of a derailment or collision due to accident cause i in segment j (#/car-mile)

MTS = Track modifier

MSP = Speed modifier

MSG = Signal modifier

T_χ = Chemical χ traffic (car-miles/year)

MTR = Traffic modifier

L_S = UAE segment length (km)

The frequency modifiers have been estimated using segment-specific characteristics and simple models developed based on expert judgement (CSC, 1987).

For each segment the traffic volume for each chemical χ was obtained by summing together the traffic for each "traffic group" that makes up the chemical category represented by the chemical χ.

Level Crossing Accidents: Accident data for level crossing accidents were segregated by protection type. The level crossing protection types were divided into - (i) gates, (ii) flashing lights and bells and other, and (iii) signs and unprotected. Knowing the number of level crossings and their protection type in a given UAE segment, the frequency of level crossing accidents in that segment involving a chemical χ was estimated by:

$$F_{L,j} = \Sigma_k \, [F_{L,k} \, N_{k,j} \, K_{k,j}] \, / \, L_S \qquad (6)$$

where, $F_{L,j}$ = Frequency of level crossing accident in segment j (# / year-km)

$F_{L,k}$ = Global accident frequency of level crossing accidents for protection type k (#/year-crossing)

$N_{k,j}$ = Number of level crossings of protection type k in segment j

$K_{k,j}$ = Probability that, given a level crossing accident in segment j, a chemical χ tankcar will be struck

= $N^* \, [f_L \, f_{\chi,L} + f_T \, f_{\chi T}]$

where, N^* = number of rail cars involved in a level crossing accident, assumed to be 1. (level crossing accidents resulting in a derailment are treated as derailments separately)

f_L = fraction of trains through segment that are local trains

f_T = fraction of trains through segment that are through trains

$f_{\chi,L}$ = fraction of local cars that are chemical χ

$f_{\chi,T}$ = fraction of through cars that are chemical χ

The expected accident frequency for a given level crossing is influenced by the volume of train traffic and volume of vehicle traffic that go by that crossing. These factors could not be taken into account because these data for all the estimated 18,000 level crossings in the Canadian rail network were not available. It was assumed that these factors are incorporated in the selection of the protection type for each level crossing.

3.1   Quantification of Fault Trees - Estimation of the Release Rate Frequencies for a Given Chemical and Segment

In the fault tree, accident frequencies for a chemical χ in a segment j are combined with post-accident severity probabilities to estimate frequencies for each release rate type using Boolean algebra. CSC (1987) provides the complete details regarding the fault tree and the post-accident probabilities used. Where supportable data were available, these probabilities were estimated separately for different chemicals and/or different types of containers. The output of the fault tree is the frequency of occurrence of a release rate l, for chemical χ in segment j ($F_{Rl,\chi,j}$).

3.2   Quantification of Event Trees - Estimation of the Hazardous Event Frequencies

Event trees are used to identify hazardous events and also to quantify them. In order to quantify the event trees a probability has to be estimated for every branch in the tree. The full approach can be seen in Figures 4 and 5. The output of the event tree is the frequency of occurrence of a hazardous event, m, due to release rate l and chemical χ in segment j ($F_{Hm,l,\chi j}$).

3.3   Frequency Assessment Results

Table 3 shows, as an example, the frequencies for a segment in a densely populated urban environment (some results associated with sections below are also shown in this table). It can be seen from this table that release rates were grouped where the consequences were expected to be the same, or where they led to the same hazardous events as identified in the event trees (e.g., a propane jet fire is possible for both SRR/NEF and SRR/WEF).

Hazardous events that are expected to lead to one or more fatalities, are estimated to occur as frequently as once every 150 years in this segment (reciprocal of the sum of the hazard frequencies multiplied by the segment length). The methods used for estimating the number of fatalities that might result from each of these events are described below.

Table 3: Frequencies, consequences and societal risk rate for hazardous events.

ALTERNATIVE A1, SEGMENT 06, 1991 POPULATION DATA

CHEMICAL	HAZARD	RELEASE RATE	HAZARD FREQUENCY (events/(yr.km))	AVERAGE FATALITIES (fatality/event)	MAXIMUM FATALITIES (fatality/event)	MINIMUM FATALITIES (fatality/event)	SOCIETAL RISK RATE (fatality/yr.km)
Ammonia	Toxic Plume	LRR/NEF	1.51×10^{-8}	66	122	29	1.00×10^{-6}
		LRR/WEF	1.29×10^{-7}	74	133	16	9.50×10^{-6}
		SRR/WEF	9.60×10^{-7}	1	2	0	7.97×10^{-7}
		SRR/NEF	6.19×10^{-5}	6	6	5	3.42×10^{-4}
Chlorine	Toxic Plume	LRR/NEF	1.75×10^{-7}	1200	3220	180	2.10×10^{-4}
		LRR/WEF	1.50×10^{-6}	2833	11300	138	4.25×10^{-3}
		SRR/WEF	1.12×10^{-5}	20	45	2	2.19×10^{-4}
		SRR/NEF	1.13×10^{-4}	56	64	50	6.31×10^{-3}
Explosive Gasoline	Blast Wave	LRR/(WEF & NEF)	1.96×10^{-9}	68	68	68	1.34×10^{-7}
	Jet Fire	SRR/(WEF & NEF)	1.77×10^{-4}	54	54	54	9.53×10^{-3}
	Pool Fire	SRR	5.01×10^{-5}	91	91	91	4.57×10^{-3}
		LRR	1.84×10^{-5}	147	147	147	2.71×10^{-3}
Oleum	Toxic Plume	LRR	2.64×10^{-6}	39	93	7	1.02×10^{-4}
		SRR/WEF	2.60×10^{-5}	0	0	0	2.50×10^{-6}
Phosphorous	Toxic Plume	SRR	1.04×10^{-4}	14	34	1	1.45×10^{-3}
		LRR	1.07×10^{-7}	20	50	1	2.14×10^{-6}
Propane	Flash Fire	SRR	3.76×10^{-5}	3	7	1	1.09×10^{-4}
		LRR	3.57×10^{-8}	20	26	9	7.13×10^{-7}
	Jet Fire	SRR/(WEF & NEF)	1.36×10^{-4}	54	54	54	7.32×10^{-3}
	Fire Ball	LRR/NEF	1.83×10^{-7}	1010	1010	1010	1.85×10^{-4}
		LRR & SRR/WEF	5.32×10^{-6}	508	508	508	2.70×10^{-3}
						TOTAL	0.0401

4.   CONSEQUENCES OF HAZARDOUS EVENTS

The consequence assessment component of this risk assessment is concerned with determining the harmful effects of hazardous events to humans with death as the measure of the harmful effect. In mathematical terms, the consequences, C_m, of a hazardous event, m, are:

$$C_m = \int [P_d(x,y)]_m \, \rho(x,y) \, dA \qquad (7)$$

where, C_m = Expected consequences (deaths/event)

$[P_d(x,y)]_m$ = Probability of death at a point (x,y) away from the location of the hazardous event (deaths/person-event)

$\rho(x,y)$ = Population density at (x,y) (persons/m^2)

A = area (m^2)

The probability of death is dependent on the amount of harmful effect, or load, that a person receives and the vulnerability of the person to the harmful effect. These quantities are very chemical dependent. The load at a point (x,y) depends on the chemical and physical properties of a hazardous substance, while, for example, human vulnerability is very much dependent on a substance's toxicological properties for toxic substances.

4.1 Population Density

All population data for this study were obtained from the Toronto Area Regional Modelling System (TARMS) information available from the Ontario Ministry of Transport and Communications (MTC). These data provide population and employment counts for the Toronto region, broken down by TARMS zone (census tracts). Projections were made for 1991 and 2011.

4.2 Load and Human Vulnerability for Each Hazard

The load that a hazardous event imposes at a point away from the event location depends on several factors as shown in Table 4. The load is dependent on the hazard.

The constant n usually has a value greater than 1.0, indicating the greater effect of concentration or thermal intensity than exposure time. For toxic substances, n is chemical dependent.

Table 4: Hazard loads.

Hazard	Load
Toxic Cloud	Concentration, exposure time; $L = c^n t$
Flammable Cloud	Flame impingement
Pool Fire	Thermal intensity, exposure; $L = I^n t$
Fireball	Thermal intensity, exposure time; $L = I^n t$
Jet Fire	Thermal intensity; $L = I$
Blast Wave	Overpressure; $L = P_0$

When comparing the thermal radiation hazards, the load for nontransient or steady fires is just the thermal radiation intensity, while for transient fires, the load is the joint effect of thermal radiation intensity and exposure time. Jet fires which burn for many minutes without changing size significantly are treated as non-transient fires. Fireballs burn out quickly and are unquestionably transient. The load due to pool fires is transient or steady depending on whether the pool is unconfined or confined. Here, they were treated as unconfined (transient) so that the fire impingement area is maximized.

For all the hazards, with the exception of flammable cloud, there is a minimum threshold load that must be exceeded for death to occur. The assumption made for flammable cloud is that persons found within a flash fire (including persons indoors) would become fatalities.

This brings us to the problem of assessing the vulnerability of humans to the various hazards considered in this risk assessment. In order for a proper assessment of risk, it is necessary to relate the intensity of the load such as thermal radiation from a fire to the degree of impact (or probability of death). The methods used in this study are:

• Empirical assessment of observed damage for confined explosions,

• Thermal radiation criteria for jet fires, and

• Probit analysis for all the other types of loads.

The most extensively used method was probit analysis. A detailed account of this method is given by Finney (1971) and its application to major hazards is described by Eisenberg et al. (1975).

The major advantages of the probit approach are (i) it is very flexible, especially in how one decides to define the load, (ii) it is easy to use and (iii) incorporates varying human sensitivity to load (i.e., a certain toxic load might kill one person but not another). This method has become a very popular hazard assessment tool over the last decade and has been used in major risk assessments (Health and Safety Executive 1981; Cremer and Warner, 1982).

The vulnerability criteria used in this study are shown in Table 5. Some general comments regarding the criteria selected for each chemical are given below.

Ammonia: Probit equation published by Eisenberg et al. (1975) was used. This equation was the most conservative of those found in the literature.

Chlorine: Probit equation published by the Health and Safety Executive of the United Kingdom was used (Purdy et al., 1987). One other equation was reviewed from Whithers and Lees (1985) who conducted an extensive study on chlorine toxicity.

Explosive: Probit equation published by Eisenberg et al. (1975) was used.

Gasoline: Probit equation for pool fires used for this study was published by Eisenberg et al. (1975) and used by Hymes (1983). The jet fire vulnerability criteria were developed from information supplied in the second Canvey report (1981) and include consideration of escape. The criteria used for confined explosions were determined from overpressure effects (lung hemorrhage) in the near field ($P_0 > 1.5$ bar) and from observed damage in the far field ($P_0 < 1.5$ bar).

Oleum: Probit equation for oleum was developed for this study from animal toxicological data for mice and rats obtained from Environment Canada (1984). Most probit equations are developed from animal data since data on humans are either scarce or unreliable. In these cases, as in this one, it is assumed that humans react to hazardous loads in a similar manner to the animals from which the probit equation was developed. The equation used here is for sulphuric acid mist which is the atmospheric product from evaporating SO_3 from an oleum spill. The equation was developed from two points, an LL_{50} and an LL_{LO}, where the LL_{LO} is assumed to be the LL_{01}.

Table 5: Vulnerability criteria.

CHEMICAL	HAZARD	VULNERABILITY CRITERIA		
Full Cars				
Ammonia	Toxic Cloud	$Y = -30.57+1.385 \ln [C^{2.75}t]$, t in minutes		
Chlorine	Toxic Cloud	$Y = -4.4+0.52 \ln [C^{2.75}t]$, t in minutes		
Explosive	Blast Wave	$Y = -77.1+6.91 \ln [P_o]$, P_o in N/m^2		
Gasoline	Pool Fire	$Y = -14.9+2.56 \ln [10^{-4} I^{4/3}t]$, t in seconds		
	Jet Fire	$I > 12.6$	kW/m^2	Pd = 1.00
		$6.5 < I < 12.6$	kW/m^2	Pd = 0.25
		$4.0 < I < 6.5$	kW/m^2	Pd = 0.05
Oleum	Toxic Cloud	$Y = -39.6+2.5 \ln [C_o^2 t]$, t in minutes		
Phosphorus	Toxic Cloud	$Y = -6.7+0.52 \ln [C_o^{2.75}t]$, t in minutes		
Propane	Flammable Cloud	Within cloud Pd = 1.00		
	Fireball	$Y = -14.9+2.56 \ln [10^{-4} I^{4/3}t]$, t in seconds		
	Jet Fire	$I > 12.6$	kW/m^2	Pd = 1.00
		$6.5 < I < 12.6$	kW/m^2	Pd = 0.25
		$4.0 < I < 6.5$	kW/m^2	Pd = 0.05
Residues				
Ammonia	Toxic Cloud	$Y = -30.57+1.385 \ln [C^{2.75}t]$, t in minutes		
Chlorine	Toxic Cloud	$Y = -4.4+0.52 \ln [C^{2.75}t]$, t in minutes		
Propane	Confined Explosion	$P_o > 5.0$	bar	Pd = 1.00
		$1.5 < P_o < 5.0$	bar	Pd = 0.75
		$0.68 < P_o < 1.5$	bar	Pd = 0.38
		$0.2 < P_o < 0.68$	bar	Pd = 0.13
Gasoline	Confined Explosion	$P_o > 5.0$	bar	Pd = 1.00
		$1.5 < P_o < 5.0$	bar	Pd = 0.75
		$0.68 < P_o < 1.5$	bar	Pd = 0.38
		$0.2 < P_o < 0.68$	bar	Pd = 0.13

Notes: (1) C has units of ppm
    (2) C_o has units of mg/m^3
    (3) Pd is Probability of death
    (4) I is thermal radiation intensity (W/m^2)
    (5) P_o is peak overpressure

Phosphorus: Probit equation developed for this study was developed from toxicological data obtained from Sax (1979) for phosphorus pentoxide which is the principal combustion product from burning phosphorus. Only an LL_{50} was found in the literature for phosphorus pentoxide. Thus to produce a probit equation the slope (k_2) was assumed to be equivalent to that of chlorine.

Propane transient thermal radiation from fireballs was assessed using the same probit equation used for gasoline pool fires. Steady thermal radiation from jet fires was assessed using the same probit equation used for gasoline jet fires. As mentioned earlier, persons trapped within a flash fire (indoors and outdoor) are assumed to become facilities. The criteria used for confined explosions were determined from overpressure effects (lung hemorrhage) in the near field $(P_0 > 1.5$ bar) and from observed damage in the far field $(P_0 < 1.5$ bar).

4.3 Estimation of Loads for Each Hazard

In order to estimate the load for each hazard at a given point various hazards assessment models were used. These models are considered to be state-of-the-art and include the latest hazard modelling techniques.

Loads Due to Toxic Cloud: For hazardous substances that disperse in the atmosphere, the load at a given point is very much dependent on the prevailing meteorology. It was assumed that the meteorology would not vary for the duration of time that the toxic hazard persisted. In addition, indoor loads vary considerably from outdoor loads. Since the majority of people at any given moment are indoors (90 percent assumed to be indoors, a gas infiltration model was used to estimate the indoor concentration-time profile from the outdoor concentration-time profile (Purdy et al., 1987).

The models used to estimate toxic loads were COBRA and INPUFF.

COBRA is a heavy gas spill and dispersion modelling system developed by Concord Scientific (Alp, 1985; Alp and Matthias, 1991) based on Colenbrander's theoretical model (1980). The model has been used to estimate toxic loads for ammonia, oleum (sulphuric acid mist), and chlorine (LRR/NEF,LRR/WEF and SRR/NEF for all chemicals).

INPUFF is a Gaussian dispersion model developed by the US EPA (1986). The model has been used to estimate toxic loads for ammonia, chlorine and oleum (SRR/WEF only for these chemicals) and for phosphorus (phosphorus pentoxide) release rates.

Loads Due to Flammable Clouds: The dispersion of propane vapour in the atmosphere was also modelled by COBRA. A flammable cloud ignition model was used to predict the probability of the cloud being ignited at a particular point in time (and space). The probability P_n that the cloud will be ignited by the nth ignition source it encounters, is equal to the ignition probability per source, k_o, multiplied by the probability that ignition has not taken place due to the n-1 ignition sources already encountered:

$$P_n = k_o (1 - k_o)^{n-1} \tag{8}$$

Here, it is assumed that an ignition source either ignites the cloud when it first sees it, or it does not ignite the cloud at all.

In the present study, every person is assumed to be an ignition source, with a 1% chance of igniting the cloud (TransCanada Pipelines, 1979).

Probability of death for an individual at a downwind distance x, $P_d(x)$, is dependent on the probability of being exposed to a flash fire, $P_e(x)$, and the probability of not surviving if exposed, P_S. As discussed above, we have taken $P_S = 1.0$. $P_d(x)$ then equals:

$$\begin{aligned} P_d(x) &= P_S \cdot P_e(x) \\ &= P_8(x) \end{aligned}$$

$P_e(x)$, in turn, is equal to the probability of non-ignition before x, multiplied by the probability of ignition between x and x_{max}, the maximum extent of the cloud:

$$P_e(x) = (1-k_o)^{n_x} \sum_{n=1}^{(n_{max}-n_x)} k_o(1-k_o)^{n-1} \tag{9}$$

$$= (1-k_o)^{n_x} - (1-k_o)^{n_{max}}$$

where n_x = the number of ignition sources (i.e., people) between x = o (the source) and x,

n_{max} = the number of ignition sources within the total extent of the cloud, between x = o and x = x_{max}.

Note that $(n_{max} - n_x)$ is the number of ignition sources that the cloud sees *downwind* of x.

In order to estimate the number of expected fatalities (i.e., the event societal risk, R_S) for a given release under a particular set of meteorological conditions, let us consider the following. The probability of the first person that the cloud encounters causing ignition of the cloud is k_0. Since it is assumed that this would be the only person to be exposed to the flash fire when the cloud ignites, k_0 is also the probability of one person becoming a fatality. The probability of the second person igniting the cloud is $k_0 (1-k_0)$, which is also the probability of two persons dying. Thus, the probability of n people becoming fatalities is $k_0 (1-k_0)^{n-1}$. The number of expected fatalities for the event will then by the sum of the probability-weighted number of fatalities for each ignition outcome involving n number of people.

$$R_s = \sum_{n=1}^{n_{max}} n k_0 (1-k_0)^{n-1} = \frac{1}{k_0} [1 - (1 + k_0 n_{max}) (1-k_0)^{n_{max}}] \quad (10)$$

Loads Due to Fireball: The fireball model used in this study was developed by Crocker and Napier (1987) and is a stationary ground level model for a stationary receptor. An exposure time of 30 seconds was used in estimating the thermal load.

The exposure time of 30 seconds is less than the duration of combustion of the fireball. This value was selected to compensate for the conservative assumptions of (i) stationary receptor and (ii) all persons assumed to be outdoors.

Loads Due to Jetfire: The jetfire model developed by Crocker and Napier (1987) was used to predict radiation intensities.

Loads Due to Pool Fires: A dynamic model of a spreading pool fire (Moorehouse and Pritchard, 1982) was used to estimate the thermal load at a vertical receptor located at the ground-level. The stationary target is assumed to receive the thermal radiation for an exposure time of 30 seconds. All people inside the pool fire were assumed to become fatalities.

Loads Due to VCE's: VCE is a deflagration computer model for vapour cloud explosions developed by Concord Scientific based on the theoretical model of Wiekema (1980). This model was initially used for predicting overpressure loads from potential propane VCE's.

Estimated overpressures outside the combusted cloud were not strong enough to cause fatalities (due to overpressure) beyond a few meters. Based on these results and the findings of Wiekema (1984), it was assumed that VCE effects would be treated the same as flash fires. In terms of fatalities, the effects of VCE's and flash fires are similar; however VCE's tend to produce more injuries.

Loads Due to Detonations and Confined Explosions: The estimation of personal damage due to exposure to the effects of a detonation or confined explosion is dealt with here. The hazardous effects of an explosion include:

- exposure to overpressure effects of the blast wave (e.g., lung hemorrhage),
- whole body translation,
- fragmentation and missiles.

Confined explosions of "residue" rail cars present an explosion hazard due to the residue of flammable material in the tank. Quantification of the explosion damage from an explosion inside the rail car involved the following assumptions:

- the flammable material (propane, gasoline) is at its upper flammable limit at the moment of ignition,
- the initial blast overpressure is the same as the explosion overpressure at the wall of tank (around 5 bar - the upper overpressure threshold),
- gasoline is represented by n-decane,
- 100% and 50% fatalities were calculated on the basis of direct exposure to overpressure effects of the blast wave; this again resulted in lung hemorrhage,
- 25% fatalities were estimated at the overpressure level likely to cause total destruction; this results in falling rubble of buildings and based on experience from earthquakes damage and bombings in wars, was considered as a reasonable estimate, and
- the 1% fatality level considered that the threshold of fatalities was similar to the limit of missiles based on overpressure.

These assumptions were used in conjunction with a set of scaled overpressure and scaled distance curves developed by Baker et al. (1977).

4.4 Consequence Assessment Results

Probability of death takes into account both toxic load and vulnerability. The next three exhibits present some of this information in graphical form. Figure 7 shows the probability of lethality as a function of distance for some of the hazardous events. As can be seen, the large chlorine release has the potential to cause death at much further distances than the other hazardous events shown. Furthermore, hazardous events not shown, such as jet fires, detonations and confined explosions have the potential to cause death only within 100 meters.

Figure 8 shows, for chlorine releases, the effects of LRR/NEF compared to SRR/NEF and the effects of outdoor compared to indoor loads. If we take the probability of lethality of 0.5 (50% of exposed persons to die) as a benchmark then the following conclusions can be made. The LRR/NEF distance is about 8 times greater than the SRR/NEF distance. The outdoor load distance is about twice the indoor load distance. If we consider the distance at which the outdoor P_d = 0.5, namely about 1.5 km, the indoor probability of death at that point is about 0.13. These results indicate that when assessing the risks due to toxic substances, consideration of indoor loads is important.

Another important question that has been addressed in this study is "how hazardous are residue tankcars?" A tankcar is considered a residue if it contains less than 2% of the maximum full load. Only chlorine, ammonia, propane and gasoline tankcars were considered to pose a threat beyond the rail line right-of-way. Figure 9 shows curves of probability of lethality with distance for these hazards. It can be seen by comparing Figures 8 and 9 that residue cars, as expected, are much less hazardous than full cars; however, residue cars still pose a considerable hazard.

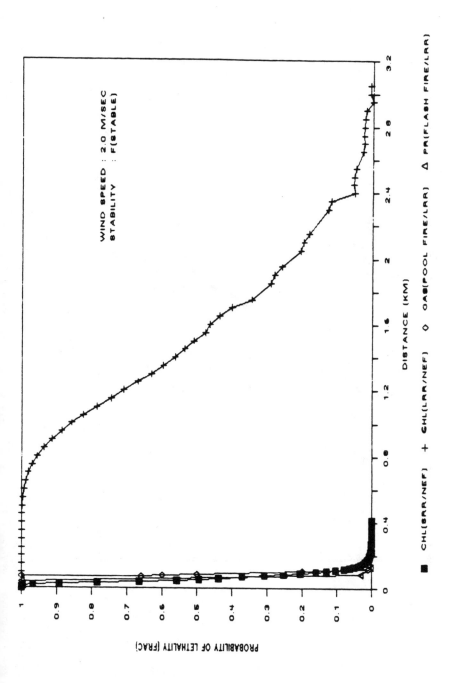

Figure 7: Probability of lethality vs. distance.

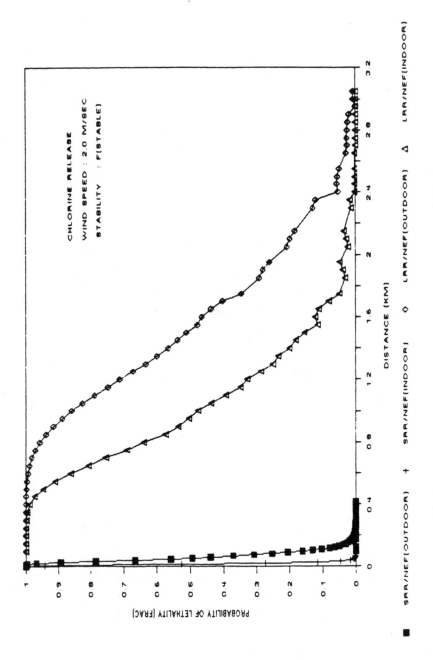

Figure 8: Probability of lethality vs. distance.

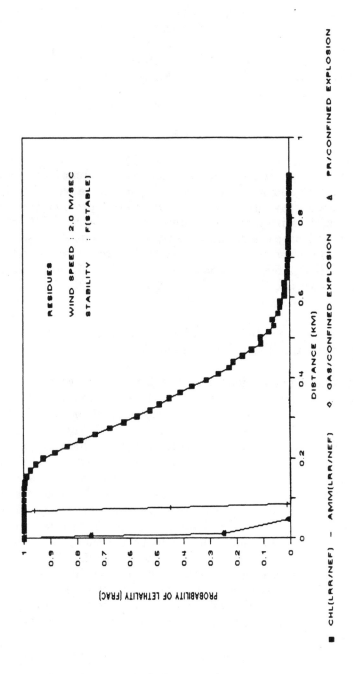

Figure 9: Probability of lethality vs. distance.

Table 3 shows the consequences in terms of average fatalities per hazardous event for a selected segment. Maximum and minimum fatalities per event are also shown for hazardous events that are influenced by meteorology, namely toxic plumes (or clouds) and flash fires. For these events the average fatalities per event, C_{ma}, was calculated by

$$C_{ma} = \sum_{i}^{72} P_{wi} \, C_{mi} \qquad (11)$$

For a given hazardous event, C_{mi} can vary by almost two orders of magnitude. In addition, fatalities per event range from 0 to 11,300. The hazardous events that are estimated to produce the highest number of fatalities are toxic plumes from large (LRR) chlorine releases and propane fireballs. This is very consistent with what has been observed in actual incidents. The toxic vapour release in Bhopal killed 2,500 persons while the BLEVE's and associated fireballs in Mexico City and San Carlos de la Rapita, Spain resulted in 452 and 210 deaths respectively.

It is now appropriate to consider the degree of conservatism (or overestimation) in the results presented.

Errors in the population density estimates are considered to be random. The criteria used to estimate human vulnerability probably err on the safe side. The probit equations used for toxicity are mainly derived from animal data. The assumption is made that humans succumb to the same toxic loads as these test animals.

The hazard loads estimated are also conservative, with some more than others, for the following reasons. In all cases stationary targets were assumed, allowing no opportunity for escape. Secondly, in the case of fires, no shielding was permitted from thermal radiation (everyone was assumed to be outdoors), nor was there assumed to be any chance of surviving a flash fire. Toxic loads are felt to be more realistically estimated because of the consideration of indoor loads. However, consideration was not given to persons in high-rise buildings who might not be influenced by heavy gas toxic or flammable clouds, increasing the level of conservatism in these estimates.

5. RISK CALCULATIONS

5.1 Societal Risk Calculations

The key parameter of importance in this study is the system societal risk which would give an indication of the relative merits of the alternative rail systems examined in this study in terms of risk to the public.

For a given hazardous event H, the societal risk rate is calculated by:

$$R_{H_{m,\chi,j,i}} = F^*_{H_{m,\chi,j,i}} C_{H_{m,\chi,j,i}} \qquad (12)$$

$$F^*_{H_{m,\chi,j,i}} = F_{H_{m,\chi,j}} P_{M_i} \qquad (13)$$

$$C_{H_{m,\chi,j,i}} = \int [P_d(x,y)]_{m,\chi,i} [\rho(x,y)]_j \, dA \qquad (14)$$

where

$R_{H_{m,\chi,j}}$ = Societal risk rate for a given hazardous event m, due to chemical χ in segment j, under meteorological condition i (deaths/year-km).

$F^*_{H_{m,\chi,j,i}}$ = Joint frequency of hazardous event m, occurring with meteorological condition i, due to chemical χ, in segment j (# events/year-km)

$F_{H_{m,\chi,j}}$ = Frequency of hazardous event m, occurring due to chemical χ, in segment j (# events/year-km)

P_{M_i} = Probability of meteorological condition i

$[P_d(x,y)]_{m,x,i}$ = Probability of death at a given point (x,y) away from the hazardous event m, due to chemical χ, meteorological condition i (deaths/person-event).

$[\rho(x,y)]_j$ = Population density at point (x,y) (Persons/m^2)

A   =   area (m^2)

$C_{Hm,\chi,j,i}$ =   Consequence of hazardous event m, due to chemical χ, in segment j, under meteorological condition i (deaths/ event).

Then, for a given segment j, the societal risk rate is:

$$R_{rj} \quad = \quad \sum_{\chi} \sum_{m} \sum_{i} R_{Hm,\chi,j,i} \tag{15}$$

The societal risk rate is a useful quantity when comparing risk between one UAE segment and another, since segments vary significantly in length. The societal risk for a segment is then:

$$R_{sj} \quad = \quad R_{rj} L_s$$

and the societal risk for an alternative is:

$$R_s \quad = \quad \sum_{j} R_{sj} \tag{16}$$

The societal risk as determined by Equation 16 has units of deaths per year, and is useful when comparing one alternative to another. However, when comparing risks from major hazards to commonly understood risks, attention must be given to low frequency, high consequence events. The most useful form of this comparison is the Risk Spectrum curve (or F-N curve).

5.2 Individual Risk Calculations

The individual risk calculations are used to determine how the level of individual risk changes as one moves away from the rail line. For a given hazardous event the individual risk rate is:

$$R_I(x,y)_{m,x,j,i} \quad = \quad F^*_{Hm,x,j,i} [P_d (x,y)]_{m,\chi,i} \tag{17}$$

$$F^*_{Hm,\chi j,i} \quad = \quad F_{Hm,\chi j} P_{Mi} \tag{18}$$

where $R_I(x,y)_{m,x,j,i}$ = Individual risk for hazardous event m, due to chemical χ, in segment j under meteorological condition i.(death/year-km-person).

Since accidents that lead to hazardous events can occur anywhere along a UAE segment and since the impact zone is usually much smaller than the length of the segment, one cannot simply multiply the individual risk rate by segment length to obtain individual risk along the segment. The approach adopted here is to allow all hazardous events to initiate at one location. Individual risk rates are then summed for all hazards for that point, thereby obtaining the individual risk rate - distance relationships for all eight wind directions.

The next step in the approach, which has been developed for this study and has not been published by others in the literature, is to integrate the risk rate/distance relationship along its interaction length for a stationary receptor away from the rail line.

6. DISCUSSION OF RESULTS

6.1 Societal Risk for Rail System Alternatives

We shall now present some of the key risk results of the study and put them in perspective by providing comparisons with other more familiar risks.

The key parameter of importance in this study is the system societal risk which would give an indication of the relative merits of the alternative rail systems examined in this study in terms of risk to the public. The system societal risk estimates for the base case (Alternative A1) and 1991 population densities is about 4.1 fatalities per year for the entire Greater Toronto Area (GTA). This can be compared with societal risks due to some other hazardous events (see Table 6).

Societal risk comparisons similar to those made in Table 6 are frequently used in risk assessments. However, when making such comparisons, one ought to be cautious. Although the comparisons made provide some value, a more meaningful comparison could be made with a proper basis of comparison: the exposed populations for each activity. This extension in comparison was not carried through in this study because exposed populations for the common activities shown in Table 6 were not readily available. The entire Canadian population can be used in several, but not all cases, since phenomena such as earth movements would not affect the entire population.

Table 6: Societal risk due to various hazardous events.

	Societal Risk (fatalities per year)
Motor Vehicle Accidents	4,238
Falls	1,829
Poisoning+	665
Dwelling Fires	487
Excessive Cold	121
Cataclysmic Storms	13
Earth Movements	5
Lightning	3
Rail Transport of Dangerous Goods (TDG) in the Greater Toronto Area (baseline risks based on existing system)	4.1*

* This is estimated societal risk in "statistical" fatalities per year as determined by risk assessment. All of the other societal risk numbers are "actual" fatalities recorded Canada-wide. (Source: Statistics Canada, 1985, "Causes of Death", Publication #84203.)

+ Includes accidental poisoning due to poisonous and other substances, surgical complications and misadventures to patients.

One must make another distinction between the rail TDG risks as estimated in this study and the others in Table 6. The rail risk estimates are in terms of what is usually termed as "statistical fatalities;" they are not the actual expected number of fatalities in any given year. The average annual risk given in "statistical fatalities per year" reflect the likelihood that there might not be any such deaths over an extended period of years but there is always the possibility of an accident which could cause multiple deaths, thereby bringing the average up to the societal risk number given above.

Another way of comparing risks due to various hazardous activities or events is to compare their risk spectra on one graph. Risk spectra are curves which present a more detailed picture of the frequency and consequence of the individual events that make up the hazardous activity. An example is shown in Figure 10. Here, the ordinate is the frequency of events having a consequence of N fatalities or more, and the abscissa is the number of fatalities N. In the figure, the risk spectrum for the existing rail system is shown together with the curves for various other hazards as given in the Reactor Safety Study published by the U.S. Nuclear Regulatory Commission in 1975. As can be seen, the rail TDG risks in the Toronto Area are again relatively small compared to these hazards.

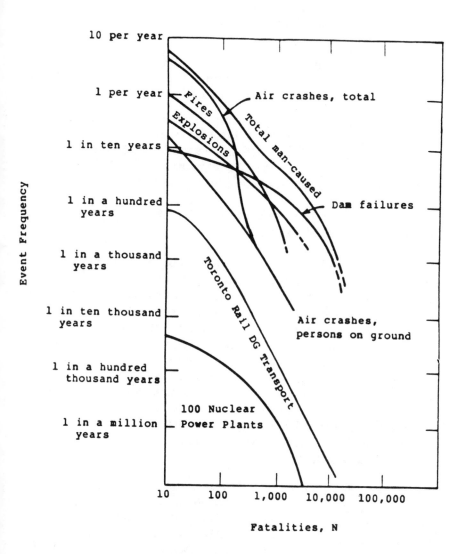

Figure 10:   Societal risk for the existing system compared to other risks.

Another factor that must be kept in mind in the interpretation of the risk estimates of this study is the uncertainties in these estimates. In general, these uncertainties are of three types:

- Bias introduced in the selection of, for example, the representative chemicals or release events for their respective groupings. The bias introduced here is on the conservative side, i.e., it will lead to higher estimates of risk than actual.

- Uncertainties due to errors in input data, such as accident frequencies, population data, meteorological data, etc. These errors would normally be random and not introduce any systematic bias in the risk estimates.

- Uncertainties due to model simplifications and assumptions in the consequence models or in the rail system model used in estimating event frequencies. The model assumptions are in general made such that the model results would "err on the safe side", i.e., lead to higher estimates of risks than actual.

The overall results of introducing these uncertainties are expected to be conservative risk estimates.

It is difficult to define the magnitude of these uncertainties without extensive sensitivity and error propagation analyses. However, a rough estimate can be provided based on professional judgement: the overall system societal risk would likely be between a high of two times and a low of one tenth of the estimate given, and probably closer to about one-third of the estimate. It should be noted, that these uncertainties would be of particular importance when absolute risks are required. When relative risks are being addressed, as in the present study, comparisons between various alternatives can be made with a lot more confidence than the above uncertainties might indicate, because the risk estimates for each alternative would have uncertainties in the same direction (e.g., risk estimates would be high) by about the same magnitude.

With this perspective in mind, let us now examine the system societal risk estimates for the various alternatives presented in Figure 11. Let us first compare the year 1991 and 2011. The differences here are strictly due to population changes. For all the alternatives, the estimated societal risk increases for the year 2011. This is an expected result since population is expected to increase from 1991 to 2011. The relative increases from one alternative to another are quite different, however, the percentage increase for the existing system (A1, (4.6-4.1)/4.1 = 12%) being much less than the percentage increase, for example, for alternative C1 (Route C - North). This is also as expected since population growth is projected to be much faster in the outlying areas than within the presently built-up areas.

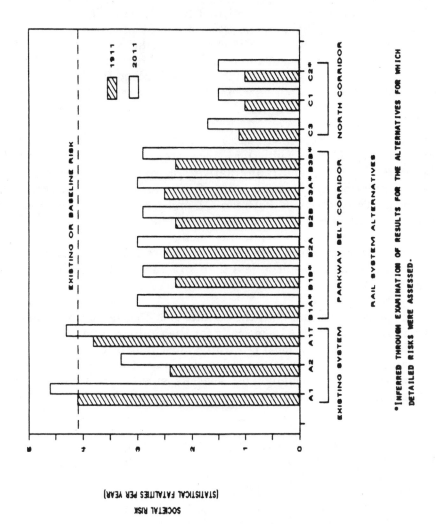

Figure 11: Comparison of societal risks for rail system alternatives.

Alternatives A2 and B2 (both B2A and B2B) represent significant societal risk reductions over the base case A1. The main reason for this is the obvious one: CP dangerous goods traffic is diverted away from the more densely populated regions to less densely populated regions. However, there is practically no difference between A2 and B2 (1991), indicating that diversion of both CN and CP dangerous goods traffic from the present CN Halton Subdivision near the Georgetown area south to the Milton cut-off produces a negligible net effect for 1991 even though the Milton cut-off route is shorter. For 2011, the projected higher population growth along the CN Halton Subdivision leads to higher societal risk estimates for A2 than for B2.

Alternatives C1 (northern route) and C3 (southern route) also represent significant reductions in system societal risk, with respect to both the A and B alternatives, mainly again due to much lower population densities, in spite of the much longer rail travel distances. The difference between C1 and C3 is not as much as would be expected due to this compensating effect.

The technology improvements (A1T), as proposed for application to the existing system, result in less than 10% reduction in the system societal risk.

One of the main reasons that the societal risk does not drop further than shown relative to A1 especially for the C alternatives is that a considerable amount of local DG traffic must remain in the densely populated areas, to service the industries located in these areas. If the traffic patterns change for reasons such as re-location of industries, the estimated 2011 societal risk might well be reduced from the 1991 level.

A striking observation that can be made from Figure 11 is that there are essentially three principal levels of societal risk for the alternatives examined. The risk reduction from the existing level for Alternatives A2 through B3B is roughly 40% and for C1 through C3 about 75%. Now, to examine how particular segments are being affected and by how much, one must look at the societal risk distribution in the study area, which is discussed in the next section.

6.2 Societal Risk Distribution in the Toronto Area

The rail system societal risk comprises the societal risk estimates for the individual rail segments which make up the system. To give an example of the distribution of the rail TDG risks in the Toronto Area, the societal risk estimates for each segment in the existing system (A1) are presented in detail in Table 7. Here, the segment societal risk in terms of fatalities per year due to rail operations on that segment is given for the subdivisions (and yards) as listed. For risk assessment purposes some of the subdivisions have been further divided into rail segments of uniform accident environment. Also shown in Table 7 is the societal risk rate along each segment in terms of fatalities/year•km.

Although the segment societal risk is more useful when comparing one alternative to the other (the sum of the segment societal risks gives the system societal risk), the societal risk rate along each segment is more meaningful when comparing one segment with another. Hence, this last parameter is presented graphically on the map in Figure 12 for the existing system (1991 population). The CP North Toronto and CN York (Thornhill area) Subdivisions have higher societal risk rates than the other subdivisions in the Toronto Area.

By examining similar maps for each of the alternatives, one can identify which areas would experience higher or lower risks relative to the base case and each other.

6.3 Individual Risk Near Rail Lines

Individual risk is a measure of the chance of a person who is at a particular distance from the rail line becoming a fatality due to TDG accidents on that rail line. The individual risk profiles near a high-traffic rail segment in the Toronto Area are shown in Figure 13. Also shown are individual risk numbers for the other hazards that were considered earlier for comparative purposes.

It should be noted here that all of these risks except that due to smoking and being in a motor vehicle can be considered to be involuntary risks. There is generally less public acceptance of involuntary risks than of voluntary risks such as smoking. The one in a million level, which is generally considered to be an acceptable level for individual risks of an involuntary nature, is reached at about 400 m from the rail line. Before making any judgements as to the acceptability of the risks near a rail line, however, the uncertainties in the risk estimates must be considered in some detail since for such judgements one must be concerned more about absolute risk than relative risk which has been the focus of the present study.

Table 7: System risk table: Alternative A1 for 1991.

Segment	Subdivision	Length (km)	Societal Risk for Segment (fatali-ties/year)	Societal Risk Rate Along Segment (fatalities/yr-km)
A1-01	CP Galt	15.6	0.046	0.0030
A1-02	CP Galt	34.5	0.171	0.0049
A1-03	CP Galt	2.4	0.030	0.0125
A1-04	CP Galt	3.9	0.068	0.0173
A1-05	CP Galt/N. Toronto	3.0	0.234	0.0780
A1-06	CP N. Toronto	9.1	0.364	0.0400
A1-07	CP Belleville	12.5	0.179	0.0143
A1-08	CP Belleville	4.6	0.026	0.0056
A1-09	CP Belleville	64.3	0.103	0.0016
A1-10	CN Oakville	38.0	0.118	0.0031
A1-11	CP Canpa	4.2	0.011	0.0026
A1-12	CN Oakville	13.4	0.020	0.0015
A1-13	CN Weston	9.4	<0.001	<0.0001
A1-14	CN Weston	5.7	0.028	0.0049
A1-15	CN Kingston	32.6	0.005	0.0002
A1-16	CP Havelock	42.8	<0.001	<0.0001
A1-17	CN Halton	65.6	0.385	0.0059
A1-18	CN Halton	7.6	0.065	0.0085
A1-19	CN Halton/York	9.0	0.021	0.0023
A1-20	CN York	7.3	0.195	0.0267
A1-21	CN York	8.0	0.044	0.0054
A1-22	CN York	23.4	0.005	0.0002
A1-23	CN Kingston	52.5	0.106	0.0020
A2-24	CP MacTier	16.4	0.006	0.0004
A1-25	CP MacTier	17.8	0.009	0.0005
A1-26	CP MacTier	9.2	0.018	0.0020
A1-27	CP MacTier/CN Weston	7.1	0.145	0.0204
A1-28	CN Newmarket	44.6	0.013	0.0003
A1-29	CN Newmarket	14.2	0.059	0.0041
A1-30	CN Bala	21.3	0.014	0.0007
A1-31	CN Bala	62.0	0.046	0.0008
A1-32	CN Bala	22.9	0.059	0.0026
A1-33	CP Owen Sound	57.9	<0.001	<0.0001
A1-34	CN Guelph	13.5	<0.001	<0.0001
	Yards			
A1-Y1	CN MacMillan		0.6776	
A1-Y2	CN Mimico		0.0015	
A1-Y3	CN Don		0.0059	
A1-Y4	CP Agincourt		0.0931	
A1-Y5	CP Lambton		0.6068	
A1-Y6	CP Obico		0.0045	
A1-Y7	CP Ray Ave.		0.0917	

Total Societal Risk for Toronto Area Rail System = 4.1 fatalities/year.

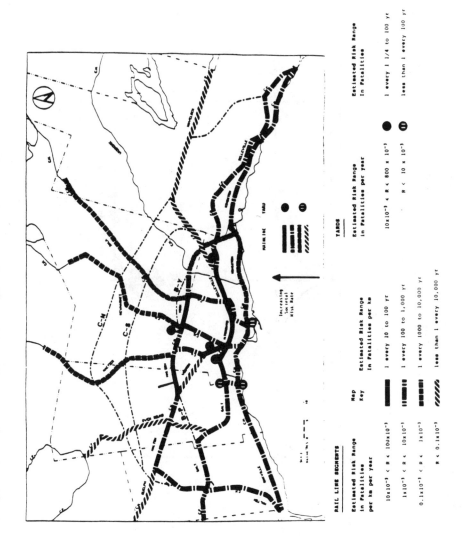

Figure 12: Societal risk distribution in the Toronto Area for DG rail transport routes in the existing system.

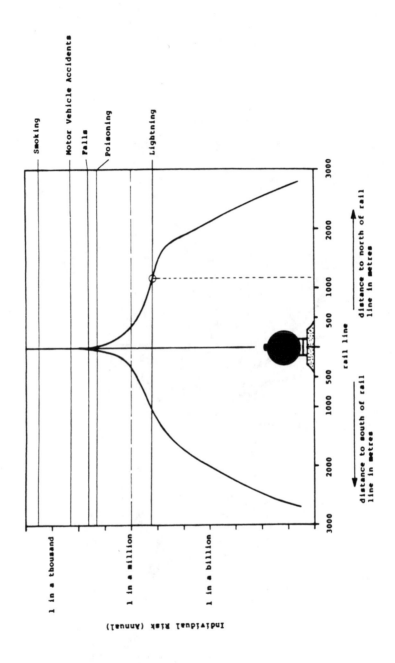

Figure 13:   Individual risk - Chance of an individual near a
                 rail line becoming a fatality (high rail traffic
                 segment in GTA, 1991).

7. SUMMARY

This paper has briefly described the approach and results of the assessment of risk to the public for the possible rail system alternatives relative to the baseline risk for the existing system. The full details of this work are documented in the major report on this study.

The study constitutes the most comprehensive of its kind to date for assessing risks due to dangerous goods rail transport in the Greater Toronto Area. It has utilized state-of-the-art consequence and frequency assessment models covering the full range of possible hazards due to DG transport. It has drawn upon a large and comprehensive volume data, such as:

- DG traffic pattern and operational strategies in the GTA,

- population densities adjacent to rail lines within the GTA,

- meteorological data for the GTA,

- track information (fixed plant) both within the GTA and Canada-wide,

- equipment information (e.g., tankcar design),

- historical accident data (Canada-wide), human vulnerability data.

Throughout the study, focus was on quantitative estimates of risk as opposed to qualitative aspects, which were the subject of all the earlier studies on risk assessments for DG rail transport in the study area.

The results of this study indicate that the baseline societal risk for the existing rail system is 4.1 statistical fatalities per year based upon 1991 population data and 4.6 for the year 2011. These levels of public risk are relatively low when compared to other societal risks.

If these levels of societal risk are considered unacceptable, significant reductions in public risk are afforded by the various alternatives investigated.

While it is anticipated that the estimated levels of societal risk for the present system will be examined with a view to assessing the acceptability of these levels, it is important to bear in mind the uncertainties associated with the estimates.

Even though the above considerations, which focus on the risk estimates in an absolute sense, are unavoidable, it must be stressed that the principal thrust of this work is on the relative risk to the public from possible alternatives to the current rail system. The results of this study which are based upon a quantitative assessment of the risks provide valuable guidance on which alternatives produce the most significant measure of risk reduction.

Acknowledgements This work was done under funding by Transport Canada. Contributions of Mr. Marcello Oliverio, Prof. Douglas Napier, and Mr. Wyman Jones during the course of this study are gratefully acknowledged.

8. REFERENCES

ALP E. (1985). COBRA: A LNG model. In: R.V. Portelli (ed.) Proceedings of the Heavy Gas (LNG/LPG) Workshop. January 29-30, Toronto, Ontario. Published by Concord Scientific Corporation. ISBN 0-920747-00-0. 294 pp.

ALP, E., & MATTHIAS (1991). COBRA. A Heavy Gas/Liquid Spill and Dispersion Modelling System. J. Loss. Prev. Process Ind., Vol. 4, April, 139-150.

BAKER, W.E., KULESZ, J.J., RICKER, R.E., WESTINE, P.S., PARR, V.B., & OLDHAM, G.A. (1977). Workbook for Predicting Pressure Wave and Fragment Effects of Exploding Propellant Tanks and Gas Storage Vessels. Report NASA CR-134906.

BURTON, I., & POST, K. (1983). The Transport of Dangerous Commodities by Rail in the Toronto Census Metropolitan Area: A Preliminary Assessment of Risk. Prepared for the Canadian Transport Commission, Railway Transport Committee.

COLENBRANDER, G.W. (1980). A Mathematical Model for the Transient Behaviour of Dense Vapour Clouds. Presented at: The 3rd International Symposium on Loss Prevention and Safety Promotion in The Process Industries. Basbe, Switzerland, September 15-19.

CREMER & WARNER (1982). Risk Analysis of Six Potentially Hazardous Industrial Objects in the Rijnmond area: A Pilot Study. Prepared by: Cremer and Warner Ltd. for the Executive Board of Rijnmond Public Authority. D. Reidel Publishing Company. Boston, Ma.

CROCKER, W.P., & NAPIER, D.H. (1987). Assessment of Mathematical Models for Fire and Explosion Hazards of Liquefied Petroleum Gases. Presented at: International Specialist Meeting on Major Hazards in the Transport and Storage of Pressure Liquified Gases. University of New Brunswick Fire Science Centre, Fredericton, New Brunswick, August 10-13.

CSC (1987). Risk Assessment for Rail Transportation of Dangerous Goods Through the Toronto Area. Concord Scientific Corporation Report CSC.J1029.1. Prepared for the Toronto Area Dangerous Goods Rail Task Force.

EISENBERG, N.A., LYNCH C.J., & BREEDING R.J. (1975). Vulnerability Model. A simulation system for assessing damage resulting from marine spills. Rockville, MD.: Enviro Control Inc. (Available from: Nat. Tech. Inf. Service Rep. AD-A015-245.)

ENVIRONMENT CANADA (1984). TIPS Manual-Sulphuric Acid and Oleum. Technical Services Branch. Ottawa, Ontario. ISBN0-662-13019-7.

ENVIRONMENT CANADA (1985). TIPS Manual-Introduction Manual. Technical Services Branch. Ottawa, Ontario. ISBN 0-662-14325-6.

FINNEY, D.J. (1971). Probit analysis. Third Edition Cambridge University Press, London.

HEALTH AND SAFETY EXECUTIVE, U.K. (1981). Canvey: A second report: A review of potential hazards from operations in the Canvey Island/Thurrock area three years after publication of the Canvey Report HMSO, London, ISBN 0-11-883459-2.

HYMES I. (1983). The Physiological and Pathological Effects of Thermal Radiation. UKAEA, Safety and Reliability Directorate Report SRD 275.

MANDA, L.J. (1975). Phase II report on full scale fire tests. Association of American Railroads, RPI-AAR Cooperative Program. Report RA-11-6-31.

MOOREHOUSE, J., & PRITCHARD, M.J. (1982). Thermal Radiation Hazards From Large Pool Fires and Fireballs. The Institution of Chemical Engineers. Symposium Series No. 71. The Assessment of Major Hazards. London, pp. 397-428.

PETERSEN, W.B., & LAVDAS, L.G. (1986). INPUFF 2.0 - A Multiple Source Gaussian Puff Dispersion Algorithm User's Guide. US Environmental Protection Agency Report No. EPA/600/8-86/024.

PURDY, G., CAMPBELL, H.S., GRINT, G.C., & SMITH, L.M. (1987). An Analysis of the Risks Arising from the Transport of Liquefied Gases in Great Britain. Presented at: The International Specialist Meeting on Major Hazards in Transport and Storage of Pressure Liquefied Gases. Fredericton, NB, August 10-13.

SAX N.I., & BRACKEN, M.C. (1979). Dangerous Properties of Industrial Materials. 5th Edition. Van Nostrand Reinhold Co. New York, NY.

TRANS CANADA PIPELINES (1979). Application for a Certificate of Authorization to Construct and LNG Terminal at Gros Cacouna, Province of Quebec. Arctic Pilot Project. Volume 4, Public Interest, Parts A and B.

WADE, P.E. & ASSOCIATES (1986). A Strategic Overview: Hazardous Goods Transportation by Rail in Toronto. Prepared for the City of Toronto Planning and Development Department.

WIEKEMA, B.J. (1980). Vapour Cloud Explosion Model. Journal of Hazardous Materials, v. 3, pp. 221-232.

WIEKEMA, B.J. (1984). Vapour Cloud Explosions - An Analysis Based on Accidents. Journal of Hazardous Materials: Part 2, v. 8, pp. 313-329.

WITHERS, R.M.J., & LEES, F.P. (1985). The Assessment of Major Hazards: The Lethal Toxicity of Chlorine: Part 1, Review of Information on Toxicity. Journal of Hazardous Materials, v. 12, pp. 231-282.

Methodology for Assessing the Public Risk from Transporting Hydrogen Sulphide

S.B. Russell
T.F. Kempe
Radioactive Materials Management
Engineering Department
Ontario Hydro
700 University Avenue
Toronto, Ontario
Canada M5G 1X6

ABSTRACT

A methodology for assessing the fatality risk to members of the public from the transportation of hydrogen sulphide under pressure by truck is developed. The model uses vehicle accident data and population densities along a transport route from Sarnia to the Bruce Heavy Water Plant in Ontario. Accident data for the transport of liquefied petroleum gas have been used to estimate the fraction of accidents that lead to the release of hydrogen sulphide. The subsequent dispersion and transport of hydrogen sulphide were estimated using a heavy gas dispersion model and the individual and collective risks to the population near the route were determined. The results from hydrogen sulphide transport in Ontario were compared with those from risk assessments for the transport of other hazardous goods near population centres in France and Ontario.

1. INTRODUCTION

Ontario Hydro uses hydrogen sulphide (H_2S) in the production of heavy water at the Bruce Heavy Water Plant (BHWP) for use in its nuclear generating stations. H_2S is a very toxic gas that can cause death after exposures of about 500 ppm in air for a few minutes.

While other physiological effects can occur at lower concentrations, the analysis in this paper has been confined to assessing the public risk of death from exposure to H_2S. Since H_2S is not produced at BHWP, it is necessary to transport the material via rail and/or truck to the site.

This paper outlines the methodology for assessing the public safety of transporting hydrogen sulphide (H_2S) from Sarnia to BHWP via truck. The risk to the public from transporting liquid H_2S under pressure has been estimated using accident statistics and release frequency data from the transport of liquefied petroleum gas (LPG) in the United States and Canada (Swoveland and Cawdery, 1986; Arnott, 1989; Saccomanno, 1990) due to the very limited data base for transport accidents involving H_2S. (The public risk of loading, transferring and unloading H_2S is beyond the scope of this paper).

The methodology for calculating the public risk from transporting H_2S was adapted from previous transportation risk assessment models (Kempe and Russell, 1987; Kempe, 1992). The H_2S transportation accident frequency, the risk to an individual along the route and the collective risk to the population have been developed for the H_2S risk assessment.

2. H_2S TOXICITY

The toxicity of H_2S was reviewed to determine the appropriate concentration values and exposure times which can lead to a fatality. The American Institute of Chemical Engineers (AICE) has recently published guidelines for estimating the quantitative risk from exposure to hazardous chemicals (AICE, 1989). The report outlines the probabilistic approach to estimating the impact on human health from exposure to hazardous chemicals, such as H_2S, using probit (PROBability unIT) equations which can be written as:

$$Pr = a + b \ln(C^m t) \qquad (1)$$

where a, b and m are constants specific to each chemical based on animal and human (where available) exposure data, C is the concentration in ppm and t is the exposure time in minutes. The probit variable is normally distributed with a mean value of 5 and a standard deviation of 1. Thus the probability of a fatality P can be found by evaluating the following equation (Abramowitz and Stegun, 1970):

$$P\,(Pr\text{-}5) = \int_{-\infty}^{Pr\text{-}5} \frac{1}{\sqrt{2\pi}}\; e^{\frac{y^2}{2}}\, dy \qquad (2)$$

For H_2S, the probit coefficients for a, b and m are -31.42, 3.008 and 1.43, respectively (AICE 1989). For example, a 30 minute exposure to H_2S at a concentration of 500 ppm gives a fatality probability of about 0.71. However, if the exposure time is reduced to 15 minutes, the fatality probability becomes 0.06. A 10 minute exposure time gives a fatality probability of 0.003. Therefore, predicting the fatality probability using probit equations can be very sensitive to concentration, exposure time and the values of the coefficients in the probit equation.

Environment Canada (1984) has recommended an H_2S concentration of 300 ppm for the IDLH (immediately dangerous to life and health) and a 30 minute exposure at 600 ppm for the LC_{LO} (lower lethal concentration). The review by Baynes (1986) showed 13 cases of accidental death and severe poisoning documented in the literature in the period from 1979 to 1985. It appears that exposures to H_2S at 200 ppm for 10 to 14 minutes will cause unconsciousness and exposures above 600 ppm will cause death in a few minutes. However, the Canadian Centre for Occupational Health and Safety (1988) reports that exposures above 500 ppm rapidly cause unconsciousness and death.

Based on the available information on H_2S toxicity and the inherent uncertainty in determining a precise time-integrated H_2S concentration that will lead to death, a simplistic approach to estimating fatalities was chosen. A concentration of 500 ppm ($7.07 \times 10^{-4} kg \cdot m^{-3}$) with an exposure period of 5 minutes was selected as the reference lethal value. Thus, all H_2S exposures with a time-integrated concentration exceeding $2.1 \times 10^{-1} kg \cdot s \cdot m^{-3}$ are assumed to cause death, and all exposures below this value do not cause death. Inserting an H_2S concentration of 500 ppm for 5 minutes into the probit equation for H_2S suggest that this exposure criteria is conservative. However, a concentration of 500 ppm is consistent with the value used in previous H_2S safety assessments (Kempe and Russell, 1987; O'Neill, 1988).

3. ACCIDENT RATES AND TRANSPORTATION DATA

The rate of accidents along the transport route from Sarnia to BHWP have been calculated for 3 transport zones: rural, suburban and urban. The vehicle accident statistics along the truck route in Ontario were taken from the Ontario Ministry of Transportation (1988a,b). For a long truck containing the H_2S, the truck accident rate Ra_i can be written as:

$$Ra_i = \frac{La}{\tau} \, Va_i \quad \text{(long truck-km)}^{-1} \qquad (3)$$

where i is the transportation zone (rural, suburban, urban), La is the long truck accident fraction (long truck accidents per vehicle accidents), τ is the long truck traffic fraction (long truck-km per vehicle-km), and Va_i is the vehicle accident rate in transportation zone i (vehicle-km)$^{-1}$.

The long truck accident fraction La was about 0.02, and the long truck traffic fraction τ was about 0.05. The truck accident statistics are listed in Table 1. The accident rate units have been simplified to km^{-1} for clarity.

Table 1: Truck accident statistics.

Zone i	Vehicle Accident Rate Va_i (vehicle-km)$^{-1}$	Truck Accident Rate Ra_i (km^{-1})
Rural	7.7×10^{-7}	3.1×10^{-7}
Suburban	1.2×10^{-6}	4.8×10^{-7}
Urban	8.2×10^{-7}	3.3×10^{-7}

The truck travel distance d from Sarnia to BHWP is about 200 km. The travel fractions and populations densities for the transport zones are listed in Table 2.

Table 2: Transportation travel fractions and population densities.

Zone i	Travel Fraction Ft_i	Population Density ρ_i (persons·km^{-2})
Rural	0.75	10
Suburban	0.05	400
Urban	0.20	1000

4. ACCIDENT SEVERITY CATEGORIES

There have not been any registered transport accidents involving H_2S in Canada upon which to base projections of the rate of accidents leading to a release of H_2S, and the number of H_2S transport accidents in the United States is very limited. However, there have been a number of transport accidents involving Liquefied Petroleum Gas (LPG) both in Canada and the United States. Since the physical properties of LPG and H_2S are similar and both are transported as liquids in fully pressurized containers, it appears reasonable to apply LPG accident statistics to H_2S transport.

Transport Canada's Transport Dangerous Goods Directorate has been compiling accident statistics for LPG (and other hazardous chemicals) since 1986. A total of 52 accidents involving trucks have been recorded over the period from 1986 to 1988 (Arnott, 1989).

In the United States, the Office of Hazardous Materials of the U.S. Department of Transport (DOT) has been maintaining a record of surface transportation incidents. An analysis of LPG accidents over the period from 1972 to 1985 by Swoveland and Cawdery (1986) has indicated a total of 112 truck transport accidents which resulted in a release of LPG. Both the Canadian and U.S. LPG data bases were used to determine the accident severity categories for H_2S transport.

The Canadian accident statistics at Transport Canada (Arnott, 1989) were examined to identify the fraction of accidents resulting in a "release" of H_2S. After reviewing the data base, it was necessary to consider an alternative approach to assess the release probability for an accident since a large fraction of transport accidents are not "dangerous occurrences" as defined in the regulations (Transport Canada, 1986) and are not reported (Learning, 1990). To estimate the probability of accidents that can release H_2S, the fault tree analysis of LPG tanker containment systems by Saccomanno (1990) was adopted for this study. The release probability for trucks, Pr, was estimated to be 0.037.

The more detailed U.S. accident statistics (Swoveland and Cawdery, 1986) were used to subdivide the release incidents into the following severity categories:

a) Total inventory release; a large fraction ($\geq$ 50%) of inventory lost due to external puncture, external heat, body/weld failure, defective/loose fittings, unknown,

b) Partial inventory release; a small fraction (< 50%) of inventory lost due to minor leaks from the vessel,

c) Total inventory release with fire and explosion.

The accident severity and the probability of accidents in each severity category are listed in Table 3.

Table 3: Accident severity categories.

Severity Category j	Severity Probability Pa_j
Total Release	0.268
Partial Release	0.688
Fire & Explosion	0.045

5.   H₂S EMISSION AND DISPERSION

5.1 Emission Rate

It is difficult to estimate the H_2S emission rate following a transport accident which leads to a release. A number of factors such as the size and location of the hole, the internal vapour pressure of the H_2S in the tanker, the size of the liquid spill and the presence or absence of fire will dictate the emission rate and behaviour of H_2S near the accident scene.

The internal pressure of the transport vessel is a function of the H_2S temperature but is approximately 2000 kPa at 25° C (Environment Canada, 1984). Therefore, a hole in the vessel could release a large fraction of the H_2S inventory over a short period of time. For example, a 0.25 m diameter hole at the bottom of the tank could have a discharge rate of about 3000 $L.s^{-1}$ and could empty a large rail tanker (80000 L) in about 20 seconds (Environment Canada, 1984). The same sized hole at the top of the tank above the liquid level could empty the contents in about 4 minutes.

Since the LPG accident data bases do not contain information on hole size, it was necessary to make some assumptions on the diameter of the hole for each of the release severity categories. After reviewing the accident data bases (Swoveland and Cawdery, 1986;

Arnott, 1989) and the behaviour of H_2S as a function of hole diameter (Environment Canada, 1984), the "total release" category was assigned an equivalent hole diameter of 0.25 m at the bottom of the tank and the "partial release" category was assigned a value of 0.01 m at the top of the tank.

The truck tank was assumed to have a capacity of 5100 gallons (US) or 19.3 m^3. Assuming the vessel is 80% full, each truck has a payload capacity, M, of about 1.5 x 10^4 kg of H_2S. The number of shipments per year, n, is approximately 24 based on an annual demand of 350 Mg (Goodwin, 1990). The release of H_2S from the tankers into the environment was calculated using the Environment Canada AQPAC model and computer code (Daggupaty, 1988a; 1988b) for the "total" and "partial" release categories. The results are given in Table 4.

Table 4: H_2S emissions to the environment.

Release Type	H2S Emission	
	Total Release	Partial Release
Puff (kg)	4.12 x 10^3	0.0
Short Term (kg·s^{-1})	3.51 x 10^{-1}	2.23 x 10^{-1}
Long Term (kg·s^{-1})	0.0	1.40 x 10^{-6}

For the "total" release scenario, 4.12 x 10^3 kg of H_2S, or about 27% of the inventory, was calculated by AQPAC to be released as a puff during the first 7 seconds. The remaining H_2S was released at a rate of 3.51 x 10^{-1} kg·s^{-1} until the tank was depleted.

For the "partial" release scenario, AQPAC calculated an H_2S release rate of 2.23 x 10^{-1} kg·s^{-1} for the first 307 minutes, and then at a much smaller rate of 1.4 x 10^{-6} kg·s^{-1} until depleted.

The "fire & explosion" severity category was treated in a different manner from the other two release scenarios since an H_2S explosion has the effect of dispersing the gases (H_2S and the combustion product SO_2) to great heights and significantly reducing the ground level concentrations. However, the resulting fireball and flying debris from the BLEVE (Boiling Liquid-Expanding Vapour Explosion) can cause considerable damage and risk to life (Swoveland and Cawdery, 1986).

Because of the uncertainty associated with modelling a BLEVE and the subsequent dispersion of H_2S and SO_2, the exposure to H_2S and SO_2 concentrations in air was not assessed for this severity category. Instead, the lethal area from the fireball and flying debris was estimated using the ground level radius r of the fireball (AICE, 1989):

$$r = 4.21 \, M^{0.325} \quad (m) \tag{4}$$

where M is the mass of H_2S in the transport vessel (kg). Using an H_2S payload capacity of 1.5×10^4 kg, the radius becomes 96 m and the lethal area becomes 2.9×10^{-2} km^2.

5.2 Atmospheric Dispersion

The release of pressurized H_2S from the transport vessel will generally be gaseous if the puncture (or hole) is above the liquid level of the tank (Environment Canada, 1984). If the puncture is near the bottom of the tank, a portion of the H_2S will flash to vapour and the remainder will spill as a liquid. Since H_2S boils at $-60.7°$ C (at 1 atmosphere), the spilled liquid will boil rapidly and release H_2S as a vapour. The emission rate is a function of the spill size and ambient conditions near the accident scene.

Since a large fraction of the inventory can be released over a short period of time, a "puff" model will reflect the initial behaviour of the released H_2S more accurately than a "plume" model.

Given the low temperature of the gas and the high molecular weight of H_2S with respect to air, the use of a heavy gas dispersion model appears to be appropriate for a realistic assessment of the behaviour of the plume. The Atmospheric Environment Service of Environment Canada has developed an air quality package of programs (AQPAC) to calculate the impact from the accidental release of toxic chemicals to the atmosphere. The AQPAC model and computer code (Daggupaty, 1988a; 1988b) has been designed to calculate both the H_2S source strength and the dispersion downwind. The initial stage of the release can be treated using a heavy gas "puff" model and the later stages of the release can be treated using a heavy gas "plume" model.

The AQPAC code was used to calculate the crosswind and downwind distances where the time-integrated H_2S concentration dropped below 2.10×10^{-1} kg·s·m^{-3} for the "total" and "partial" release scenarios and for each of the 6 Pasquill weather stability classes. The lethal plume areas are listed in Table 5 and the lethal downwind plume distances are listed in Table 6. The weather probability data Pw_k were taken from reference Canadian weather frequency data (CSA, 1987).

Table 5: Lethal H_2S plume area.

Pasquill Weather Stability Class	Lethal H_2S Plume Area A_{jk} (km^2)		
	Total	Partial	Fire & Explosion[*]
A	1.93×10^{-1}	6.57×10^{-3}	2.90×10^{-2}
B	1.40×10^{-1}	2.31×10^{-3}	2.90×10^{-2}
C	7.05×10^{-2}	1.00×10^{-3}	2.90×10^{-2}
D	1.42×10^{-1}	9.95×10^{-4}	2.90×10^{-2}
E	6.77×10^{-1}	2.87×10^{-3}	2.90×10^{-2}
F	3.65×10^{0}	8.11×10^{-3}	2.90×10^{-2}

[*]   Estimated lethal area from the exploding fireball and flying debris; assumed to be independent of weather class.

Table 6: Lethal H_2S downwind plume distance.

Pasquill Weather Stability Class	Lethal H_2S Downwind Plume Distance X_{jk} (km)		
	Total	Partial	Fire & Explosion
A	8.00×10^{-1}	1.40×10^{-1}	9.60×10^{-2}
B	7.50×10^{-1}	1.00×10^{-1}	9.60×10^{-2}
C	6.50×10^{-1}	1.00×10^{-1}	9.60×10^{-2}
D	$1.05 \times 10^{+0}$	8.00×10^{-2}	9.60×10^{-2}
E	$2.80 \times 10^{+0}$	1.40×10^{-1}	9.60×10^{-2}
F	$1.50 \times 10^{+1}$	2.20×10^{-1}	9.60×10^{-2}

From Table 6 the maximum downwind distance $Xmax$ for a lethal H_2S plume was calculated to be 15 km for the "total" release category under class F weather conditions with a wind speed of 2 m·s^{-1}. The average lethal plume width was approximately 240 m.

6.   RISK ASSESSMENT MODEL

The following risk assessment model was developed to calculate the public risk from transporting H_2S from Sarnia to the BHWP in

Ontario. The model can be used for the truck transportation or rail transportation, and depends on the travel distances, accident statistics, severity categories and population data.

6.1 Accident Frequency

The accident frequency f_{ijk} is defined here as the probability per year of an H_2S transportation accident occurring along the route to BHWP that can release H_2S such that the time-integrated concentration in air is sufficient to cause death to an exposed individual near the accident scene (see Figure 1). It also includes those accidents that can lead to public fatalities as a result of fire and explosion. The frequency is dependent on the number of H_2S shipments, the distance in each transportation zone (rural, suburban, urban), the accident rate, the fraction of accidents that can release H_2S to the environment, the fraction of accidents in each severity category and the fraction of time that a particular weather stability class occurs.

The frequency f_{ijk} of an accident in transportation zone i for accident severity category j during weather stability class k can be written as:

$$f_{ijk} = n \, d \, Ft_i \, Ra_i \, Pr \, Pa_j \, Pw_k \quad (a^{-1}) \qquad (5)$$

where i is the transportation zone (rural, suburban, urban), j is the accident severity category (with release of H_2S), k is the weather stability class, n is the number of shipments per year (a^{-1}), d is the one-way travel distance per shipment (km), Ft_i is the fraction of travel in zone i, Ra_i is the accident rate in transportation zone i (km^{-1}), Pr is the probability that an accident will lead to a release of H_2S, Pa_j is the probability that an accident leads to a release in severity category j and Pw_k is the probability of weather stability class k.

Since the probabilities of accidents in transportation zone i, accident severity category j and weather stability class k are mutually exclusive, their frequencies can be summed to give the total accident frequency f_{tot}:

$$f_{tot} = \sum_{i=1}^{3} \sum_{j=1}^{3} \sum_{k=1}^{6} f_{ijk} \quad (a^{-1}) \qquad (6)$$

where f_{ijk} is the frequency of an accident in transportation zone i for accident severity category j during weather stability class k (a^{-1}).

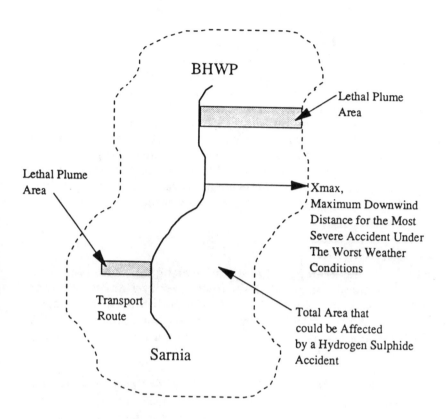

Figure 1: H_2S transportation route and plume areas.

6.2 Individual Risk - Along The Route

The individual risk along the route has been defined as the fatality risk r_{jk} to an individual located along the transportation route due to an accident involving the release of H_2S. While the individual risk can be defined for each major segment of the route, for simplicity, the risk has been defined for the entire route and is not dependent on the transportation zone i. The individual fatality risk can be written as follows:

$$r_{jk} = \sum_{i=1}^{3} f_{ijk} \; E_{jk} \quad (a^{-1}) \qquad (7)$$

where E_{jk} is the individual exposure factor for accident severity category j and weather stability class k. The individual risk is based on the accident frequency f_{ijk} summed over the transportation zones i and the exposure factor E_{jk}. The exposure factor can be defined in terms of the dimensions, relative to the total route length, of the H_2S plume whose time-integrated concentration can lead to a fatality for an accident in severity category j and weather class k. (The reduced exposure due to indoor shielding effects has not been credited). E_{jk} corresponds to the probability that, given an accident occurs somewhere along the route, an individual at any particular point along the route will be exposed to a lethal time-integrated concentration. For the severity category involving fire and explosion, the exposure factor considers the "plume area" as the estimated lethal area from the exploding fireball and flying debris. The individual exposure factor E_{jk} can be written as follows:

$$E_{jk} = \frac{L_{jk}}{d} \tag{8}$$

where L_{jk} is the average lethal distance along the route (km) for a lethal H_2S plume for accident severity category j during weather stability class k, and d is the one-way travel distance per shipment (km). L_{jk} has been calculated by considering the dimensions of the H_2S plume and the assumption that all wind directions are equally probable.

For the "Total Release" and "Partial Release" accident severity categories, where the wind direction can affect the exposure, the average lethal distance along the route L_{jk} can be written as (see Appendix A):

$$L_{jk} = \frac{2}{\pi}\left[X_{jk} \ln\left(\tan\left(\frac{\phi^*_{jk}}{2} + \frac{\pi}{4}\right)\right) + \frac{w_{jk}}{2} \ln\left(\frac{1}{\tan\left(\frac{\phi^*_{jk}}{2}\right)}\right)\right] \tag{9}$$

where X_{jk} is the downwind distance for a lethal H_2S plume (km), w_{jk} is the average width of the lethal H_2S plume (km) and ϕ^*_{jk} is the wind angle at which the downwind corner of the assumed rectangular, lethal H_2S plume crosses the transport route (see Appendix A):

$$\phi^*_{jk} = \tan^{-1}\left(\frac{w_{jk}}{2X_{jk}}\right) \tag{10}$$

For the "Fire & Explosion" severity category, where the wind direction is not a factor, the average lethal distance along the route

L_{jk} becomes:

$$L_{jk} = 2\ X_{jk}\ (km) \tag{11}$$

where the factor 2 accounts for accidents that can happen as the transport vehicle approaches and recedes from the individual.

Finally, the total individual risk R_{ind} can be calculated by summing the individual fatality risk over the accident severity category j and the weather stability class k:

$$R_{ind} = \sum_{j=1}^{3} \sum_{k=1}^{6} r_{jk}\ (a^{-1}) \tag{12}$$

where r_{jk} is the individual fatality risk for accident severity category j during weather stability class k (a^{-1}).

The average individual fatality risk R_{av} to persons residing within a distance $Xmax$ of the transport route can be estimated using the total collective risk and the total population near the route (see Section 6.4):

$$R_{av} = \frac{R_{col}}{N_{TOT}}\ (a^{-1}) \tag{13}$$

where R_{col} is the total collective risk (persons$\cdot$a^{-1}) and N_{TOT} is the total population along the H_2S transportation route (persons).

6.3 Individual Risk - Off The Route

The individual risk off the route has been defined in a manner similar to that of the previous section. At a distance y perpendicular to the route, the fatality risk to that individual is $r_{jk}(y)$ due to an accident involving the release of H_2S. While the individual risk can be defined for each major segment of the route, for simplicity, the risk has been defined for the entire route and is not dependent on the transportation zone i. The individual fatality risk off the route can be written as follows:

$$r_{jk}\ (y) = \sum_{i=1}^{3} f_{ijk}\ E_{jk}\ (y)\ (a^{-1}) \tag{14}$$

where $E_{jk}(y)$ is the individual exposure factor for accident severity category j and weather stability class k at a location y off the route. The individual risk is based on the accident frequency f_{ijk} summed over the transportation zones i and the exposure factor $E_{jk}(y)$ which corresponds to the probability that, given an accident occurs somewhere along the route, an individual at a location y perpendicular to the route will be exposed a lethal time-integrated concentration. For the severity category involving fire and explosion, the exposure factor considers the "plume area" as the estimated lethal area from the exploding fireball and flying debris.

The individual exposure factor $E_{jk}(y)$ takes into consideration the length of the route S_{jk} where an accident has the potential to affect a person at y and the average value of the wind angle θ_{jk} along this section of the route that directs the lethal plume towards the individual (see Appendix B). The exposure factor off the route can be written as follows:

$$E_{jk}(y) = \frac{S_{jk}(y)}{d} \frac{\theta_{jk}(y)}{2\pi} \tag{15}$$

where $S_{jk}(y)$ is the length of the route (km) for a lethal H_2S plume for accident of severity category j during weather stability class k to affect a person at y, d is the one-way travel distance per shipment (km) and $\theta_{jk}(y)$ is the average value of the wind angle that can affect a person at y. $S_{jk}(y)$ has been calculated by considering the dimensions of the H_2S plume, the off-route distance y and the assumption that all wind directions are equally probable.

The length of the route $S_{jk}(y)$ where an accident has the potential to affect a person at y can be written as:

$$S_{jk}(y) = 2\sqrt{X^2_{jk} - y^2} \ \ (km) \tag{16}$$

$$= 2X^*_{jk}(y)$$

where $x_{jk}^*(y)$ is half the length of the route for a lethal H_2S plume to affect a person at y (km).

For the "Total Release" and "Partial Release" accident severity categories, where the wind direction can affect the exposure, the average wind angle $\theta_{jk}(y)$ can be written as (see Appendix B):

$$\theta_{jk}(y) = \frac{2}{X^*_{jk}(y)} \int_0^{x^*_{jk}(y)} \alpha_{jk}(y,x) \, dx \tag{17}$$

where $2\alpha_{jk}(y,x)$ is the wind angle that can affect a person at y given an H_2S accident at x. The integral in the preceding equation is solved using numerical integration.

For off-route distances between half the plume width and the downwind distance of the plume, that is $w_{jk}/2 \le y \le X_{jk}$, the angle $\alpha_{jk}(y,x)$ can be written as:

$$\alpha_{jk}\ (y,x)\ =\ \text{sin-1}\left[\frac{w_{jk}}{2y}\ \sin\ \left(\text{tan}^{-1}\left(\frac{y}{x}\right)\right)\right],\ 0 \le x \le x^*_{jk} \qquad (18)$$

For off-route distances between zero and half the plume width, that is $0 \le y \le w_{jk}/2$, the angle $\alpha_{jk}(y,x)$ can be written as:

$$\alpha_{jk}\ (y,x)\ =\frac{\pi}{2}\ ,\ 0 \le x \le x'_{jk}\ (y)$$

$$=\ \text{sin}^{-1}\left[\frac{w_{jk}}{2y}\ \sin\ \left(\text{tan}^{-1}\left(\frac{y}{x}\right)\right)\right],\ x'_{jk}\ (y) < x \le x^*_{jk}\ (y) \qquad (19)$$

where $x_{jk}^*(y)$ is half the length of the route for a lethal H_2S plume to affect a person at y (km), w_{jk} is the average width of the lethal H_2S plume (km). Distance $x_{jk}'(y)$ is the distance along the route where the H_2S plume is over the location y for a wind angle $\alpha_{jk} = \pi/2$. It can be written as (see Appendix B):

$$x'_{jk}\ (y)\ =\sqrt{\left(\frac{w_{jk}}{2}\right)^2 - y^2}\ \ \ \ (\text{km}) \qquad (20)$$

For the "Accident & Explosion" severity category, where the wind direction is not a factor, the average value of the wind angle $\theta_{jk}(y)$ becomes:

$$\theta_{jk}\ (y)\ =\ 2\pi \qquad (21)$$

The total individual risk off the route $R_{ind}(y)$ can be calculated by summing the individual fatality risk (at location y) over the accident severity category j and the weather stability class k:

$$R_{ind}\ (y)\ =\ \sum_{j=1}^{3}\ \sum_{k=1}^{6}\ r_{jk}(y) \quad (a^{-1}) \qquad (22)$$

where $r_{jk}(y)$ is the individual fatality risk at location y for accident severity category j during weather stability class k (a^{-1}).

6.4 Collective Risk

The collective impact on the population from a transportation accident involving H_2S can be calculated by multiplying the accident probability p_{ijk} with the potential number of people N_{ijk} that

---

could be killed by an accident in each transportation zone, severity category and weather stability class. It can be written as follows:

$$C_{ijk} = f_{ijk} N_{ijk} \quad \text{(persons a}^{-1}) \tag{23}$$

where f_{ijk} is the frequency of an accident in transportation zone i for accident severity category j during weather stability class k (a^{-1}) and N_{ijk} is the number of people that could be killed in an accident in transportation zone i for accident severity category j during weather stability class k (persons):

$$N_{ijk} = \rho_i A_{jk} \quad \text{(persons)} \tag{24}$$

where ρ_i is the population density in transportation zone i (persons·km^{-2}) and A_{jk} is the area of H$_2$S plume whose time-integrated concentration is greater than or equal to the level required to cause death for accident severity category j and weather stability class k (km^2).

The total collective risk R_{col} to the population along the route can be calculated by summing the collective risk c_{ijk} over the transport zone i, the accident severity category j and the weather stability class k:

$$R_{col} = \sum_{i=1}^{3} \sum_{j=1}^{3} \sum_{k=1}^{6} C_{ijk} \quad \text{(persons a}^{-1}) \tag{25}$$

where c_{ijk} is the collective risk in transportation zone i for accident severity category j during weather stability class k (persons·a^{-1}). The total population N_{TOT} along the H$_2$S route which is used in the calculation of average individual risk is based on the maximum downwind distance of a lethal H$_2$S plume for the worst severity category and weather condition, $Xmax$, and the travel distance and population density along the route. It can be approximated by the following:

$$N_{TOT} = \sum_{i=1}^{3} \rho_i Ft_i 2[d + 2Xmax] Xmax \quad \text{(persons)} \tag{26}$$

7. RESULTS AND DISCUSSION

The frequency of a transport accident leading to a release of H$_2$S in the rural, suburban and urban zones is given in Table 7. The results indicate that a truck accident involving H$_2$S emissions can be expected about once every 17,000 years. The highest accident frequency was predicted to occur in the rural zone. This was not unexpected since the accident rates among the three zones are similar but the travel distance in the rural zone is much larger than the urban or suburban zones.

Table 7: Accident frequency in the transportation zones.

Zone i	Accident Frequency (a^{-1})
Rural	4.1×10^{-5}
Suburban	4.3×10^{-6}
Urban	1.2×10^{-5}
Total	5.7×10^{-5}

The most probable accident scenario was a partial release in the rural zone with Pasquill weather stability class D. The frequency of this event occurring was 1.6×10^{-5} a^{-1}. The number of fatalities expected for this scenario was calculated to be 1.0×10^{-2} persons.

The risk to an individual residing along the transportation route has been calculated as a function of accident severity category and weather stability class. Since an individual can reside anywhere along the route, the risk is not a function of transportation zone. The individual exposure factor to persons along the route is listed in Table 8. The individual risks are listed in Table 9.

Table 8: Individual exposure factor along the route.

Pasquill Weather Stability Class	Individual Exposure Factor E_{ik}		
	Total	Partial	Fire & Explosion[*]
A	1.4×10^{-3}	2.6×10^{-4}	9.6×10^{-4}
B	1.1×10^{-3}	1.4×10^{-4}	9.6×10^{-4}
C	7.2×10^{-4}	7.5×10^{-5}	9.6×10^{-4}
D	9.5×10^{-4}	8.4×10^{-5}	9.6×10^{-4}
E	1.9×10^{-3}	1.4×10^{-4}	9.6×10^{-4}
F	2.5×10^{-3}	2.4×10^{-4}	9.6×10^{-4}

The accident scenario with the largest individual risk along the route was a Total release with Pasquill weather stability class D. The risk was $8.2 \times 10^{-9} a^{-1}$. The total individual risk along the route R_{ind} for truck transport of H_2S was $2.7 \times 10^{-8} a^{-1}$, which is small. For comparison, the individual fatality risk from lightning is more than 10 times larger at $5 \times 10^{-7} a^{-1}$ (United States Nuclear Regulatory Commission, 1975).

Table 9: Individual risk along the route.

Pasquill Weather Stability Class	Individual Fatality Risk r_{jk} (a^{-1})		
	Total	Partial	Fire & Explosion*
A	2.1×10^{-10}	1.0×10^{-10}	2.5×10^{-11}
B	1.0×10^{-9}	3.3×10^{-10}	1.5×10^{-10}
C	1.1×10^{-9}	2.9×10^{-10}	2.5×10^{-10}
D	8.2×10^{-9}	1.9×10^{-9}	1.4×10^{-9}
E	2.9×10^{-9}	5.5×10^{-10}	2.5×10^{-10}
F	6.6×10^{-9}	1.6×10^{-9}	4.2×10^{-10}

The total individual fatality risk as a function of distance off the route $R_{ind}(y)$ is illustrated in Figure 2. The risk was estimated to decrease from a maximum value of $2.7 \times 10^{-8} a^{-1}$ at $y = 0$ km, to 2.8×10^{-9} a^{-1} at $y = 2.5$ km. The risk then decreased slowly with distance y until the maximum plume distance of 15 km was reached. Only the Total Release category under Class F weather conditions was estimated to affect persons at distances beyond 2.8 km (see Table 6).

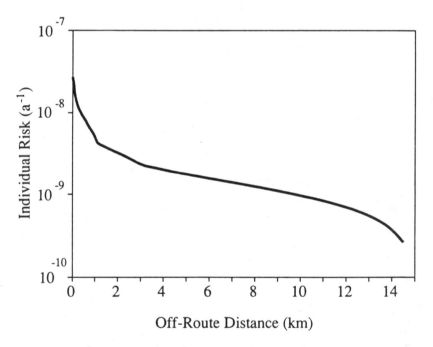

Figure 2: Total off-route individual risk from H_2S accidents.

The collective risk to the population along the route was calculated for the three transport zones. The results are listed in Table 10. For each transportation zone, the collective risk to the population is small.

Table 10: Collective risk in the transportation zones.

Zone i	Collective Fatality Risk (persons·a^{-1})
Rural	8.8 x 10^{-5}
Suburban	3.6 x 10^{-4}
Urban	2.5 x 10^{-3}
Total	3.0 x 10^{-3}

The average individual risk to persons residing within the maximum downwind distance of the plume (15 km) along the route was estimated from the total collective risk and population near the route. The average individual risk R_{av} was 1.9×10^{-9} a^{-1} using a total population N_{TOT} of 1.6×10^{6} persons, which is very small.

A summary of the total accident frequency, total individual risk, average individual risk and total collective risk for the transportation of H_2S are given in Table 11.

Table 11: Summary of H_2S transportation results.

Category	Transportation Results
Accident Frequency f_{tot} (a^{-1})	5.7 x 10^{-5}
Individual Risk R_{ind} (a^{-1})	2.7 x 10^{-8}
Average Individual Risk R_{av} (a^{-1})	1.9 x 10^{-9}
Collective Risk R_{col} (person·a^{-1})	3.0 x 10^{-3}

The collective risk on the population from an H_2S transport accident is calculated by multiplying the accident frequency with the potential number of people that could be affected by an accident in each transportation zone, severity category and weather stability class. The results can be given graphically by plotting the downward cumulative frequency with the potential number of fatalities along the route (see Figure 3).

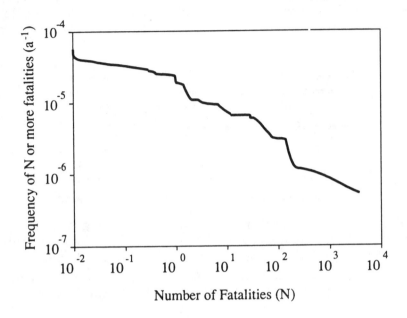

Figure 3: Societal risk from H$_2$S accidents.

The curve indicates a general trend of decreasing frequency as the number of potential fatalities increases. For H$_2$S transport via truck, the frequency of exceeding 1 fatality was calculated to be about 2×10^{-5} a^{-1}. The frequency of a more "catastrophic event" having 10^3 fatalities was about 8×10^{-7} a^{-1}. These collective risk results and the individual risks can be compared with those from other risk assessments for the transport of hazardous goods near population centres (Concord, 1987; Hubert and Pages, 1989; Walton et al., 1990).

The French study by Hubert and Pages (1989) examined the transport of motor fuel, LPG, ammonia and chlorine along two routes through the city of Lyon. Their analysis showed that the frequency of exceeding 1 fatality was about 10^{-2} a^{-1}, which is about a factor of 10^2 greater than the results in the present H$_2$S study. The frequency of exceeding 10^3 fatalities was between 10^{-5} and 10^{-6} a^{-1}, depending on the choice of route, which is similar to the H$_2$S analysis. The maximum individual risk along the transport route near Lyon was estimated to be 2×10^{-5} a^{-1}, which is several orders of magnitude greater than the H$_2$S risk analysis from Sarnia to BHWP.

The study by Walton et al. (1990) investigated the public risk from transporting ammonia from Sarnia to Lakeview generating station in Toronto. The probability of exceeding 1 fatality was about 3×10^{-5} a^{-1} and the frequency of exceeding 100 fatalities was about $2 \times 10^{-6} a^{-1}$. The transportation study by the Concord Scientific Corporation (1987) of dangerous goods through the Toronto area found that the frequency of exceeding 10^3 fatalities was between 10^{-8} and 10^{-6} a^{-1}, depending on the choice of route and section of travel near the city. The fatality risk to an individual also depended on the travel segment but was approximately 10^{-4} a^{-1} near the rail line and dropped rapidly to about 10^{-6} a^{-1} at 400 m from the line. The collective risk to the population along the route was calculated to be 4 persons$\cdot a^{-1}$ compared with 3×10^{-3} persons$\cdot a^{-1}$ for the H_2S study. These existing transport risks are significantly greater than the risks from the proposed H_2S transport to BHWP, however, the quantity of hazardous goods transported through the Toronto area is much greater than the amount of H_2S being transported from Sarnia to BHWP.

The areas in this risk assessment model with the largest parameter uncertainty are the estimates of the H_2S emission during an accident which results in a release, and the impact of H_2S on exposed individuals. Consequently, better estimates of public risk can be made with an improved understanding of those transport accidents that can lead to H_2S emissions and improved knowledge of H_2S toxicity on humans.

8.   CONCLUSIONS

The methodology for assessing the public risk from transporting hydrogen sulphide under pressure by truck has been developed. The model is simple and makes extensive use of existing accident data and release probabilities from pressurized transport containers. Individual fatality risks can be estimated to persons on-route and off-route. Collective risks to the public can also be estimated. The assessment model can be applied to both rail and truck modes of transport.

In the example calculation, the frequency of a transport accident leading to a release of H_2S was estimated to be about $5.7 \times 10^{-5} a^{-1}$ along the route. The fatality risk to an individual residing along the H_2S transportation route was conservatively estimated to be 2.7 x

10^{-8} a^{-1}. This risk was estimated to decrease by an order of magnitude at a distance of about 2.5 km off the route.

The individual risk along the route was nearly 2 orders of magnitude less than the proposed early fatality risk safety goal of 10^{-6}a^{-1} recommended for Ontario Hydro's nuclear generating stations (King, 1990). The average individual risk to the population that could be affected by an H_2S accident, that is, those individuals living within 15 km of the transport route, was 1.9×10^{-8}a^{-1}. The fatality risks from transporting H_2S were compared with the risks associated with transporting other hazardous goods and found to be small. Consequently, the public fatality risk of transporting H_2S from Sarnia to BHWP is considered to be acceptably small.

9. REFERENCES

ABRAMOWITZ, M., & STEGUN, I.A. (1970). Handbook of Mathematical Functions with Formulas, Graphs and Mathematical Tables. Dover Publications Inc. New York, NY.

AMERICAN INSTITUTE OF CHEMICAL ENGINEERS. (1989). Guidelines For Chemical Process Quantitative Risk Analysis. Center for Chemical Process Safety. TP155.5.G76. New York, NY.

ARNOTT, N. (1989). Personal Communication. Transport Canada, Transport Dangerous Goods Directorate, Risk Management Branch. Ottawa, ON.

BAYNES, C.J. (1986). Probabilistic Consequence Assessment Of Hydrogen Sulphide Releases From A Heavy Water Plant: Addendum and Summary. Research report prepared by Monserco Limited for the Atomic Energy Control Board. Project Number 85.2.1, Report INFO-0102-5. Ottawa, ON.

CANADIAN CENTRE FOR OCCUPATIONAL HEALTH AND SAFETY (1988). CHEMINFO, Hydrogen Sulphide. Hamilton, ON.

CANADIAN STANDARDS ASSOCIATION. (1987). Guidelines For Calculating Derived Release Limits For Radioactive Material In Airborne And Liquid Effluents For Normal Operation Of Nuclear Facilities. Report No. CAN/CSA-N288.1-M87, Toronto, ON.

CONCORD SCIENTIFIC CORPORATION. (1987). Risk Assessment For Rail Transportation Of Dangerous Goods Through The Toronto Area. Report CSC.J1029.1. Prepared for Toronto Area Dangerous Goods Rail Task Force. Toronto, ON.

DAGGUPATY, S.M. (1988a). AQPAC User's Reference Manual, Version 5. Draft Report, April 1988. Atmospheric Environment Service, Environment Canada. Downsview, ON.

DAGGUPATY, S.M. (1988b). Response To Accidental Release Of Toxic Chemicals Into The Atmosphere Using - AQPAC. In Natural and Man-Made Hazards, (M.I.El-Sabh and T.S. Murty, editors). D. Reidel Publishing Company, pp. 599-608.

ENVIRONMENT CANADA. (1984, July). Hydrogen Sulphide, Environmental And Technical Information For Problem Spills (EnviroTIPS). Environment Canada, Technical Services Branch, Environmental Protection Programs Directorate, Environmental Protection Service. Ottawa, ON.

GOODWIN, S.R. (1990). H_2S Supply Study For The Bruce Heavy Water Plant. Ontario Hydro, Design and Development Division - Generation Report No. 89089. Toronto, ON.

HUBERT, P., & PAGES, P. (1989). Risk management for hazardous materials transportation: a local study in Lyons. Risk Analysis, 9, 4, pp. 445-451.

KEMPE, T.F., & Russell, S.B. (1987). Hydrogen Sulphide - Preliminary Transportation Safety Review. Ontario Hydro, Design and Development Division - Generation Report No. 87053. Toronto, ON.

KEMPE, T.F. (1992). Used Fuel Transportation Assessment, Public Radiological Safety Assessment. Ontario Hydro, Design and Development Division - Generation Report No. 92014. Toronto, ON.

KING, F.K. (1990). Risk-Based Safety Goals For Ontario Hydro Nuclear Generation Stations. Ontario Hydro, Design and Development Division - Generation Report No. 89412. Toronto, ON.

LEARNING, D.A. (1990). Personal Communication. Transport Canada, Transport Dangerous Goods Directorate. Ottawa, ON.

O'NEILL, M.K. (1988). Bruce Heavy Water Plant - An Assessment
Of Public Risk From Acute H_2S Releases. Ontario Hydro,
Radioactivity Management and Environmental Protection Report
RMEP-IR-07131-3. Toronto, ON.

ONTARIO MINISTRY OF TRANSPORTATION. (1988a). '87
Ontario Road Safety - Annual Report.

ONTARIO MINISTRY OF TRANSPORTATION. (1988b).
Provincial Highways traffic Volumes: King's Highways,
Secondary Highways and Tertiary Roads - 1987.

SACCOMANNO, F.F. (1990). Release probabilities of chlorine and
LPG from rail and truck bulk tankers. Report prepared for Stone &
Webster Canada Limited. In Ontario Hydro Lakeview Thermal
Generating Station Rehabilitation Dust and Combustion Systems
Flue Gas Conditioning: Safety Evaluation Phase 1 Report, by R.
Walton, M. Oliverio & N. Hyatt. Report prepared by Stone &
Webster Canada Limited for Ontario Hydro.

SWOVELAND, C., & CAWDERY, J. (1986). LPG Transport R&D
Risk-Benefit Analysis. Transport Canada Report No. TP 7458E.
Ottawa, ON.

TRANSPORT CANADA. (1986, Jan.). Transportation of
Dangerous Goods Regulations. Ottawa, ON.

UNITED STATES NUCLEAR REGULATORY COMMISSION
(1975). Reactor Safety Study: An Assessment of Accident Risks in
U.S. Commercial Nuclear Power Plants, Executive Summary.
WASH-1400, NUREG-75/014. Springfield, VA.

WALTON, R., OLIVERIO, M., & HYATT, N. (1990). Ontario
Hydro Lakeview Thermal Generating Station Rehabilitation Dust
and Combustion Systems Flue Gas Conditioning: Safety Evaluation
Phase 1 Report. Report prepared by Stone & Webster Canada
Limited for Ontario Hydro.

APPENDIX A:
DERIVATION OF AVERAGE LETHAL DISTANCE L_{jk}

A.1 INTRODUCTION

The individual exposure factor E_{jk} for an H_2S accident along the transportation route considers the dimensions of the H_2S plume and the impact of the wind direction on the path of the H_2S plume. The average lethal distance along the route L can be estimated with the help of Figure A-1. For simplicity, the subscripts j and k for severity category and weather class have been dropped. Thus, we can write:

$$L = L_{jk}$$
$$X_m = X_{jk}$$
$$W = W_{jk} \qquad\qquad (A.1)$$
$$\phi^* = \phi^*_{jk}$$

If a release incident occurs at some location along the route, then the parameters of interest are the lethal downwind H_2S plume distance and width, plus the wind angle ϕ that will allow the H_2S plume to reach a person at a distance x along the route from the release point. The lethal H_2S plume area can be approximated by a rectangle with length X_m and width w, as illustrated in Figure A-1.

A.2 CALCULATION OF AVERAGE LETHAL DISTANCE

From Figure A-1, the lethal distance along the route x depends on the wind angle ϕ, and can be written in the first quadrant ($0 \leq \phi \leq \pi/2$) as follows:

$$x(\phi) = \frac{X_m}{\cos(\phi)} \quad , \quad 0 \leq \phi \leq \phi^*$$

$$ \qquad\qquad (A.2)$$

$$= \frac{w}{2\sin(\phi)} \quad , \quad \phi^* \leq \phi \leq \frac{\pi}{2}$$

where w is the average width of the lethal H_2S plume (km) and X_m is the maximum downwind distance of the plume (km) in the direction of the wind ϕ. From symmetry, the behaviour of x in the other three quadrants from $\pi/2 \le \phi \le 2\pi$ is similar to the above equation.

For angles between 0 and $\pi/2$, the region is subdivided into $0 \le \phi \le \phi^*$ and $\phi^* \le \phi \le \pi/2$. Angle ϕ^* is the wind angle where the downwind corner of the assumed rectangular, lethal H_2S plume crosses the transport route. It is written as:

$$\phi^* = \tan^{-1}\left(\frac{w}{2X_m}\right) \tag{A.3}$$

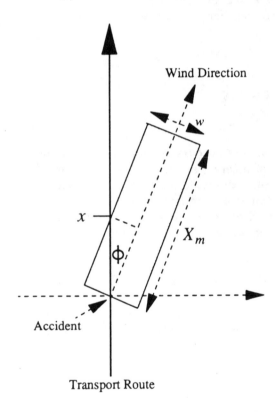

Figure A-1: Lethal distance x along the route for an H_2S release.

The average value of the lethal distance along the route can be determined by integrating x over all wind angles, or 2π. The average lethal distance L can be written as:

$$L = \frac{1}{2\pi} \int_0^{2\pi} x(\phi) \, d\phi$$

(A.4)

$$= \frac{1}{2\pi} 4 \int_0^{\frac{\pi}{2}} x(\phi) \, d\phi$$

Taking into account the change in the expression for the lethal distance x at angle ϕ^*, the average lethal distance becomes:

$$L = \frac{1}{2\pi} 4 \int_0^{\frac{\pi}{2}} x(\phi) \, d\phi$$

(A.5)

$$= \frac{2}{\pi} \left[\int_0^{\phi^*} \frac{X_m}{\cos(\phi)} \, d\phi + \int_{\phi^*}^{\frac{\pi}{2}} \frac{w}{2\sin(\phi)} \, d\phi \right]$$

The solution to the above trigonometric integrals can be found in standard tables (CRC, 1979) and gives:

$$L = \frac{2}{\pi} \left[X_m \ln\left(\tan\left(\frac{\phi^*}{2} + \frac{\pi}{4} \right) \right) + \frac{w}{2} \ln\left(\frac{1}{\tan\left(\frac{\phi^*}{2}\right)} \right) \right]$$

(A.6)

A.3 REFERENCE

CHEMICAL RUBBER COMPANY (CRC). (1979). CRC Handbook of Chemistry and Physics, 60th Edition. The Chemical Rubber Company, CRC Press, Inc. Boca Raton, Florida.

APPENDIX B:
DERIVATION OF THE OFF-ROUTE EXPOSURE FACTOR $E_{jk}(y)$

B.1 INTRODUCTION

The off-route individual exposure factor $E_{jk}(y)$ for an H_2S accident on the transportation route considers the length of the route S_{jk} where an accident has the potential to affect a person at an off-route location y, and the average value of the wind angle θ_{jk} along this section of the route that directs the lethal plume towards the individual.

For simplicity, the subscripts j and k for severity category and weather class will be dropped. Thus, we can write:

$$
\begin{aligned}
X_m &= X_{jk} \\
w &= w_{jk} \\
S(y) &= S_{jk}(y) \\
x^*(y) &= x^*_{jk}(y) \\
\theta(y) &= \theta_{jk}(y) \\
\alpha(y,x) &= \alpha_{jk}(y,x) \\
x'(y) &= x'_{jk}(y)
\end{aligned}
\tag{B.1}
$$

The lethal H_2S plume area can be approximated by a rectangle with length X_m and width w, as illustrated in Figure B-1. The individual exposure factor $E(y)$ takes into consideration the length of the route $S(y)$ where an accident has the potential to affect a person at y and the average value of the wind angle $\theta(y)$ along this section of the route that directs the lethal plume towards the individual. The exposure factor off the route can be written as follows:

$$
E(y) = \frac{S(y)}{d} \frac{\theta(y)}{2\pi}
\tag{B.2}
$$

where $S(y)$ is the length of the route for a lethal H_2S plume that can affect a person at y, d is the one-way travel distance per shipment and $\theta(y)$ is the average value of the wind angle that can affect a person at y. From Figure B-1, it is clear that the length of the route $S(y)$ where an accident has the potential to affect a person at an off-route distance y is given by:

$$S(y) = 2\sqrt{X^2_m - y^2} \quad \text{(km)} \qquad\qquad \text{(B.3)}$$

$$= 2x^*$$

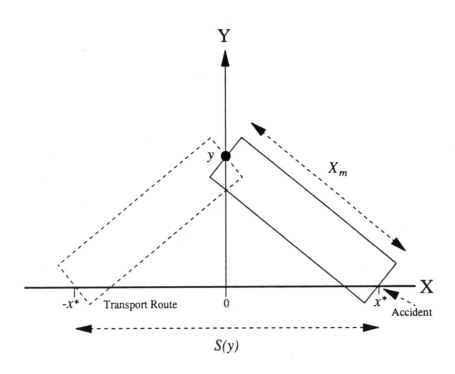

Figure B-1: Length of route $S(y)$ where an H_2S accident can affect a person at off-route location y.

B.2 CALCULATION OF AVERAGE WIND ANGLE θ

For the "Total Release" and "Partial Release" accident severity categories, where the wind direction can affect the exposure, it is necessary to take into consideration the effect of the wind angle in order to determine the exposure factor to a person located at y. While the length of the route that can affect a person at y is limited to $S(y) = 2 x^*(y)$, only a fraction of wind angles will direct the plume towards the person at y. This fraction depends upon the width of the plume w, the location of the accident x, and the location of the person y.

As illustrated in Figure B-2, an accident releasing H_2S is assumed to occur at position x along the route. For a person located at an off-route distance y which is greater than $w/2$ and less than X_m, the wind angle that can cause the lethal plume to pass over y is given by 2α.

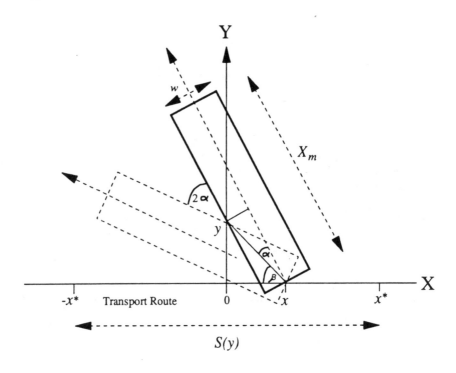

Figure B-2: Wind angle 2α where an H_2S accident can affect a person at off-route location y.

From Figure B-2, if we define z as the straight-line distance between x and y, the following expressions can be made:

$$\sin\alpha = \frac{w/2}{z}$$

$$\sin\beta = \frac{y}{z} \tag{B.4}$$

$$\tan\beta = \frac{y}{x}$$

Therefore, for off-route distances between half the plume width and the downwind distance of the plume, that is $w/2 \leq y \leq X_m$, the angle $\alpha(y,x)$ can be written as:

$$\alpha\ (y,x) = \sin^{-1}\left[\frac{w}{2y}\sin\left(\tan^{-1}\left(\frac{y}{x}\right)\right)\right], 0 \leq x \leq x^* \qquad (B.5)$$

For off-route distances y that are less than half the plume width, special consideration must be given to the location of the accident x since the plume can be over y for an angle $\alpha = \pi/2$ for small values of x. Therefore, for off-route distances between zero and half the plume width, that is $0 \leq y \leq w/2$, the angle $\alpha(y,x)$ can be written as:

$$\alpha\ (y,x) = \frac{\pi}{2} \qquad , 0 \leq x \leq x^{'}\ (y)$$

$$= \sin^{-1}\left[\frac{w}{2y}\sin\left(\tan^{-1}\left(\frac{y}{x}\right)\right)\right], x^{'}\ (y) < x \leq x^*\ (y) \qquad (B.6)$$

where $x^*(y)$ is half the length of the route for a lethal H_2S plume to affect a person at y, w is the average width of the lethal H_2S plume. Distance $x^{'}\ (y)$ is the distance along the route where the H_2S plume is over the location y for a wind angle $\alpha = \pi/2$. It can be written as:

$$x^{'}\ (y) = \sqrt{\left(\frac{w}{2}\right)^2 - y^2}\ \text{(km)} \qquad (B.7)$$

The average wind angle $\theta(y)$ is the average value of angle $2\alpha(y,x)$ over the route length $0 \leq x \leq x^*(y)$. It can be written as:

$$\theta\ (y) = \frac{2}{x^*(y)} \int_0^{x^*(y)} \alpha(y,x)\ dx \qquad (B.8)$$

The above integral can be solved using numerical integration using the appropriate expression for $\alpha(y,x)$ depending upon the location of y, that is $0 \leq y \leq w/2$, or $w/2 \leq y \leq X_m$.

For off-route distances greater than X_m, the exposure factor $E(y)=0$.

As the off-route distance y approaches zero, the numerical expression for the exposure factor $E(y)$ approaches the analytical expression given earlier in the text.

Techniques for Risk Assessment of Ships Carrying Hazardous Cargo in Port Areas

J.R. Spouge
Technica Ltd.
7/12 Tavistock Square
London WC1H 9LT

ABSTRACT

This paper describes recent developments in the techniques for assessing the risks from hazardous cargo shipments in port areas. It is based on work by Technica as part of a major study of the risks from the transport of bulk dangerous substances throughout Great Britain for the Health and Safety Commission.

The study covered the risks of major hazard accidents affecting people ashore from bulk shipments of dangerous substances including liquefied gases, flammable liquid petroleum products and chemicals. For each significant hazard, it estimated the accident frequencies and modelled their consequences using the software package SAFETI, in order to determine individual and societal risks.

The paper provides illustrative examples of frequency calculations, and presents some of the estimated individual and societal risk results for a single port. It summarises the risk criteria which were used to assess the results, and presents a cost-benefit analysis of an example risk reduction measure.

1. INTRODUCTION

The transport of hazardous cargoes by sea creates risk problems which are quite different to other transportation modes. Ships may carry over 100,000 tonnes of flammable liquid, or 20,000 tonnes of liquefied gas, and the potential effect zone in the event of a spill could be very large. On the other hand, ships spend most of their time at sea where the only people exposed to the hazard are the ship's crew. The main risk of major accidents affecting the public is when the ship comes into port to load and unload. Analysis of the risks from such an operation has some similarities to the analysis of risks from fixed installations, and some similarities to other transport modes. Recent developments incorporate techniques from both disciplines, and form the subject of this paper.

The paper is drawn from a major new risk assessment of shipping operations which has recently been published by the Health and Safety Commission (HSC) as part of their study of the risks of transporting dangerous substances throughout Great Britain (HSC, 1991). The study was organised by the Sub-Committee on Major Hazards of Transport (MHT) of the HSC's Advisory Committee on Dangerous Substances (ACDS). It was carried out by Technica Ltd.

There are in total 88 terminals in 42 ports handling bulk hazardous cargoes in Great Britain. A detailed analysis of all of them would have been impracticable, so three ports were analysed in detail and a simplified method was developed for the others. For brevity, this paper concentrates on the analysis of one of the three ports which were analysed in detail. This is the port of Felixstowe, which has a significant trade in liquefied gas and flammable liquids, and provides a good illustration of the techniques which were used.

The background to this type of study was given in an earlier paper (Spouge, 1990), together with a definition of terms and a simple introduction to the techniques involved. The present paper describes selected analysis techniques in more detail, and assumes that the reader is familiar with the basic concepts of risk analysis. Further details on other analysis techniques and the national risk calculations are given in the main study report (HSC, 1991).

The overall methodology for the analysis is illustrated in Figure 1. It follows the usual structure for quantified risk assessment (QRA). This paper's approach to the subject is similar.

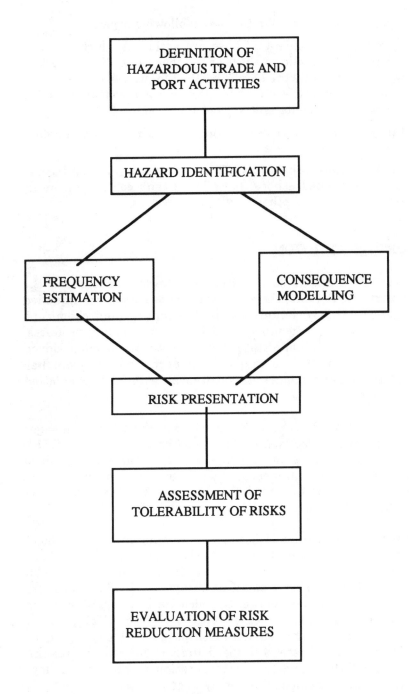

Figure 1: Port risk assessment.

The scope of the study was limited to the following aspects:

- "Major hazard" accidents. These are accidents which threaten members of the public on-shore or in other vessels in the port, or crew members while the ship is berthed in port. Accidents which only affect the ship's crew while at sea are considered to be occupational accidents and are not included.

- Fatality risks. The risks of injuries, environmental damage and property damage are not calculated.

- Bulk dangerous substances. This includes liquids and gases which are pumped ashore. It excludes tank containers, Ro-Ro vehicles, drums and other small packages.

2. PORT DESCRIPTION

Felixstowe (Figure 2) is part of Harwich Haven and is one of the busiest ports in Britain. Its main hazardous trades are liquefied petroleum gas (LPG), liquid petroleum products and chemicals. There are also significant quantities of petroleum products, chemicals and ammonium nitrate moving past it to the other Haven ports. In addition, it is a major ferry port, and therefore has potential to combine dangerous substances with high population concentrations.

In 1983, the LPG trade through Felixstowe was the subject of a major risk assessment exercise by the Health and Safety Executive (HSE). A marine traffic management study was also carried out. Since then most of the remedial measures which were recommended have been adopted, and the LPG trade has been significantly reduced. As a result, the port already has high safety standards.

Approximately 11000 ships visit Harwich harbour each year, 5000 of which come to Felixstowe. Harwich harbour traffic includes about 2600 ferry visits per year, some of which may carry up to 2000 passengers. There is also a small passenger ferry which operates between Harwich and Felixstowe. Harwich Harbour Board operates an advanced radar-based Vessel Traffic System (VTS), which provides information and advice to ships as well as helping to enforce regulations. There is a speed limit of 8 knots within the harbour. The VTS allows the port operation to continue in fog, although special restrictions then apply to gas carriers.

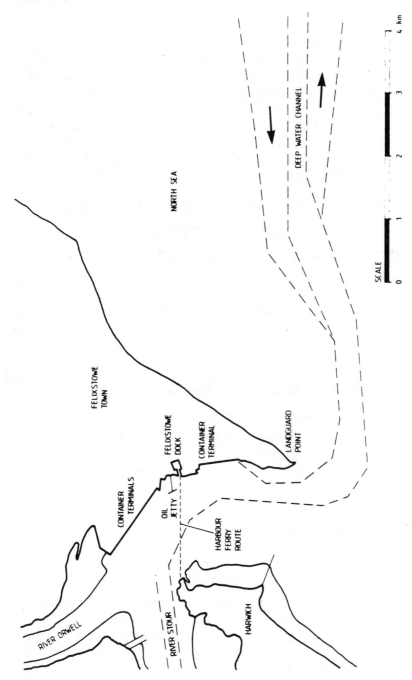

Figure 2: Port of Felixstowe.

The hazardous trades all use Felixstowe Oil Jetty, and are summarised in Table 1. LPG is imported and exported by both large fully-refrigerated and small semi-refrigerated gas carriers. The small vessels have cylindrical pressure tanks, but carry the cargo fully refrigerated. Their cargo capacity is typically 3700m³. The large gas carriers have prismatic tanks and a total capacity of typically 23,000m³ (Figure 3). The cargo is carried fully refrigerated at atmospheric pressure. An articulated hard arm with quick-release and emergency shut-down (ESD) is used for ship-to-shore transfer.

Table 1: Felixstowe bulk dangerous substances traffic.

MATERIAL	QUANTITY (tonnes/yr)	VISITS (cargoes/yr)	SHIPMENTS (tonnes)
LPG (Large gas carriers)	35000	5	7000
LPG (Small gas carriers)	21000	14	1500
Low-flash petroleum products	60000	40	1500
High-flash petroleum products	60000	40	1500
Low-flash chemicals	48000	32	1500
High-flash chemicals	72000	48	1500

Petroleum products and chemicals are imported in small coastal tankers of typically 2000 dwt, using flexible hoses for transfer. For reasons of commercial confidentiality, only very approximate shipment figures for broad categories of materials could be used in the study.

3.   HAZARD IDENTIFICATION

Hazard identification involves identifying all relevant types of accidents for the analysis. This has two purposes:

- Classification of accidents. Possible accidents can be identified from historical accident records and from previous risk analyses. An exhaustive list of individual accident scenarios would be impractical. Instead, a hazard classification which could fit all known scenarios is chosen.

- Preliminary screening. Many types of accidents can be excluded as they have no potential to cause fatalities among the public.

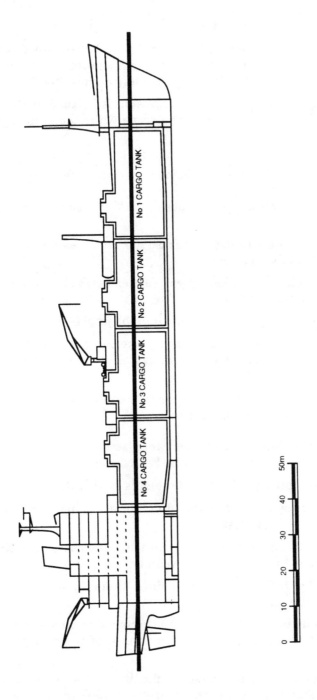

Figure 3: Large refrigerated gas carrier.

The main groups of people at risk in Felixstowe are:

- Residents and workers in Felixstowe port, Felixstowe Town and Harwich Town. Since the Oil Jetty is 250m long, only spills or explosions of LPG are likely to reach any of these people.

- Ship crew and jetty staff while the ship is berthed at the jetty. (The crew are excluded from the scope while the ship is under way in the port.)

- People on other vessels. In particular, this includes large numbers of people on ferries which pass close to the jetty.

The categories of accident which pose a fatality risk to these people are:

- Collision - where a tanker or gas carrier under way in the port is struck by another vessel.

- Striking - where a tanker or gas carrier berthed at the jetty is struck by a passing vessel.

- Impact - where a gas carrier runs into the jetty head.

- Fire/explosion - where a fire on a tanker or gas carrier spreads to involve the cargo, or where an explosion is initiated directly in the cargo tanks.

- Cargo transfer failure - where cargo is spilled while a tanker or gas carrier is loading or unloading.

The following accident categories were screened out as having negligible major hazard potential in Felixstowe:

- Grounding - since the harbour bottom is mud and soft rock.

- Foundering/capsize - this would create a salvage problem but not a major fatality risk.

- Structural failure - historical experience indicates that this is only a major hazard on larger vessels.

- Striking by falling aircraft - since there is no significant airport in the region.

- Spontaneous tank failure - this has occurred in on-shore storage tanks, but there has been no experience of it on marine gas carriers.

- Domino accidents from on-shore facilities - the nearest major hazard installation is a large refrigerated LPG tank 800m from the jetty head.

Neglecting these hazards allows the analysis to concentrate on the types of accidents which are likely to dominate the risk. Once a first estimate is obtained, it is possible to go back and include hazards which were previously screened out, to provide a quantitative check of their significance.

4.   FREQUENCY ESTIMATION

4.1 General Approach

The frequencies which are required for the study are in effect likelihoods per year of an accidental release or explosion of hazardous cargo in the port.

For collisions, strikings, impacts and gas carrier fires the frequencies of cargo release were estimated as the product of two components:

- The frequency of the accident (i.e., collision, striking, impact or fire) for the ship, whether or not it results in release of cargo. These are relatively minor accidents and are similar for all ship types, and hence occur sufficiently often to give reliable frequencies from historical data.

- The conditional probability of a release of cargo given that the accident (collision, striking, impact or fire) has occurred. These are specific to the hazardous cargo ship, and must be theoretical estimates since such events are so rare.

Tanker explosion frequencies were estimated directly from historical data, and cargo transfer failures from historical data and an event tree analysis.

The techniques are illustrated here with the example of LPG releases due to striking, which proved to be one of the main risks in the port.

4.2 Striking Frequencies

Harwich Harbour Board had good records of accidents which had occurred at Felixstowe. Over a 12.4 year period up to 1987 there had been 3 incidents of a tanker at the Oil Jetty being struck by a passing ship, and 2 others where the jetty head was struck by a passing ship while no tanker was present. Approximately 9100 ships per year berth upstream of the Oil Jetty (i.e., 18,200 passing movements), so the striking frequency for a ship of length equal to the jetty head was estimated as:

$$\frac{5}{18200 \times 12.4} = 2.2 \times 10^{-5} \text{ per passing movement}$$

Expressing the frequency in this form allows it to be compared with striking frequencies in other ports with different traffic levels. For example, the estimated generic striking frequency for British ports on wide estuaries is 4.0×10^{-6} per passing movement (HSC, 1991). The difference may be attributed to unusually exposed position of the Oil Jetty at Felixstowe.

This frequency may then be applied to the large refrigerated gas carrier trade, using the following data:

- Ship length - 166m
- Jetty head length - 140m
- Time at berth per visit - 14 hours
- Visit frequency - 5 per year
- Passing movements - 18,200 per year

The resulting striking frequency is:

$$2.2 \times 10^{-5} \times \frac{166}{140} \times \frac{14 \times 5}{24 \times 365} \times 18{,}200 = 3.8 \times 10^{-3} \text{ per year}$$

This is consistent with the gas carriers never having been struck by another ship while at the Oil Jetty.

4.3 Release Probability

The above striking frequency is based on the vessel being struck while at the jetty, whether loaded or not. For a release of gas to result, the following conditions must be satisfied:

- The gas carrier must be struck in way of the cargo tanks. A probability of 0.65 is used, equal to the proportion of the ship length occupied by cargo tanks.
- The tank must be loaded. A probability of 0.5 is assumed.
- The collision must be at an oblique angle, not a glancing blow. Based on the courses of passing vessels, a probability of 0.38 was estimated.

- The striking ship must have sufficient energy to breach the cargo tank. This energy is proportional to the striking ship's displacement and the square of its speed, and is also affected by the strength of the gas carrier's cargo tanks and side structure, the strength of the jetty, the geometry and strength of the striking ship's bow and the precise angle of impact. This aspect has been studied in detail previously, and a detailed investigation was outside the scope of the present study. The probability of having sufficient energy was estimated to be on average 0.31 for large gas carriers and 0.85 for small gas carriers. These were based on the full harbour speeds of the striking ships (around 8 knots off the Oil Jetty, rising to 12 - 15 knots at Cork Spit). A further reduction of 50% was applied to account for speed reduction at the last moment.

The calculation is illustrated as an event tree in Figure 4. The overall release probability for the large gas carriers is 0.019 per striking event.

4.4 Failure Cases

A striking could produce a range of hole sizes and locations, but for simplicity just two failure cases are modelled:

- Cold leak through a hole the size of the loading pipe connection, located at the bottom of a prismatic tank or at the mid-height of a cylindrical or spherical tank.

- Cold rupture in which the tank's contents are released almost instantaneously if pressurised, or within about 5 minutes if refrigerated.

It is assumed that 90% of such releases are leaks and 10% are ruptures. This has been a common assumption in previous marine risk analyses, based on a wish to be cautious where uncertainty is large. However, it is now believed that cold rupture of a gas carrier's tank is very unlikely, and modelling using two sizes of leak would be preferable. The sensitivity of the results to the rupture probability was evaluated and was found to be significant (see Section 6.5).

4.5 Other Frequencies

Using techniques similar to those outlined above for striking, frequencies were estimated for all causes of release of LPG or flammable liquids from vessels at the Oil Jetty. The components of the total release frequency from the large gas carriers are shown in Figure 5.

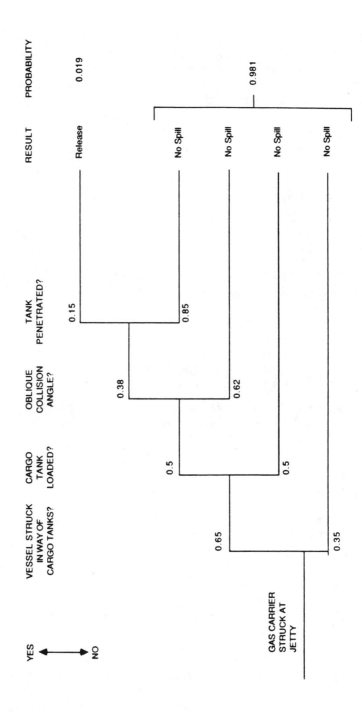

Figure 4: Event tree for gas carrier striking.

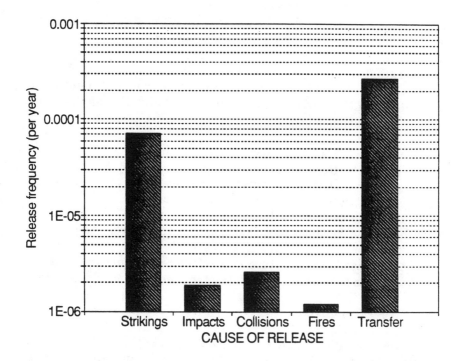

Figure 5: Release frequencies.

5. CONSEQUENCE MODELLING

5.1 General Approach

Consequence modelling was performed using Technica's software package SAFETI (Software for the Assessment of Flammable, Explosive and Toxic Impacts) (Ale & Whitehouse, 1986). This is a highly efficient computational tool which was developed for the analysis of on-shore chemical installations, but is equally applicable to ships in ports. Its approaches are outlined below.

5.2 Cold Releases of Liquefied Gas

5.2.1 Release model

For a cold leak from a pressurised tank or loading arm, the liquefied gas will be driven out in a continuous stream by the internal pressure. The steady-state flashing release rate is calculated by SAFETI, and the final conditions are used as input to the dispersion model. The momentum of the jet is assumed to be dissipated by the ship's secondary containment.

For a catastrophic cold rupture of a pressurised tank, SAFETI assumes an isentropic hemispherical expansion, and determines the initial dilution for the dispersion model, and also any rain-out which forms a liquid pool.

For a cold leak from a refrigerated tank, the flow will at first be driven by the hydrostatic head, and subsequently by the heating of the cargo by the sea water. A constant flow rate equal to 75% of the initial hydrostatically-driven release rate is assumed, and the entire contents of the tank are released at this rate. SAFETI then calculates the spreading of the pool on the water and the evaporation rate.

For a catastrophic cold rupture of a refrigerated tank, the contents are assumed to be released onto the water over a period of 5 minutes.

5.2.2 Dispersion Model

The dense gas dispersion model used by SAFETI is based on a standard "box" model (Cox & Carpenter, 1980), but with numerous enhancements to the initial cloud development to account for various entrainment regimes (Figure 6) (Emerson, 1986). An instantaneous pressurised release is represented by a cylindrical cloud which slumps radially under gravity to a "pancake" shape while advecting with the wind. A continuous pressurised release or the gas evaporating from a pool of refrigerated material is represented by a plume of rectangular cross-section which is blown down-wind and spreads laterally under gravity.

The underlying surface is assumed to be level. As a result, releases over water or near to a gently-sloping shore are correctly modelled, while travel over hilly ground or high harbour walls may be somewhat over-estimated.

Local meteorological conditions were represented in the Felixstowe study by 6 combinations of wind speed and atmospheric stability and 12 wind directions.

The surrounding population is represented by the numbers of people in each 100m grid square who may be affected by the release. Two distributions are used: one for day-time and the other for night-time. The night-time distribution is derived from census data; the day-time one uses additional data on local industries' employees.

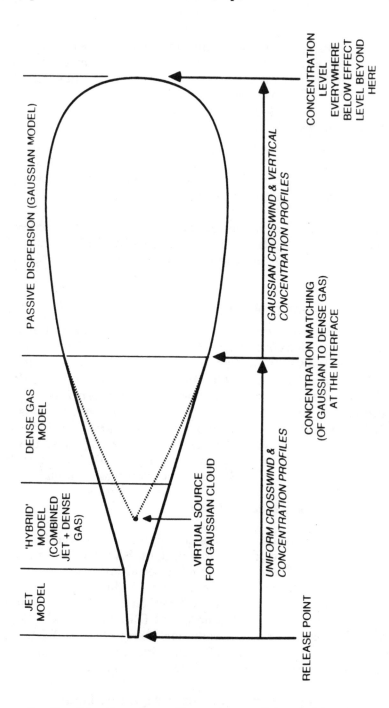

Figure 6: Combination of gas dispersion models used in the
SAFETI Program.

Corresponding ignition source distributions are used to model the probabilities and source strengths of ignition from people, motor vehicles, industries, transmission lines, trains, ships etc. These are all necessarily based on judgement.

5.2.3 Flash Fires

The possibility of ignition at each time step during the cloud dispersion by any of the ignition sources within the current lower flammable limit (LFL) is taken into account. On 90% of occasions, an ignited cloud is assumed to form a flash fire.

Whenever a flash fire is ignited, all outdoor population within the current LFL of the gas cloud is assumed to be killed, and 30% of those indoors. It is assumed that no-one beyond the LFL of the cloud is killed.

5.2.4 Vapour Cloud Explosions

A vapour cloud explosion (VCE) is assumed to be the result in 10% of ignition cases. This has been a common assumption in previous marine risk analyses, although experimental evidence indicates that VCEs are very unlikely. In the present study VCEs are predicted to cause hardly any more fatalities than flash fires, so uncertainty in this value is not important. However, if VCEs were predicted to cause domino effects elsewhere, further refinement of the VCE probability would be appropriate. SAFETI uses the TNO correlation model for VCEs (Opschoor, 1979). This has been modified to represent the HSE explosive overpressure probit.

5.3 Hot Releases of Liquefied Gas

If a fire causes rupture of a tank containing pressurised flammable liquefied gas this is modelled as a fireball. SAFETI determines the thermal radiation from the mass of material involved. This may be several hundred tonnes, and the available radiation impact models are not validated for such events. Therefore a judgmental modification of the Eisenberg radiation probit (Eisenberg et al., 1975) is used. In the region exposed to radiation energy above 500kJ/m^2, 75% of people outdoors and 25% of people indoors are assumed to be killed, either by radiation or secondary ignition effects.

A hot rupture of a tank containing refrigerated liquefied gas was initially modelled as a fireball too, on the basis that considerable pressure could build up inside the tank before it failed in a fire. However, this is uncertain, and a pool fire may be a more appropriate outcome for refrigerated material. Sensitivity tests showed that the overall difference in the risks between these options was negligible.

5.4 Cold Releases of Flammable Liquid

The release of flammable liquid from a damaged tanker will be driven by hydrostatic head. A constant flow rate equal to 75% of the initial rate is assumed, and taken to continue until hydrostatic equilibrium is reached.

For a full-bore transfer spill, the flow will emerge at the pumping rate, assuming that back-flow is negligible. For a leak during transfer, outflow at 10% of this rate is assumed.

The released liquid is considered to spread under gravity until it reaches an average thickness of 5mm on the water, and then to drift under the influence of wind and tide. The shape of the pool follows the shape of the harbour. The pool is allowed to drift for up to 20 minutes. After this time, it is assumed that the spill is contained and ignition sources controlled.

Only the population on the water or along the water's edge need to be modelled in this case, since these are the only people at risk if the pool is ignited. For low-flash products, it is assumed that 50% of people within 6m of the edge of the pool and 100% of people on small craft within the pool are killed if it ignites. Similar assumptions are made for other people at risk and for high-flash products which are less likely to be ignited.

5.5 Explosions of Flammable Liquid

The probability distribution of crew fatalities in a cargo explosion on a tanker is given in Table 2, based on an analysis of tanker total losses due to explosions in port during 1977 - 1986.

Table 2: Fatalities in flammable liquid explosions.

PERCENTAGE OF CREW KILLED	PROBABILITY GIVEN CARGO EXPLOSION
15	0.37
40	0.11
65	0.04
100	0.04

Fatalities due to overpressure and fragments have been estimated for a representative event and have shown, in general, to be insignificant.

5.6 Collisions Between Ferries and Tankers

The consequences which might result from a tanker exploding when struck by a ferry are:

- Blast damage on the ferry as well as on the tanker and jetty. This is likely to puncture the hull at the bow and break the windows on the bridge and around the forward part of the ferry.

- Pool fire around the jetty, tanker and ferry bow (if a loaded tank is breached as well). Since the ferry sides are so high, this is unlikely to endanger the ferry passengers unless they have to evacuate the vessel for some other reason.

- Fire spreading to the ferry from the tanker. Since the outside of the ferry is mostly steel, the ferry itself would only burn if this failed in the heat or if burning liquid was thrown through the blast damage. Such a fire should be controlled, but there is a possibility of it causing explosions in fuel tanks on the vehicle deck, escalating rapidly and forcing evacuation of the ferry.

- Capsize of the ferry. The ferry is designed to withstand flooding resulting from striking another ship, but extra flooding may take place through explosion damage near the waterline. Combined with other adverse circumstances such as rough weather or overloading, flooding may occur on the vehicle deck. There may also be fire-fighting water on this deck if the scuppers are blocked. These circumstances may lead to a rapid capsize of the ferry, possibly within 20 minutes.

A simple manual analysis of these effects was carried out for Felixstowe. The risk was found to be significant, and improvement of this approach would be desirable.

6.   RISK PRESENTATION

6.1 Risk Calculations

For LPG releases, the fatalities are calculated by SAFETI for each failure case and each of typically:

- 6 wind speed/atmospheric stability combinations.
- 12 wind directions.
- Day-time and night-time population distributions.
- Over 100 possible ignition sources.
- Over 50 possible time steps when ignition might occur.
- 2 possible ignited outcomes (flash fire and VCE).

These fatalities are expressed both as probabilities at each point on a 100m grid and as total expected numbers. The probabilities of each combination are combined with the failure case frequencies, and SAFETI automatically computes the measures of individual and societal risk.

For the explosions which result from collisions between ferries and tankers, or for the pool fires which result from spills of flammable liquid around the ship, equivalent calculations are performed by hand. These are necessarily for fewer combinations of parameters.

6.2 Forms of Risk Presentation

The calculated risks are presented in the following forms:

Individual risk - the likelihood of death per year for an individual continuously at a given location, assuming an average vulnerability to the hazards and allowing for realistic probabilities of being sheltered indoors. Escape action is not modelled.

- FN curves - measures of societal risk showing the relationship between numbers fatalities and frequency (i.e., likelihood per year) of accidents. The FN curves are actually cumulative plots, showing the frequency F of accidents involving N or more fatalities.

- Annual fatality rates - simpler measures of societal risk, equal to the expected long-term average number of fatalities per year (i.e., expectation values). These give no indication of the numbers of fatalities in individual accidents.

- Mean fatalities per event - the average number of fatalities in given accident types. These supply the missing information in the fatality rates. They are equal to the fatality rate divided by the event frequency.

6.3 Individual Risks

The individual risks calculated for the LPG trade are shown in Figure 7. The contours show characteristic circular shapes for events at the jetty and parallel lines for events in the channel. They are only distorted slightly by the wind rose and by ignition sources such as those concentrated on the main road into Felixstowe docks. The affected area on-shore is almost exclusively the port industrial land, although nearby housing is affected by some low probability events (i.e., the catastrophic gas releases which travel furthest). Some gas clouds travel over 1 km towards Harwich before reaching their lower flammable limit, but this is against the prevailing wind, and so their frequency is relatively low.

6.4 Societal Risks

The societal risk results for each failure case, expressed as annual fatality rates, are given in Table 3. The largest contributions come from tanker transfer spills ignited by passing small craft and from collisions between tankers and Ro-Ro ferries. However, the latter have only been modelled crudely. The results are shown as FN curves in Figure 8.

The overall long-term average number of fatalities in Felixstowe from bulk hazardous cargo shipments is found to be only 0.019 per year, i.e., roughly one fatality expected for every 50 years of port operation. This is entirely consistent with the fact that no such accidents have occurred in Felixstowe. In the whole of Great Britain, there have only been 2 fatalities from bulk hazardous cargo shipments in the period 1969 - 1988.

6.5 Sensitivity Tests

Sensitivity tests were used to illustrate the effects of some suggested changes to the gas carrier consequence modelling. These were:

• Elimination of hot ruptures on refrigerated gas carriers.

• Reduction of VCE probability from 0.1 to 0.025.

• Reduction of cold rupture probability from 0.1 to 0.01.

• Reductions of various probabilities in the event trees for transfer spills.

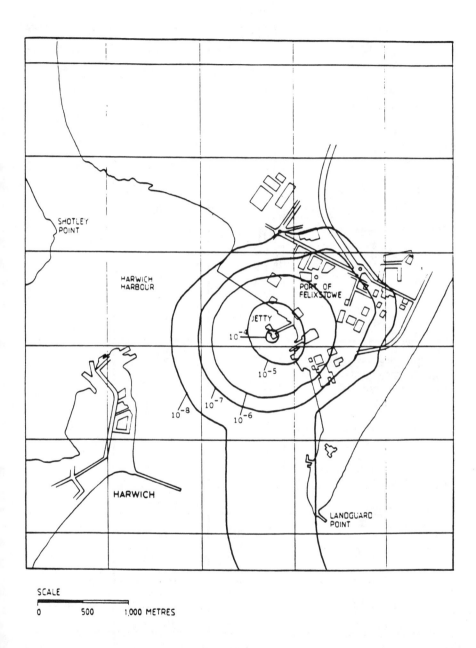

Figure 7: Felixstowe individual risks for gas carriers.

Table 3: Annual fatality rates for Felixstowe.

EVENT	ANNUAL FATALITY RATE	PERCENTAGE OF TRADE TOTAL	MEAN FATALITIES PER EVENT
Large gas carriers			
Cold leak at jetty	1.8×10^{-4}	11.1	2.6
Cold leak at jetty with ferry	9.0×10^{-5}	5.6	6.9
Cold rupture at jetty	3.5×10^{-4}	21.6	46
Cold rupture at jetty with ferry	1.7×10^{-4}	10.5	121
Hot rupture at jetty	3.8×10^{-4}	23.4	317
3 min full-bore transfer spill	1.1×10^{-4}	6.8	2.4
15 min full-bore transfer spill	9.0×10^{-5}	5.6	5.0
10 min transfer leak	2.5×10^{-4}	15.4	1.3
TOTAL (Large gas carriers)	1.6×10^{-3}	100.0	
Small gas carriers			
Cold leak at jetty	1.44×10^{-3}	34.4	5.3
Cold leak at jetty with ferry	6.9×10^{-4}	16.4	13
Cold rupture at jetty	8.1×10^{-4}	19.4	27
Cold rupture at jetty with ferry	5.8×10^{-4}	13.8	96
Hot rupture at jetty	3.0×10^{-4}	7.2	316
3 min full-bore transfer spill	1.5×10^{-4}	3.6	1.2
15 min full-bore transfer spill	9.5×10^{-5}	2.3	1.9
10 min transfer leak	1.2×10^{-4}	2.9	0.2
TOTAL (Small gas carriers)	4.2×10^{-3}	100.0	
Ferry-tanker collision			
Blast damage only	2.2×10^{-3}	16.8	11
Fire on the ferry which is controlled	1.7×10^{-3}	12.8	18
Fire forcing evacuation at jetty	7.2×10^{-4}	5.5	39
Fire forcing evacuation in channel	1.6×10^{-4}	1.2	88
Fire causing capsize at jetty	7.2×10^{-4}	5.4	342
Fire causing capsize in channel	1.5×10^{-4}	1.2	732
TOTAL (ferry-tanker collision)	5.6×10^{-3}	42.9	
Tanker transfer spill			
Ignited at jetty	1.8×10^{-3}	14.0	1
Ignited by small craft	4.2×10^{-3}	32.0	7
TOTAL (Transfer spill)	6.0×10^{-3}	46.0	
Tanker Explosion without ferry	1.5×10^{-3}	11.1	1
TOTAL (Tankers)	1.3×10^{-2}	100.0	
TOTAL (All Trades)	1.9×10^{-2}		

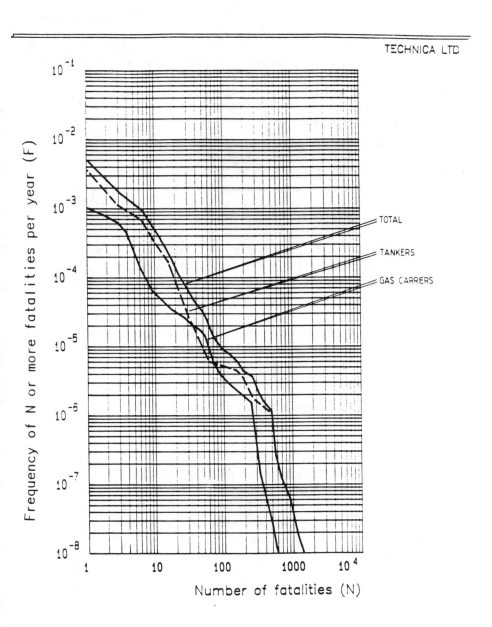

Figure 8: Felixstowe societal risk curves.

Table 4: Sensitivity test annual fatality rates.

CASE	ANNUAL FATALITY RATE	FRACTION OF BASE CASE
Base Case	4.75×10^{-3}	1.00
No refrigerated ship hot ruptures	4.60×10^{-3}	0.97
VCE probability 0.025	4.41×10^{-3}	0.93
Cold rupture probability 0.01	2.41×10^{-3}	0.51
Transfer spill development	4.47×10^{-3}	0.94

The effects on the fatality rate results are given in Table 4. Changing the cold rupture probability by a factor of 10 causes a factor of 2 change in the fatality rate. The other tested parameters cause less than a 7% change.

The reason why these changes are relatively small is that the gas carriers only provide 31% of the total fatality rate at Felixstowe. Thus the sensitivities would be about 3 times greater for gas carriers alone.

These results were included in an uncertainty analysis, which estimated uncertainties of a factor of 6 - 7 (higher or lower) in the fatality rate results for the gas carrier and tanker trades.

7. ASSESSMENT USING RISK CRITERIA

7.1 General Approach

Risk criteria are used to indicate whether the estimated risks are tolerable or whether remedial measures are needed, and to rank risk sources in order of priority for testing remedial measures. They only provide an indication, because many factors which are relevant to the decision cannot be included in numerical criteria.

The criteria used here have been proposed by a joint working party of the Subcommittee of ACDS with members from industry, HSE and Technica. Individual and societal risk criteria are used in combination; both of which must be satisfied.

7.2 Individual Risk Criteria

The proposed criteria for individual risk of death for members of the public are based on previous criteria used by the HSE (1987):

- Negligible criterion 10^{-6} per year.
- Maximum tolerable criterion 10^{-4} per year.

These divide individual risks into three zones:

- Negligible - below the negligible criterion.
- ALARP - between the two criteria. These risks are considered tolerable if they have been made as low as reasonably practicable on cost-benefit grounds.
- Intolerable - above the maximum tolerable criterion.

The risks are calculated for individuals at specific locations, assuming:

- Realistic proportions of time present at the location.
- Realistic proportions of time spent indoors.
- Average susceptibility to the hazards.

The criteria apply to the most exposed members of the public. These may be residents who happen to be continuously present at the location, or workers who may be present up to one third of the time. The criteria do not apply to workers involved in the particular hazardous trade being assessed, but do apply to workers employed by other companies nearby.

7.3 Societal Risk Criteria

The proposed local societal risk criteria for a port are in the form of lines on an FN plot. Two of the lines are fixed, and indicate overall tolerable levels for any size of port. They are based on the judgements made about the Canvey Island industrial complex (HSE, 1978), where about half the risks came from marine activities.

The third line recognises that the tolerability of a port's societal risks depends in part on its size and on the benefits against which the risks must be set. Thus risks which are tolerable in a complex as large as Canvey might possibly be unjustifiable if they arose from a small trade which has little value (in terms of jobs, tax revenues etc.) to the society. Unfortunately, at present the "value added" by a port can only be approximated by the mass of bulk hazardous cargoes shipped through the port. This in effect assumes that the value added per tonne shipped is the same for all these materials.

The balance between such possibly unjustifiable risks and tonnage shipped is also undefined at present, so the third line is called a "scrutiny level" rather than a criterion. Above it, further examination of the overall risks and benefits of the port would be necessary, and the hazardous trade might possibly be found to be unjustifiable. Below it, in the ALARP region, only the marginal costs and benefits of remedial measures need be examined.

Thus the local societal risk criteria for a port are:

	Intercept with N = 1	Slope
Negligible criterion	10^{-4} per year	-1
Tolerable criterion	10^{-1} per year	-1
Scrutiny level	3.2×10^{-8} per tonne per year but not outside 10^{-4} to 10^{-1}	-1

These divide the FN space into four zones:

- Negligible region - below the negligible criterion.

- ALARP region - between the negligible criterion and the scrutiny level.

- Possibly unjustifiable region - between the scrutiny level and the maximum tolerable criterion. This region does not exist if the port shipments are more than 3.1×10^6 tonnes per year.

- Intolerable region - above the maximum tolerable criterion.

The societal risks are calculated including all workers and members of the public. They are modelled as present and as indoors or outdoors for realistic proportions of the time, and given a realistic distribution of susceptibility to the hazards.

7.4 Assessment of Felixstowe

The highest individual risk at Felixstowe other than on the ship or jetty is 10^{-5} per year (Figure 7). Workers here are only likely to be present for at most 30% of the time, so their actual individual risk is only 3.3×10^{-6} per year, which is low in the ALARP region.

The most exposed residents are near the 10^{-7} contour. Even allowing for additional risks from the on-shore LPG storage, this is probably still negligible by the criteria used here.

The societal risks are compared with the criteria in Figure 9, and are shown to be high in the ALARP region.

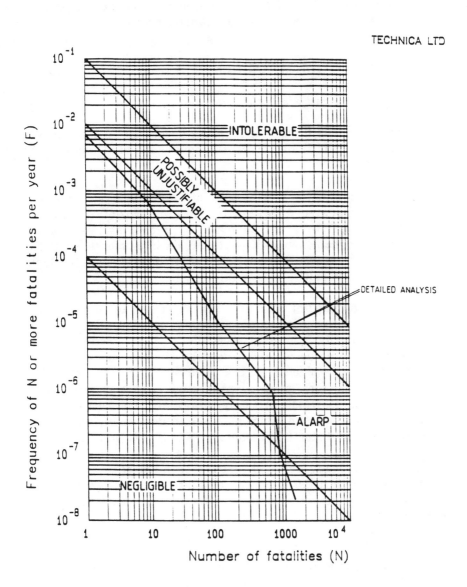

Figure 9: Felixstowe societal risk tolerability overall.

Thus Felixstowe is in the ALARP region mainly because of societal risk considerations. Its risks are therefore considered tolerable provided that they are as low as reasonably practicable.

8.   RISK REDUCTION MEASURES

8.1   Cost-Benefit Analysis Technique

In order to show whether or not the risks in Felixstowe are ALARP, several possible additional measures were evaluated using cost-benefit analysis.

In cost-benefit analysis, the costs of implementing remedial measures are compared with their benefits in terms of the costs of accidents averted. In order to make a port's risks as low as reasonably practicable, measures should be implemented if their costs are not "grossly disproportionate" to their benefits. Such measures are then called "reasonably practicable".

The benefits of remedial measures are expressed in terms of the risk-factored costs of accidents avoided. A provisional monetary value of £2 million per statistical fatality is used. This is equivalent to the value of £0.5m used by the UK Department of Transport for road accidents, with a somewhat arbitrary factor of 4 increase to allow for "gross disproportion". This ensures that remedial measures which are only marginally non-cost-effective are still implemented, and hence allows a prudent safety margin on the uncertain and distasteful monetary valuation of human life.

In the present study the risk estimates have only considered fatalities, although it has been argued that a complete cost-benefit analysis should take other accident costs into account. These would include:

- Injuries
- Ships damaged and totally lost
- Cargo lost
- Jetties and terminal buildings damaged
- Third party property damage
- Business interruptions
- Port closure costs

- Environmental damage

- Environmental clean-up costs

Information on some of these aspects (especially business interruption costs) is difficult to obtain and a detailed estimate was outside the scope of the study. Nevertheless, an approximate estimate was made, based on accident reports. This showed that the total cost for many accidents is three times as high as the cost of fatalities alone.

8.2 Example Analysis

As an example, one possible risk reduction measure at Felixstowe would be to re-schedule or re-route small craft away from the Oil Jetty while tankers are present. This could reduce the risk by 4.2 x 10^{-3} per year (Table 3).

If an expenditure of £2m was justified to avert a statistical fatality, then this measure would be reasonably practicable if it cost less than:

4.2 x 10^{-3} fatalities per year x £2m per fatality = 8400 per year

Since most of this risk comes from relatively disciplined craft such as the harbour ferry, the cost of enforcing such a restriction is not expected to be any greater than this sum, and hence the measure is found to be reasonably practicable.

8.3 Conclusions About Remedial Measures

A general conclusion from the cost-benefit analysis is that most measures involving purchases of additional hardware are not reasonably practicable. Measures which involve improved safety management, such as training, communications, procedures, emergency response and water-use planning (as in the above example) appear much more effective. This is partly because these measures have received less attention in previous risk assessments and partly because they are relatively cheap. However, estimates of the risk reductions from improved safety management are even less certain than for hardware measures. The present study did not set out to model these factors, so more specific studies of them are desirable.

Another general conclusion is that several measures are on the borderline of reasonable practicability, but only appear reasonably practicable if non-fatality costs are included in the cost-benefit analysis.

9. REFERENCES

ALE, B.J.M., & WHITEHOUSE, R.J. (1986). Computer Based System for Risk Analysis of Process Plants. In Heavy Gas and Risk Assessment III, S. Hartwig (Ed.), D. Reidel, Dordrecht, The Netherlands.

COX, R.A., & CARPENTER, R.J. (1980). Further Development of a Dense Vapour Cloud Model for Hazard Analysis. In Heavy Gas and Risk Assessment, S. Hartwig (Ed.), D. Reidel, Dordrecht, The Netherlands.

EISENBERG, N.A., LYNCH, C.J., & BREEDING, R.J. (1975). Vulnerability Model: A Simulation System for Assessing Damage Resulting from Marine Spills, US Coast Guard Report USCG-D- 136-75.

EMERSON, M.C. (1986). Dense Cloud Behaviour in Momentum Jet Dispersion. IMA Conference on Mathematics in Major Risk Assessment, Oxford, July 1986.

HEALTH & SAFETY COMMISSION (HSC) (1991). Major Hazard Aspects of the Transport of Dangerous Substances, Advisory Committee on Dangerous Substances, HMSO.

HEALTH & SAFETY EXECUTIVE (HSE) (1978). Canvey: An Investigation of Potential Hazards from Operations in the Canvey Island/Thurock Area, HMSO.

HEALTH & SAFETY EXECUTIVE (HSE) (1987). The Tolerability of Risk from Nuclear Power Stations, HMSO.

NATIONAL PORTS COUNCIL (NPC) (1976). Analysis of Marine Incidents in Ports and Harbours, September 1976.

OPSCHOOR, G. (1979). Methods for the Calculation of the Physical Effects of the Escape of Dangerous Material, TNO.

SPOUGE, J.R., (1990). The Use of Risk Assessment for Ships Carrying Hazardous Cargo in Port Areas. Conference on Safety at Sea and in the Air - Taking Stock Together, Royal Aeronautical Society, London, November 1990.

Risk Assessment of Transportation for the Canadian Nuclear Fuel Waste Management Program

T.F. Kempe
L. Grondin
Ontario Hydro
Toronto, Ontario
Canada M5G 1X6

ABSTRACT

The calculation of public risk arising from a conceptual future program of large-scale transport of used fuel to a disposal centre is described. The data required for the assessment are examined, in particular the data for estimation of the frequencies of low-probability, high consequence accidents. Radiological risks are summarized and compared with those arising from conventional injuries in accidents.

1. INTRODUCTION

A nuclear generating station such as Pickering produces about 25,000 used fuel bundles per year. At present this used fuel is stored in water pools at the station sites, but under the Corporate Plan for used fuel management, Ontario Hydro plans to dispose of the used fuel when a disposal facility becomes available (Ontario Hydro, 1991). For planning purposes, the in-service date for disposal is assumed to be 2025.

In 1978 the Federal Government of Canada agreed on a joint program with the Province of Ontario to assure the safe and permanent disposal of nuclear fuel waste from power reactors.

The reference disposal concept, developed by Atomic Energy of Canada Limited under the Canadian Nuclear Fuel Waste Management Program (CNFWMP), is for the fuel to be shipped to a disposal facility located somewhere in the Canadian Shield in Ontario. As part of the CNFWMP, Ontario Hydro is developing the technology for a large-scale used fuel transportation system. This paper describes the assessment of the impact on the public of the transportation component of the disposal system, considering both normal and accident conditions.

This paper is a condensed account of the assessment. The full assessment will be available for public review during Federal Environmental Assessment Review Office (FEARO) hearings, now expected to take place in 1994-95.

2.   TRANSPORTATION SYSTEM DESCRIPTION

Until such time as a disposal facility is available, fuel will continue to be stored at Ontario Hydro's CANDU nuclear generating stations. These stations are located on the Great Lakes, and are accessible by road, rail and water. All three modes of transportation are being considered in the concept assessment.

Approximately 180 000 used CANDU fuel bundles, or about 3 600 tonnes of irradiated uranium, would be shipped to the disposal facility each year.

A detailed transportation system description for road, and conceptual systems for rail and water, have been developed as a basis for analysis of logistics, costs and environmental impacts. The road system is based on an existing tractor/trailer/cask system, designed by Ontario Hydro (Ribbans, 1988). A Type B(U) design approval certificate for this cask was received from the Atomic Energy Control Board in July 1987, licensing it to carry up to 192 CANDU fuel bundles at ten years' cooling following discharge from a reactor. The cask, illustrated in Figure 1, is almost cubical in shape. It is made of solid stainless steel, and has a stainless-steel-sheathed redwood impact limiter bolted to the lid, protecting the seal area from impact and fire. Transport is with a dry (air-filled) cavity. The fuel is contained in two irradiated fuel storage-transportation modules.

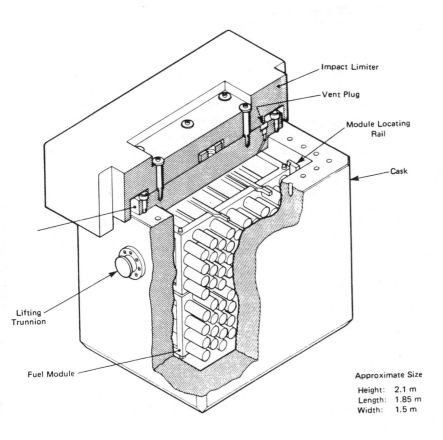

Impact Limiter

Vent Plug

Module Locating
Rail

Cask

Lifting
Trunnion

Fuel Module

Approximate Size

Height: 2.1 m
Length: 1.85 m
Width: 1.5 m

Figure 1: Ontario Hydro used fuel transportation cask.

Based on the existing road cask, a larger rail cask has been
developed to the concept stage, accepting six fuel modules. One
cask would be carried on each of 10 railcars in a special train.

The water mode of transport would use either road casks or rail
casks, depending upon the land mode of transport selected to
interface with the water system. These two options are referred to
in the results as water-road and water-rail. The conceptual barge
system would carry multiple road or rail casks below decks. The
concept is an integrated push-type barge, with the tug fitting in a
notch at the back.

3.   TRANSPORT PACKAGING OF RADIOACTIVE
MATERIALS REGULATIONS

Transport of radioactive material in Canada is regulated by
Transport Canada, via the Transport of Dangerous Goods
Regulations, and by the Atomic Energy Control Board, via the
Transport Packaging of Radioactive Materials Regulations
(AECB, 1983). The AECB's packaging regulations are based on
the model regulations prepared by the International Atomic
Energy Agency (IAEA, 1979). Among the requirements of the
regulations are specified the following:

- radioactive contents limits for different types of package;

- approval by AECB of packages for large quantities of
  radioactive material and packages containing fissile
  material;

- allowable external radiation levels;

- allowable external surface contamination;

- limits on leakage of radioactivity in normal conditions; and

- requirements for retention of shielding and containment of
  radioactive material in accident conditions.

The range of potential hazard from radioactive material is very
wide. To cover this range, two approaches are used in prescribing
package design requirements:

- defined limits on package contents, for packages containing
  relatively small quantities of radioactive materials (Type A
  packages and low specific activity materials), and

- design requirements, such as limits on external radiation
  levels and allowable releases of radioactive materials, for
  packages for which no contents limits are specified (Type B
  packages).

The design of a Type A package is required to provide adequate
shielding and containment under normal conditions of transport, including rough handling and adverse weather conditions.
The design of a Type B package is required, in addition, to
withstand accident conditions comprising severe impact,
specified as a 9 m drop onto an unyielding target plus a 1 m drop

onto a punch, followed by a half-hour exposure to an enveloping 800°C fire. The accident conditions which a Type B package must survive are intended to give a reproducible representation of real accidents. For the used fuel, the quantity and type of radionuclides present require that a Type B package is used.

Experience with use of the regulations world-wide shows that the regulatory prescriptions provide a high degree of safety.

4. ASSESSMENT METHODOLOGY

4.1 General

Although there is a considerable body of literature on the safety and risk assessment of transport of radioactive materials, at present there is no detailed methodology or code recommended by IAEA. However, a number of IAEA activities have focused on transport risk assessment. A current IAEA Coordinated Research Programme (CRP), titled "Development of Risk Assessment Techniques Related to the Safe Transport of Radioactive Materials" has the aim of publishing a document containing guidance material. As part of the CRP, the participating Member States are evaluating the US code RADTRAN4, developed by Sandia National Laboratories, for wider international use.

The present assessment uses the code INTERTRAN, developed during an earlier IAEA CRP, to calculate radiation doses to the public in normal conditions. An Ontario Hydro code, TADS, has been developed to calculate doses in accident conditions, and is based on models similar to those of RADTRAN4.

The radioactive material carried in the cask can cause radiation exposure of members of the public by two main pathways. In normal conditions, the material is tightly contained within the cask, and the only radiation exposure is due to gamma rays and neutrons penetrating through the cask walls and travelling through the air to the exposed person. The cask wall thickness is designed to limit the amount of radiation to below the limits specified in the AECB transport regulations. In this assessment, the small doses remaining are calculated for the population surrounding the transport routes. The cask is represented as a point source of radiation, quantified by means of the transport index. The transport index is the external radiation dose rate at one metre from the package surface. It is used in categorizing, labelling, and controlling transport of radioactive material. Doses at greater than one metre are calculated from the transport index using an inverse square relationship between dose rate and distance from the package. Doses to groups beside the route, in other vehicles, and at shipment stops are calculated.

The other way in which radiation exposure is received is by internal exposure after radioactive material has been inhaled or ingested. This applies only if radioactive material escapes from the cask. This would be small quantities, and would essentially only happen in a transport accident, as shown in Figure 3. Released material can also settle on the ground and give off gamma rays and neutrons which would lead to enhanced external doses.

If material is incorporated in the body, it follows the normal metabolic processes for that particular element. It may remain in the lung, or move to bone, or to the liver or other organs, or all of these. As an example, plutonium, present in the used fuel, is an element which tends to stay in the lung, giving off concentrated alpha radiation, and producing high doses from a small amount. Cesium, on the other hand, tends to move to bone, where it gives off energetic gamma rays, which deposit energy in the whole body.

Out of approximately 150 radionuclides present in the used fuel, the twenty most significant (representing about 99% of the potential risk) were selected for inclusion in the assessment. These radionuclides were identified by examination of the inventory of each radionuclide in 10-year-cooled fuel (Tait et al., 1989) together with the dose conversion factor, i.e., the dose from unit quantities inhaled or deposited on the ground, taken from tabulated values generally used in the nuclear industry (Johnson and Dunford, 1983; Holford, 1989). Individual and collective risks are calculated making use of a linear dose-risk relationship (ICRP, 1991).

4.2 Accident Severity Categorization

The used fuel transportation cask is designed to withstand accident conditions at least as severe as those specified in the IAEA Regulations and the AECB's TPRMR. These conditions have been estimated to encompass up to 99.9% of transportation accidents (Wilmot, 1981; McClure, 1981). Extended testing and analysis on the Ontario Hydro used fuel transportation cask has shown that the cask would maintain containment in accident conditions more severe than the regulatory conditions. The probability of any release occurring is therefore very small, and it is difficult to obtain an estimate of the frequency of occurrence of releases from statistical data. A simplified form of fault tree analysis was therefore used to estimate the probability of each severity category, for each mode. This methodology is commonly used to estimate the probability of rare events where little or no historical data are available for those specific events.

The range of accident conditions was divided into a number of accident severity categories, described further below.

4.3 Calculation of Risk from Accident Conditions

The scheme for calculation of the consequences of an accident in a particular severity category is shown in Figure 2. First, the seal damage and damage to the used fuel bundles are quantified for each severity category. The amount of radioactive material released from the cask is then calculated for each category.

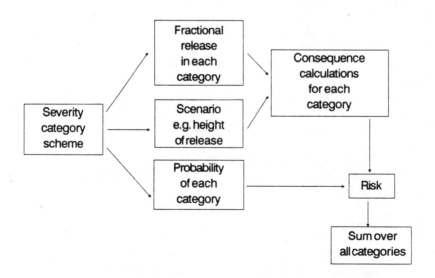

Figure 2: Scheme for risk calculations.

To complete the assessment, the probability of an accident occurring in each severity category is estimated, together with the probability that the accident would occur in a particular Pasquill stability class, and in a particular population density zone (i.e., rural, suburban or urban).

The consequence calculations are similar to those carried out for fixed facilities such as nuclear reactors and radioactive waste treatment facilities.

The transport and dispersion of radioactivity released from the cask was modelled using the Gaussian plume model for a short-term release and for a prolonged release. The short-term release model was used for releases taking place over a very short time period, i.e., those occurring due to fuel and seal-area damage caused by impact, while the prolonged release model was used for releases taking place over a period of more than a few minutes, due to heating of the fuel and thermal degradation and distortion of the seal and seal-area.

For the calculation of long-term doses, it was assumed that, following exposure for 24 h to the external dose rate from initially-deposited radionuclides, the contamination was cleaned-up, if necessary, and the exposure then continued for a further period of 50 a. Account was taken of radioactive decay and weathering of surfaces. With the exception of cleanup of ground deposits, the results of mitigating actions are not included in the TADS calculations. In the discussion below, only short-term, or acute, doses are used. Inclusion of long-term exposure would add about 60% to the maximum short-term doses. Long-term collective doses could be considerably higher than short-term, reflecting the lesser impact of cleanup, however, control measures are likely to be the most important factor.

For simplicity, only adult doses are discussed in this paper. Doses to infants were included in the assessment, and were less than a factor two higher.

4.4 Pathways of Exposure to Radiation

Figure 3 shows the exposure pathways by which material released from the cask reaches members of the public. Following release, the plume of released radioactivity disperses downwind, with exposure of the public taking place via inhalation and exposure to

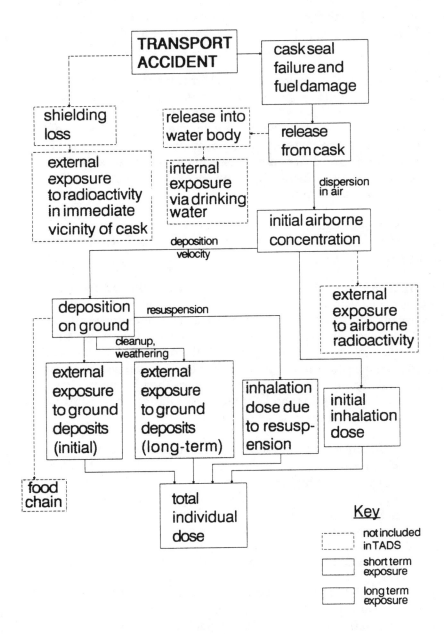

Figure 3: Exposure pathways in accident conditions.

radioactivity deposited from the plume. In the long term, after the airborne radioactivity is dispersed, exposure takes place via inhalation of material resuspended from ground deposits, and exposure to the ground deposits.

In preliminary calculations, the most important pathways were shown to be inhalation, groundshine in the short and long term, and inhalation following resuspension of deposits. External exposure to the material remaining in the cask, and external exposure to airborne material were not important. The food chain can give significant doses, but control measures are likely to be the most important factor (this pathway has been examined separately in a more qualitative way).

The consequence calculations are similar to those carried out for fixed facilities such as nuclear reactors and radioactive waste treatment facilities.

The transport and dispersion of radioactivity released from the cask was modelled using the Gaussian plume model for a short-term release and for a prolonged release. The short-term release model was used for releases taking place over a very short time period, i.e., those occurring due to fuel and seal-area damage caused by impact, while the prolonged release model was used for releases taking place over a period of more than a few minutes, due to heating of the fuel and thermal degradation and distortion of the seal and seal-area.

For the calculation of long-term doses, it was assumed that, following exposure for 24 h to the external dose rate from initially-deposited radionuclides, the contamination was cleaned-up, if necessary, and the exposure then continued for a further period of 50 a. Account was taken of radioactive decay and weathering of surfaces. With the exception of cleanup of ground deposits, the results of mitigating actions are not included in the TADS calculations. In the discussion below, only short-term, or acute, doses are used. Inclusion of long-term exposure would add about 60% to the maximum short-term doses. Long-term collective doses could be considerably higher than short-term, reflecting the lesser impact of cleanup, however, control measures are likely to be the most important factor.

For simplicity, only adult doses are discussed in this paper. Doses to infants were included in the assessment, and were less than a factor two higher.

4.5 Conventional Risks

Risks from used fuel transport from conventional causes were also examined. These impacts are related to accident rate increases due to increased traffic, and injuries and fatalities from conventional causes in traffic accidents involving the used fuel shipments.

5. ROUTE DATA

The NFWMP will not begin selecting a disposal site until some time after the ongoing disposal concept assessment has been completed, and a decision on acceptability has been made. Since the disposal concept is based on geological formations ('plutons') which occur through out the Canadian Shield, and since Ontario is the major source of used fuel for disposal, it is assumed that the disposal centre would eventually be located somewhere within the Ontario portion of the Shield. Figure 4 shows the study area.

This is a very large area (approximately 670 000 km^2), which results in a wide range of values for the assessment parameters. However, the generic nature of the assessment implies that the environment characterization approach should not prejudge the transportation modes and routes. The parametric approach chosen was based on the establishment of generalized 'reference routes' from the source of the used fuel to the geographical centroids of three regions in the Ontario Shield: southern, central and northern.

Road routes used only existing Class A roads. The rail routes were selected from existing Canadian National Railway (CN) main tracks. Because of the geographical location of the Great Lakes water bodies in Ontario, a water reference route to the southern region was not considered. The water routes to the central and northern regions were constructed based on standard sailing routes, using normal shipping lanes in Canadian Waters where possible.

Figure 4: Study area.

For each of the three regions, the 'reference route' was defined as a composite route having characteristics constructed as the distance-weighted average for a number of existing possible routes. This approach provides a set of parameters representative of the transportation conditions in the study area.

Table 1 summarises some of the data used. Similar sets of data were compiled for each mode.

Table 1: Data required for assessment of normal conditions: road mode, southern destination.

Parameter	Zone		
	Urban	Suburban	Rural
Population density (p.km^{-2})	922	370	19
Fraction of travel in zone	0.024	0.05	0.926
Shipment speed (km.h^{-1})	15	50	85
Number of one-way vehicles per hour	153	215	75
One-way distance (km)	400		
Number of shipments	938		
Stop time per 24 hours (h)	6.7		
Number of people exposed at stops	25		
Exposure distance (m)	20		
Pedestrian density factor	6		
Fraction of urban travel on city streets	0.76		
Fraction of rural and suburban travel on freeways	0.07		
Fraction of travel during rush hours	0.15		
Persons per vehicle	2		
Transport index	4.4		

5.1 Population Density Data

Statistics Canada defines urban and rural areas according to concentration of population and population density. These two divisions are too broad for the purposes of the used fuel transportation assessment. New population density ranges were therefore defined, representing urban, suburban and rural zones. The population data were then characterized by the fraction of each route falling into each of the classes. The average density found in each zone, for the reference routes, was used in the calculation. For road, an estimate was made of the household density within 1600 m of the highway, compared with that in the township as a whole, and this factor was applied to the data obtained from the Ontario Municipal Directory, to correct for the fact that some rural areas show higher than average population densities along the highways.

5.2   Accident Data for All Modes

Overall accident rates for the road mode were taken from
Ontario Ministry of Transportation and Communications data,
for each highway segment. The number of accidents occurring
in each highway segment was divided by the traffic volume data
for that segment; this gives the accident rate for the segment. The
distance weighted average for each population zone was used in
the calculations. For rail, the number of accidents for each rail
segment was obtained from the CN Accident Prevention
Department. A reportable accident is defined as any accident
resulting in material damages in excess of $750. This was
multiplied by the number of railcars involved. The number of
railcar accidents was then divided by the total railcar km
travelled to give the railcar accident rate per railcar-km. The
railcar traffic was obtained by multiplying the train traffic,
from CN by an average number of railcars per train, from
Statistics Canada. This average number was 7 for a passenger
train, and 70 for a freight train (passenger train railcars and
freight railcars were added). For water, the number of tug-barge
accidents in a given section (1975-1985) was obtained from the
Canadian Coast Guard. The corresponding tug-barge km was
estimated using traffic data from the Welland Canal and the
Sarnia Coast Guard Traffic Centre, and assuming that the
transit distances were the same as those travelled on the
reference routes.

Accident rates are summarized in Table 2. It can be seen that
there is wide variation among the route segments. In the case of
water transport, the range represents the difference between open
water and river/canal segments.

Table 2: Accident rates.

Mode	Range of accident rates, per 10^6 vehicle-km
Road	0.9 - 1.5
Rail	0.2 - 1.9
Water	0.8 - 18.9

6. ACCIDENT SEVERITY DATA

For the analysis of accident conditions, the spectrum of possible
accidents was divided into a number of categories according to
the severity of the impact and thermal environment experienced
by the shipment. The accident severity categorization scheme
used in the assessment is shown in Figure 5. The first category
consists of those accidents which are not severe enough to affect
the integrity of the cask, and for which the radiological
consequences are therefore bounded by the doses corresponding to
the regulatory release limits. The other categories were chosen to
represent a spectrum of accident conditions for which the release
from the used fuel transportation cask would vary from minimal
up to the most severe credible. The parameters affecting the
quantity of radioactive material released from the cask (impact
severity and fuel temperature), and the characterization of the
accidents, (release duration, effective height of release, number
of casks involved) were described for use in fault tree analysis.

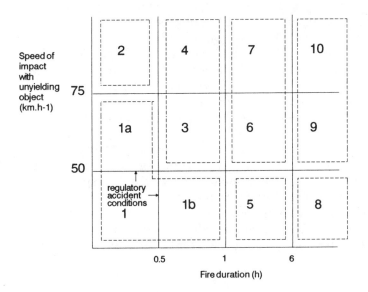

Figure 5: Accident severity categories.

6.1 Accident Severity Probabilities

A simplified form of fault tree analysis was used to estimate the probability of an accident in each category. A sample fault tree is given in Figure 6, showing the combinations of events which result in an accident in Severity Category 10. The event probabilities (e.g. probability of a collision occurring in a particular speed range) were taken from the literature, as discussed briefly below. Many conservative simplifying assumptions were made, e.g. as to orientation of the cask at the time of impact.

Probabilities for truck collisions at various speeds, and the probability of a subsequent fire, were taken from Sandia work (Dennis et al., 1978). Collisions with vehicles other than heavy trucks (>11 te) were not considered to have the potential to damage the cask. The fractions of accidents involving collision with a rock face or bridge support were taken from the Ontario MTC Annual Road Safety Report (MTC, 1986). For rail, all crossing accidents where the locomotive collides with another vehicle were considered to be in Severity Category 1. This was because, in general, the vehicle impacted by the shipment would be of much less weight. The main impact would also be taken by the locomotive. Collision accidents were broken down using data from Cook et al. (1985). For the water mode, the distribution of severities was derived from accident data available from the Canadian Coast Guard, and from other data. Due to the low speeds involved, impact at an equivalent impact speed greater than 75 km.h^{-1} was considered incredible. Only fires involving tankers were considered to have the potential to damage the cask.

The resulting accident probabilities are given in Table 3. The zero fractional occurrence of impact-damage-only accidents (i.e., Severity 2) in the water mode reflects the low speeds, as mentioned above. The fire-damage-only accident categories, Severity 5 and 8, have a higher fractional occurrence than the combination impact-fire damage accidents, Severity 3/4 and 6/7, identifying long fires as an important contributor to cask failure frequency.

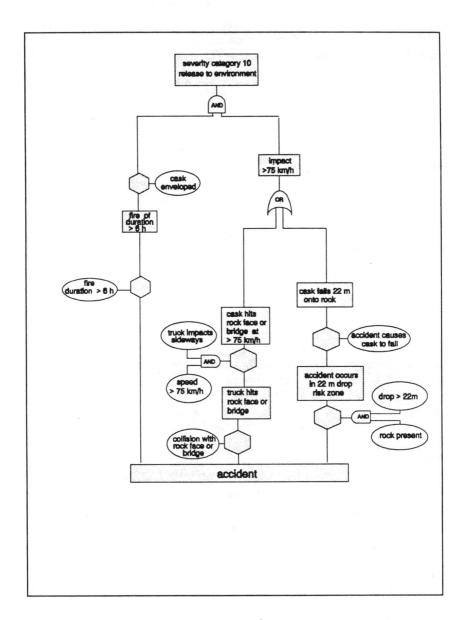

Figure 6:   Example of fault tree, Severity Category 10 (fire
duration > 6h, impact > 75 km.h^{-1})

Table 3: Fraction of accidents in each severity category

Severity	Fractional Occurrence, Given an Accident		
Category	Road	Rail	Water
1	0.99998	0.99988	0.99999
2	10-5	10-4	0
3/4	10-7	10-6	10-8
6/7	10-8	10-7	10-7
5	10-5	10-5	10-6
8	0	10-5	10-5
9/10	0	10-7	10-6

6.2 Release from Cask

To reach the environment, the radionuclides in the fuel must escape three barriers:

a) the fuel matrix,

b) the fuel cladding, and

c) the cask.

Models, based on those used for reactor accident analysis, were developed for release from the fuel matrix. Different release mechanisms are applicable to each of the nuclide groups, i.e. gases, semi-volatiles, and non-volatile fission products and actinides. The fractional release from the cask in each severity category was calculated using Oak Ridge models (Lorenz, 1980; Lorenz, 1979), together with recent Canadian data on oxidation (Hunt et. al., 1986). Following Wilmot and McClure (1981), it was then assumed that the particulates, also ^{106}Ru, ^{134}Cs, ^{137}Cs and ^{129}I at temperatures below their volatilization temperatures, would be retained within the cask with an efficiency of 0.95.

Fuel cladding damage assumptions were based on data from impact tests on CANDU fuel.

The boundaries for cask failure (i.e., impact equivalent to 75 km.h^{-1}, or fire duration greater than 1 h), were derived using the results of the cask design program, together with review of the literature on fuel cask design and testing. Potential cask failure modes envisaged were as follows:

a) Loss of integrity of the elastomeric lid, vent or drain seals due to thermal degradation in severe thermal conditions.

b) Loss of lid bolt tension, leading to seal bypass leakage, following a severe impact.

Puncture of the 270 mm stainless steel cask body or lid was not considered credible. This is supported by the rail coupler impact test carried out by Ontario Hydro, in which the cask survived impact at 104 km.h^{-1} by a locomotive coupler with only superficial scratches. Similarly, loss of all lid bolts leading to ejection of gross quantities of fuel, was not considered credible.

For the most severe accidents, involving an impact and an extended fire, fractions of 2×10^{-3} of the cesium in the fuel, and 10^{-6} of non-volatile nuclides, were estimated to be released from the cask.

7. RESULTS OF ASSESSMENT

7.1 Collective Risk

Collective risk is the total risk to members of the exposed population, equivalent to societal risk.

The annual collective dose calculated for normal conditions ranges from 0.01 to 0.1 person-Sv, the largest collective dose being for shipment to the northern region by the water mode. Although wide variations are seen among the cases, the absolute numbers are not high enough to justify drawing a distinction between the modes and destinations based on this measure.

The additional road traffic constitutes a maximum increase of 0.23% to the average road traffic. On the least travelled segment, the increase in traffic was 2.5%. For rail, the maximum increase was 5.7% to the average traffic, or 25% of the traffic on the least-travelled segment. For water, the traffic increase was 0.4%. The traffic was also compared with that from typical mining and lumbering operations. The used fuel traffic was around 15% of that for these operations. The small increases in traffic found are expected to have minimal effect on the accident probability on the reference routes.

The total expected numbers of accidents per year, calculated using accident rates derived for each route, as described in Section 5, are shown in Table 4. Also shown is the number of hypothetical accidents causing a release. This is the number of accidents in Severity Categories 2 - 10, and is derived from the total number of accidents using the fractional occurrences of Table 3. The largest number of total accidents is for the road mode, and this mode also gives the highest probability of an accident causing a release. For road, the number of accidents occurring in the urban zone is similar for the three destinations, however, the total number (in all zones) is greater for the northern destination because of the greater distance. All of the shipments have their origin in southern Ontario. For rail, the relatively high number of accidents in travel to the southern destination reflects a high accident rate on the particular routes used in defining the reference routes.

Table 4: Total annual expected number of accidents involving a used fuel shipment, and probability of a release accident.

Mode	Destination	Expected Number of Accidents Per Year	Probability of Release Accident Per Year
Road	Southern	0.42	8.4×10^{-6}
	Central	0.85	1.7×10^{-5}
	Northern	1.79	3.6×10^{-5}
Rail	Southern	0.02	2.4×10^{-6}
	Central	0.006	7.6×10^{-7}
	Northern	0.009	1.1×10^{-6}
Water-road	Central	$0.47^a/0.08^b$	$9.5 \times 10^{-6}/8.9 \times 10^{-7}$
	Northern	0.39/0.08	$7.9 \times 10^{-6}/8.4 \times 10^{-7}$
Water-rail	Central	$0.04^c/0.08$	$4.8 \times 10^{-7}/8.7 \times 10^{-7}$
	Northern	0.04/0.08	$5.5 \times 10^{-6}/8.6 \times 10^{-7}$

[a] Road   [b] Water   [c] Rail

From the accident calculations, the collective doses in accident conditions were obtained in the form of a cumulative frequency distribution curve for each case (comparable to an F-N curve). This gives the annual probability of a particular dose being reached or exceeded, and shows how the probability falls off for higher doses. An annotated curve is shown in Figure 7. The frequency distributions were used to derive the collective dose corresponding to a particular frequency, and the 90th percentile dose, so that these doses could be examined as well as the 'worst-case' dose.

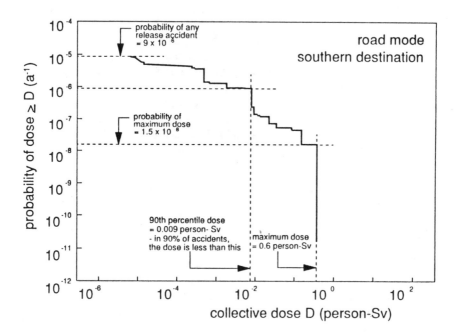

Figure 7: Typical Cumulative Frequency Distribution Curve.

The maximum collective doses were for accidents in the most severe category occurring in stable weather conditions, in the most highly populated zone. The collective dose for this set of circumstances is about 10 person-Sv for the rail and water cases. For road, there are no accidents in the most severe categories, and the maximum collective dose is less than 1 person-Sv. The probability of these mode-specific 'worst cases' varies widely, from 10^{-12} per year for the rail mode to 10^{-8} per year for the road and water modes. These probabilities are very small. For more 'credible' probabilities of around 10^{-6} per year, the collective dose ranges from 0.01 person-Sv for the road and rail modes to 1 person-Sv for the water mode.

The data were also used to give the collective dose from a 'typical' release accident. The collective dose which is not exceeded in 90 percent of these release accidents (the 90th percentile dose) was obtained by reading off the dose for a probability of one-tenth of the total probability of a release accident, as shown in Figure 7. This collective dose ranges from 0.1 person-Sv for the road mode to 1 person-Sv for the rail-water mode.

The average person-Sv per year, or the total expected collective dose, represents the sum of over all the scenarios of the dose multiplied by the corresponding probability. The maximum contribution to this total for the road mode was from accidents in Severity Category 2. For rail, the maximum contribution was from accident Severity Category 2, followed by Severity 8. For the water mode, Severity 8 gave the largest contribution, followed by Severity 9/10. The major contribution to risk is therefore seen to arise from the more probable, lower consequence events.

The effect of ionizing radiation depositing energy in tissue is to increase the probability of the individual developing cancer. It is thought that the radiation dose acts as a multiplier on the natural cancer rate. In addition, if reproductive tissues are irradiated, hereditary effects may be induced in succeeding generations. Using the latest risk factors recommended by the International Commission on Radiological Protection (ICRP, 1991), the expected collective doses in normal and accident conditions were multiplied by 5×10^{-2} per Sv to give the number of fatal cancers expected.

The resulting figures are summarized in Table 5, where they are compared with the number of fatalities and injuries arising from conventional causes.

The consequences from conventional causes for members of the public involved in an accident with a used fuel transport were also examined. Based on MTC data (MTC, 1986) 0.6% of accidents result in loss of life. For rail, the average numbers of persons killed, or injured, per train accident were 0.4 and 1.33 respectively (CN, 1986). These statistics cover the period 1974-1985 and include all public fatalities and injuries associated with railway operation. More recent data may show a decrease in these numbers. Water accident statistics for 1966-1985 show an average of 0.006 deaths and 0.003 injuries per accident. Applying these to the numbers of accidents, shown in Table 4, above, and multiplying by two to account for the return journey, the consequences were as shown in Table 5 (only the average for the three destinations is shown in this table).

Table 5: Comparison of radiological and conventional risks from the used fuel transportation program.

Mode	Number of cancers induced per year		Injuries from conventional causes, per year	Fatalities from conventional causes, per year
	Normal conditions	Accident conditions		
Road	0.003	1×10^{-9}	0.5	0.01
Rail	0.0003	2×10^{-9}	0.02	0.01
Water -road	0.001[a]/0.003[b]	5×10^{-10}/8×10^{-8}	0.3/0.0005	0.005/0.001
Water -rail	0.00005[c]/0.003	5×10^{-10}/9×10^{-8}	0.1/0.0005	0.03/0.001

[a] Road   [b] Water [c] Rail

7.2 Individual Doses

Calculated radiation doses to the public for normal transportation are small. The maximum annual individual dose are summarized in Table 6. The highest figure, 0.39 mSv, is well below the present regulatory limit of 5 mSv, and below the proposed new limit of 1 mSv (AECB, 1991). This figure is for persons exposed to all the shipments at a truck stop (i.e., exposed for 938 h per year), and could be controlled in practice by monitoring, use of alternative truck stops, and choice of parking location.

Although this paper does not address occupational risk it may be noted that occupational doses are limited by distance from the cask to the cab (10m), and by the working time spent in the cab, to a value well below the occupational dose limit.

The maximum dose of 0.39 mSv.a^{-1} may also be put in perspective by comparison with the dose due to natural background radiation in Ontario, 3 mSv.a^{-1} (Neil, 1988).

Table 6: Summary of maximum individual dose in normal transportation conditions.

Mode	Destination	Dose, mSv
Road	All	0.39[a]
Rail	All	0.0003[b]
Water	All	0.05[c]

[a] Persons continuously present at a truck stop used by the shipments (per year).
[b] Persons living beside the rail link (per year).
[c] Persons following a shipment through a canal (per trip).

The maximum individual doses in accident conditions were analyzed in a similar way to the collective doses. The results are summarized in Table 7 for each mode, together with the conditions in which they are found. The location with respect to population zone is not a factor in individual dose since, for this generic assessment, it is assumed that people are present in the vicinity of the accident, wherever it occurs.

Table 7: Summary of maximum individual doses due to transportation accidents.

Mode	Maximum Individual Dose (mSv)			Annual Probability of Worst Case	Conditions of Worst Case
	90th Percentile	Probability of 10^{-6}	Worst Case		
Road	3	9	9	3×10^{-6}	Severity 2, 3/4, 6/7; Pasquill Class F
Rail	30	2	28	4×10^{-7}	Severity 2, 3/4, 9/10; Pasquill Class F
Water	30	8	55	2×10^{-8}	Severity 2, 3/4, 9/10; Pasquill Class F

While higher than the limit for members of the public, the individual doses corresponding to a probability of 10^{-6}, 2 - 9 mSv, are less than the annual limit for occupationally-exposed persons. They are also well below the threshold for non-stochastic effects (e.g., vomiting, early death from bone marrow cell depletion) of 500 mSv (ICRP, 1984). Even the 'worst case' dose, 55 mSv, is below this level. No acute effects would therefore be expected.

It may also be noted that the figure of 2 - 9 mSv is less than the upper Protective Action Level, 100 mSv, given in the Technical Bases of the Ontario Provincial Nuclear Emergency Plan, at which members of the public would be 'automatically' evacuated, although they are of the same order as the lower Protection Action Level, 10 mSv, at which evacuation would be considered (Province of Ontario, 1984).

The calculation of individual risk requires the summation of risk to an individual located at any particular point along the highway from accidents occurring at different distances. This is a complex calculation, and has not been attempted here. An upper bound to the individual risk from accidents, for comparison with the individual risk criterion of 10^{-6} a^{-1} proposed for radioactive waste disposal (AECB, 1987), may be found by taking the expected collective dose impact, recognizing that this represents the probabilistic dose summed over all individuals potentially exposed along the transportation route. The maximum expected collective dose is 1.8×10^{-6} person-Sv a^{-1} (water-rail mode, transport to the Northern destination). Multiplying by a risk coefficient of 5×10^{-2} Sv^{-1}, the risk becomes 9×10^{-8} a^{-1} for this case. Clearly, since the summation encompasses all individuals, including those at highest risk near the source, the risk to any one individual is far less than this, and is well below the limit of 10^{-6} a^{-1}.

8. CONCLUSIONS

The used fuel shipments result in very low estimated radiological impact on the public in normal conditions. The actual routes chosen should be surveyed to identify critical groups, and, in the early stages of the program, limited health physics surveillance would be useful in ensuring doses were as low as reasonably achievable.

The probability of an accident resulting in a release of radioactivity is very small. Acute effects or any detectable increase in cancer incidence are not expected following even a severe accident. The annual risk to any individual is well within acceptability criteria.

It was concluded that, because of the high standard of safety in the design and testing of the cask, the risk of detriment to any individual from the radiological aspect of the transportation program is negligible. The radiological risk from accidents is very much smaller than that from normal operation, which in turn is similar or less than the risk from conventional causes in accidents.

9. REFERENCES

ATOMIC ENERGY CONTROL BOARD (AECB) (1983). Transport Packaging of Radioactive Materials Regulations. SOR/83-740, 29 September, and subsequent amendments.

ATOMIC ENERGY CONTROL BOARD (AECB) (1987). Regulatory objectives, requirements and guidelines for the disposal of radioactive wastes - long-term aspects. Regulatory Policy Statement R-104, 5 June.

ATOMIC ENERGY CONTROL BOARD (AECB) (1991). Proposed amendments to the AEC regulations for reduced radiation dose limits based on the 1991 ICRP Recommendations. Consultative Document C-122.

CANADIAN NATIONAL RAILWAY (CN) (1986). Safety Backgrounder (1974-1985).

COOK, M.C., MILES, J.C., & SHEARS, M. (1985). A study of flask transport impact hazards. Seminar on the resistance to impact of spent magnox fuel transport flasks. Institution of Mechanical Engineers, 30 April - 1 May.

DENNIS, A.W., FOLEY, J.T., HARTMAN, W.F., & LARSON, D.W. (1978). Severities of transportation accidents involving large packages. SAND-77-0001.

HOLFORD, R.M. (1989). Supplement to dose conversion factors for air, water, soil and building materials. AECL-9825-1.

HUNT, C.E.L., IGLESIAS, F.C, COX, D.S, KELLER, N.A., BARRAND, R.D., MITCHELL, J.R., & O'CONNOR, R.F. (1986). Fission product release during UO_2 oxidation. International Conference on CANDU Fuel, CRNL, October 6-8.

INTERNATIONAL ATOMIC ENERGY AGENCY (IAEA) (1979). Regulations for the safe transport of radioactive materials. 1973 Revised Edition (as Amended). Safety Series No.6, IAEA, Vienna.

INTERNATIONAL COMMISSION ON RADIOLOGICAL PROTECTION (ICRP) (1984). Protection of the public in the event of major radiation accidents: principles for planning. ICRP Publication 40, Pergamon Press.

INTERNATIONAL COMMISSION ON RADIOLOGICAL PROTECTION (ICRP) (1991). Recommendations of the International Commission on Radiological Protection. ICRP Publication No. 60, Pergamon Press.

JOHNSON, J.R., & DUNFORD, D.W. (1983). Dose conversion factors for intakes of selected radionuclides by infants and adults. AECL-7919.

LORENZ, R.A., COLLINS, J.L., & MALINAUKAS, A.P. (1979). Fission product source terms for the light water reactor loss-of-coolant accident. Nuclear Technology , 46, p 404, Mid-December.

LORENZ, R.A., COLLINS, J.L., MALINAUKAS, A.P., KIRKLAND, O.L., & TOWNS, R.L. (1980). Fission product release from highly irradiated LWR fuel. NUREG/CR-0722, ORNL/NUREG/TM-287/R2.

MCCLURE, J.D. (1981). The probability of spent fuel transportation accidents. SAND80-1721.

NEIL, B.C.J. (1988). Annual summary and assessment of environmental radiological data for 1987. Ontario Hydro Report SSD-AR-87-1.

ONTARIO HYDRO (1991). Radioactive materials management at Ontario Hydro: the plan for used fuel.

ONTARIO MINISTRY OF TRANSPORT AND COMMUNICATIONS (MTC) (1986). Ontario Road Safety Annual Report for 1985.

PROVINCE OF ONTARIO (1985). Nuclear emergency plan.

RIBBANS, D.J. (1988) Road cask for the transportation of CANDU irradiated fuel. International Conference on Transportation for the Nuclear Industry, Stratford-on-Avon, UK, 23-25 May.

TAIT, J.C., GOULD, I.C. & WILKIN, G.B. (1989). Derivation of initial radionuclide inventories for the safety assessment of the disposal of used CANDU fuel. AECL-9881.

WILMOT, E.L. (1981). Transportation accident scenarios for commercial spent fuel. SAND-80-2124.

Chapter 2: Analysis of Dangerous Goods Accidents and Releases

Historical Analysis of Dangerous Goods Spills in Alberta

S.P. Hammond
W.W. Smith
Dangerous Goods Control Division
Alberta Public Safety Services
Edmonton, Alberta

ABSTRACT

Alberta Public Safety Services is the provincial department concerned with the transportation of dangerous goods on Alberta highways, the enforcement of the Canadian legislation, and the response to dangerous goods incidents on the highways. Traditional vehicle accident reporting methods do not necessarily reflect the true cause of the accident, and an analysis of incidents occurring in the province during 1991 was done to assess the human error factor in dangerous goods spills in the province. Much of the analysis is based on post accident investigation, but there appears to be sufficient data to point to human error, opposed to mechanical failure, as a significant factor in dangerous goods incidents. Industry response to suggestions for improvements has been positive, and has reduced incident frequency in some areas.

1. INTRODUCTION

Alberta Public Safety Services is the provincial government agency responsible for the preparation and implementation of the emergency planning requirements for municipalities in the province and the administration of the Transportation of Dangerous Goods legislation (Transportation of Dangerous Goods Control Act, 1982) in Alberta.

To meet operational objectives APSS formed the Coordination and Information Centre (CIC) within the Dangerous Goods Control Division in late 1984, training personnel to undertake three roles:

1. to provide a 24 hour source of compliance information to industry, the enforcement agencies and the general public,

2. to provide emergency response forces with an immediate source of initial emergency response information, access to the provincial government support mechanisms, and

3. to function as the provincial government emergency response centre for all natural and man made emergencies and disasters.

The Coordination and Information Centre or (CIC) operates on a 24 hour basis and responds to approximately 300 emergencies and disasters annually, of which an average of 250 relate to dangerous goods incidents, usually in the transport phase.

Under established protocols, the local police are to notify the CIC of a dangerous occurrence, which results in the activation of the appropriate provincial government response. This action is taken in support of the municipal authorities, all of whom are required by law to have in place a municipal emergency plan (Alberta Public Safety Services Act, 1985). APSS will ensure that all appropriate government departments are notified, and, in general, an APSS dangerous goods inspector is sent to the scene to act as a resource and to conduct a compliance investigation (Alberta Public Safety Services, 1991). Significant data is gathered at the time of the notification, and further data is generated by the inspector at the scene and from his reports. An additional source of information is the 30 day report filed with Transport Canada.

The data gathered by the CIC since the implementation of the legislation in Alberta has been maintained on an APSS database, and is probably the most comprehensive regional information system related to road mode incidents involving dangerous goods in Canada today.

In previous analyses, we have estimated that the reported road mode dangerous goods incident trend shows an annual increase of between 9 and 16%, dependent on whether the recently regulated products (diesel and sulphur) are included. Figure 1 shows the impact of the amendment regulating these products, (published in August 1989) on the 12 month moving average. There are some seasonal variations and the data is quite irregular; however the impact of the regulatory amendments is quite clear.

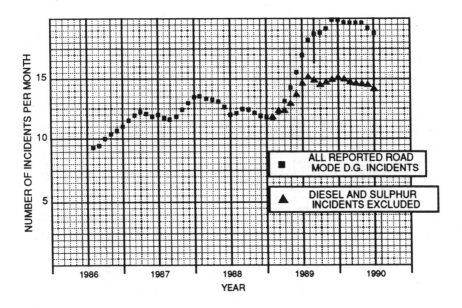

Figure 1:   Reported road mode dangerous goods incidents twelve
            month moving averages: 1986-1990 (Hammond, 1991).

Figure 2 shows the calculated trends in relation to monthly
dangerous goods incidents.

As part of the ongoing analysis at APSS, incidents were reviewed to
determine the root cause. In going beyond the traditional causes,
such as rollovers, single motor vehicle accidents etc., APSS has
attempted to break these causes down into two main categories,
human error or mechanical failure.

In doing so we have gone beyond statistically defined trends, and
calculated incident rates, to a more subjective analysis by looking
at each incident and assessing the fundamental causes of the
mishap from inspection and incident reports.

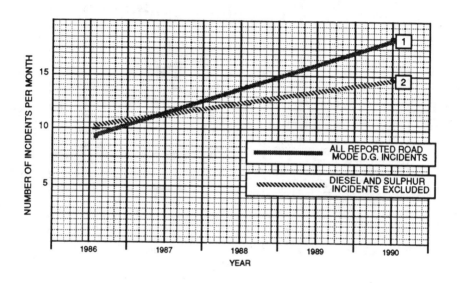

Figure 2:   Reported road mode dangerous goods incidents monthly
trends: 1986-1990 (Based on 12 month moving averages)
(Hammond, 1991).

This review is based purely on the opinions and analyses of the
APSS staff, and for the purposes of the review is confined only to the
road mode incidents occurring in Alberta during 1991.

2.   SCOPE OF THIS REVIEW

This summary considers only regulated road-mode dangerous
goods incidents that occurred in Alberta during 1991. Incidents are
complicated events and their description in simple terms is at best
ambiguous. There are all kinds of incidents, so an immense
variety of circumstances are possible.

Our concern in this area is based on the trend that most regulatory action in the area of dangerous goods is aimed at improving the mechanical aspects of spill prevention, i.e. improving the standards for vehicles, tanks, containers and packaging. But no matter how well we design and implement these standards, accidents occur, and from our review, occur primarily as a result of human error.

3.   DEFINITIONS

Before we begin it might be useful to define the terminology used in the review:

Incident   is a reportable road mode dangerous occurrence as defined in the Transportation of Dangerous Goods Act and regulations.

Human error   means a lack of skill, an attitude that certain operations are unimportant, a general lack of alertness or the failure to apply sufficient attention to a situation that it deserves, leading to an incident.

Mechanical failure   means a mechanical condition leading to the loss of control of a vehicle and thus an incident, or a mechanical condition in a tank, container or packaging, not normally detectable during routine transportation activities, leading to an incident

4.   TYPES OF INCIDENTS

Table 1 deals with general types of incidents, and the fundamental events that made up the incidents under review.

The "incident type" is a basic descriptor from the APSS databases, and in most instances is self explanatory. Where an incident type has been characterized as "unknown", it usually refers to that most difficult of all incidents that APSS has to investigate - the "unknown chemical spill at the side of the road".

Our review of the 181 regulated dangerous goods road-mode incidents revealed that spills accounted for 53.6% of the incidents.

Table 1: Incident types.

INCIDENT TYPE	Jan	Feb	Mar	Apr	May	Jun	Jul	Aug	Sep	Oct	Nov	Dec	Total	%
Rollover	7	5		6	5	6	6	4	3	6	6	4	58	32.1
Leak	1		1		1		2					2	7	3.9
Unknown				1				1					2	1.1
Fire		1				1	2	4	1	1		1	11	6.1
Eq. Failure								3					3	1.6
MVA	1								1		1		3	1.6
Spills	11	11	10	5	14	3	17	6	1	11	7	1	97	53.6
Total	20	17	11	12	20	10	27	18	6	18	14	8	181	

5. ACTIVITY AT THE TIME OF THE INCIDENT

Table 2 provides a breakdown of the spills by types.

21.6% of the spills occurred during the unloading process, compared to 14.4% during the loading operations. The remaining 64% of the spills occurring in 1991 occurred during the transportation cycle.

5.1 Loading and Unloading

Loading and unloading operations combined for a total of 36% of the incidents in 1991.

The most significant overall loss occurred during one LPG loading operation, where the driver made one cargo tank connection and moved his vehicle forward to make another connection. He drove over the connected piping rupturing the line. The escaping LPG ignited resulting in the destruction of the transport unit and the loading racks. The estimated loss totalled more than $750,000.00.

This is the type of incident which may be classified as human error. The error may have been due to a slip or a momentary lapse of attention, to poor training or instruction, or a deliberate decision not to follow instructions or legal requirements. (In Alberta there is a legal requirement that the cargo tank be blocked to prevent movement prior to commencing any loading or unloading of LPG.)

Movement of transport units with product lines attached resulted in spills on four other occasions during 1991.

Other examples of human error resulting in incidents and categorized are:

- Eight movements of transport units with an opened product discharge valve.
- Five cases of the driver overfilling the tank.
- Three cases of carrier leaving the immediate site during the loading or unloading process resulting in product overflow.
- Two cases of the carrier off-loading into the wrong tank.
- A consignee provided the carrier with the wrong tank capacity resulting in an overflow spill.
- One carrier made the wrong connection to a filled tank.

Table 2: Activities leading to spills.

	Jan	Feb	Mar	Apr	May	Jun	Jul	Aug	Sep	Oct	Nov	Dec	Total	%
Loading	1	4	2		3		6			5			21	21.6
Unloading	3		1	1	2		2	2	1	1	1		14	14.4
Other	7	7	7	4	9	3	9	4		5	6	1	62	64.0
Total	11	11	10	5	14	3	17	6	1	11	7	1	97	

- A carrier made an improper hose connection resulting in the discharge line disconnecting once produce transfer started.

- In one case the carrier reversed the pump flow resulting in a spill from the cargo tank which was supposed to be unloading.

Most of these are examples of routine tasks which went awry, usually arising from the failure to carry out a simple routine step. The intention to carry out the task was correct. The operator knew what was required, had carried out the tasks in the past, and is presumed to be capable of carrying out the activity but has a momentary lapse of attention.

5.2 Load Security

Fourteen other spills are directly attributed to insecure loading practices. In most cases when load securement procedures were questioned, the easy reply was "it's industry practice."

Eight spills were the result of punctures to the container: from other pieces of a consignment, nails in the transport unit floor, to piercing the drum with a forklift tine. (A rationale for drums of paint falling off a transport unit deck? The operator suggested it was because the deck was icy.)

6. CAUSES

Table 3 outlines, in simple terms, the causes of road mode incidents during 1991. These causes are derived from the database coding, and it required further analysis by APSS to characterize them as "human error" or "mechanical failure".

Human error can be attributed to 86 road mode incidents during 1991. Post rollover incident investigations and a positive reception by upper company management has provided positive returns.

6.1 The Environment

Alberta's winter weather conditions contributed to six incidents. It is ice-bound five months of the year. Because of this fact, to attribute an incident to icy road conditions is at best borderline, in our opinion.

Table 3: Causes.

INCIDENT TYPE	Jan	Feb	Mar	Apr	May	Jun	Jul	Aug	Sep	Oct	Nov	Dec	Total	%
Environment	1	2		1	1		1				6	1	13	7.2
Human Error	12	7	4	6	8	4	15	7	5	11	4	3	86	47.5
Insecure	2	1	2	1	1	2	3	1			1		14	7.7
Eq. Failure	4	4	5	4	6		5	3		3	1	4	39	21.5
Unknown	1	2			3	3	2	5	1	4	2		23	12.7
Vandals		1											1	0.6
Packaging					1	1	1						3	1.7
Other								2					2	1.1
Total	20	17	11	12	20	10	27	18	6	18	14	8	181	

Although bad driving conditions may have existed, they may have had nothing to do with causing the incident. The really safe driver would size up all conditions and allow for them in his driving. He can travel successfully over very bad roads and in vile weather. He knows enough, not to drive when conditions are worse than he can handle. Most road and weather conditions would not be sufficient, by themselves, to cause an incident.

Training and experience should prevent mistakes in knowledge-based and skill-based behavior. Company instructions should prevent mistakes in rule-based behavior. Companies cannot lay down detailed instructions to cover all winter driving conditions, but can issue sufficient guidelines that permits drivers to exercise judgement and discretion in conjunction with his skills and experience to ensure the safe transport of the dangerous goods.

Most Alberta trucking companies have an internal procedure which permits the operator to suspend operations until conditions improve. For this reason the drivers in the six incidents attributed to icy road conditions could be categorized as human errors due to inadequate training or a deliberate decision not to follow instructions.

One APSS post accident investigation determined that an LPG rollover was caused by the operator driving during a whiteout blizzard. This type of action is not a momentary lapse of attention, but an apparent deliberate disregard for safety.

6.2 The Road

One of the most common causes of a dangerous goods incident is the "rollover". Road conditions and driving techniques are factors in causing accidents of this type. Examples of human errors resulting in rollovers are:

- Three following too closely violations

- Three stop sign violations

- Nine speeding violations. To determine the appropriate speed the driver must understand the operational limits of his vehicle, his abilities and the roadway configuration.

- One rollover caused by driver fatigue

- A loaded refuelling truck backing into a construction excavation site. (An Alberta Occupational Health and Safety General Safety Regulation prohibits the movement of equipment on a construction work site without the assistance of a signaller to guide the equipment operator).

We suggest that these incidents are as a result of human error stemming from a deliberate decision of not following instructions or legal requirement. One could also argue that some of these incidents arise from a lapse of attention or poor training or instruction.

Two other rollover factors noted are roadway drop-offs and offramp/onramp incidents. Certaintly the sharp shoulders on many of the roads in the province are the source of many vehicles veering from the highway, once the driving or following wheels leave the road.

One incident investigation resulted in a trucking firm prohibiting the use of Highway 831, a fairly narrow highway with sharp drop-offs on each side. This firm now incurs greater cost due to a longer distance travelled and then additional time increase due to a forced stop at a previously bypassed Vehicle Inspection Station (V.I.S.). (At the V.I.S. the units are subject to weight examinations and, on occasion, a Commercial Vehicle Roadside Inspection. This roadside inspection is a monitoring tool to measure the firm's preventive maintenance program).

The company directive is enforceable because the Edmonton Area Manager resides along Highway 831 and has resulted in no further company rollovers occurring on Highway 831, or on the new route.

One highway offramp incident was determined to have been caused by speeding around the curve, creating a liquid surge in the tank. The result was the unit left the overpass and came to rest on the major highway it just departed.

Liquids transported in bulk require special driver skills because of the high center of gravity and liquid movement. The high center of gravity makes the unit top heavy and easier to rollover. Liquid surges in smooth bore tanks can also contribute to a rollover. The elevation of offramps, when combined with a liquid surge and a transport unit's high center of gravity can cause a rollover even at the posted ramp speeds.

As a sideline Alberta had one case of a transport unit striking a moose. The interesting point is no accidents were attributed to evasive action being taken to avoid the phantom deer or approaching phantom car.

7. SOME SOLUTIONS

Alberta Public Safety Service, Dangerous Goods Control, in cooperation with a trucking company and the major consignor assisted in the development of the trucking company's policy and procedure manual. This manual covers such points as:

*   ensuring Material Safety Data Sheets for all products the firm transported are carried in the cab of the units

*   detailed maps of all consignees unloading areas and emergency equipment locations. Further, the company now requires all operators to be familiar with the unloading procedures of their clients. The company will no longer permit unloading unless the consignee has approved the unloading procedure to be followed

*   the firm has also hired a safety coordinator and is now a member of the Alberta Trucking Association Safety Committee

*   the hiring practices have been expanded to include a background check

*   operator skills and knowledge are now field tested prior to full-time employment being offered

*   each time an acid cargo tank is cleaned and purged an interior examination of the tank will be conducted

This procedure was recommended as a cost saving plan to reduce emergency response, and unit down-time costs, and resulted from a road mode leak being discovered by a highway patrol member and an APSS-DG tank interior examination of the repairs to the leaking cargo tank. The examination revealed the tank had developed cracks in the tank in the area of the landing gear.

Post rollover incident investigations and a well designed procedures review by upper company management has had positive results.

8. MECHANICAL FITNESS

Another Alberta trucking firm had two anhydrous ammonia trailer separations during a two week period. An interesting point is that the separations occurred within two weeks of each other at the same general highway location, and both incidents at approximately 1400 hours.

An APSS post incident investigation revealed that the firm used single jaw fifth wheels and teflon plates. The fifth wheel manufacturer did not recommend the use of a teflon plate in place of regular greasing of the fifth wheel components.

The company has removed all single jaw fifth wheels and replaced them with double jaw fifth wheels. The use of the teflon plates has been also been discontinued, and the firm has not had any similar types of incidents since this change was made.

A review of the same company's dangerous goods incidents from December 1990 to the end of 1991 also revealed six minor molten sulphur spills. The company officials were receptive to the suggestion of concentrating on these types of incidents. This focus should provide the firm with quick results in reducing molten sulphur spills.

9.   TRAINING - A VITAL ELEMENT

Vital elements in preventing motor vehicle accidents and dangerous goods incidents are the driver's skill and knowledge, both of which can be increased through appropriate training.

Driver performance in the operation of vehicles transporting dangerous goods can be improved by company training programs and supervision by safety coordinators.

Driver attitude is affected by company philosophy and a safe working environment. A employer who actively promotes and rewards good dangerous goods handling and transportation practices can establish safety as an employee responsibility. Properly conducted safety meetings can build good rapport between the employees and management. Management's commitment of encouraging driver feedback can bring better agreement on organizational goals and how to achieve them.

Driver attitude can also be improved by written examinations and road testing before the issuance of a Commercial Driver's License, permitting the operation of a transport unit hauling dangerous goods or a bulk liquid tank.

The concept of Commercial Vehicle Driver's License endorsements for dangerous goods and bulk liquid tank vehicle are recommended by Alberta Public Safety Services. These Driver's License endorsement programs could reduce incidents caused by human error.

10. CONCLUSION

Human error is a significant factor contributing to dangerous goods spills. The errors may range from poor training, poor site or transport procedures to poor decision making on the part of the operators.

Where management has focused on better planning, training and procedures, a decrease in the accident rates has been noted.

We believe that equal focus should be given in regulatory action to address the human factors, and the mechanical standards for vehicles, packaging and containers.

By working with the industry in post accident investigations, regulatory agencies can highlight areas of concern, and in conjunction with industry, implement simple policy or procedural changes or standards to avoid a reoccurrence.

11. REFERENCES

ALBERTA PUBLIC SAFETY ACT (1985). Statutes of Alberta, Chapter P-10.5, Edmonton: Queen's Printer for Alberta.

ALBERTA PUBLIC SAFETY SERVICES (1991). Government of Alberta Support Plan for Dangerous Goods Incidents, Edmonton, AB.

HAMMOND, S.P. (1991). Severity Ratings and Incident Trend Analysis, Proceedings, ER91, Calgary, AB.

TRANSPORTATION OF DANGEROUS GOODS CONTROL ACT (1982). Statutes of Alberta, Chapter T-6.5., Edmonton: Queen's Printer for Alberta.

Characteristics of Motor Carriers of Hazardous Materials

Leon N. Moses
Ian Savage
Department of Economics &
The Transportation Center
Northwestern University
2003 Sheridan Road
Evanston, Illinois 60208
USA

ABSTRACT

This paper investigates whether trucking firms that haul hazardous materials differ from firms who do not haul these goods. It employs a database of 75,000 federal government safety audits of United States motor carriers. We find that hazardous materials firms are five times larger, in terms of annual fleet miles, than non-hazardous materials haulers, and are more likely to be general commodity carriers. Based on Poisson regression analysis, firms that carry hazardous materials exclusively have an accident rate 11% higher than comparable firms that do not carry these commodities, and a rate of fatalities and serious injuries that is 22% higher. Firms that carry hazardous materials in combination with general freight, have an accident rate that is 18% higher and a fatality and injury rate that is 24% higher.

Among hazardous materials carriers, accident rates decline with firm size. Private carriers are safer than for-hire carriers. Haulers of gases in packages and liquids in tanks have the highest accident rates. Carriers of hazardous wastes have the lowest accident rates. Firms classified as "unsatisfactory" in safety audits by the federal government have an accident rate 50% worse than other firms, though these accidents do not result in a higher incidence of fatalities and injuries.

1. INTRODUCTION

This paper investigates whether the trucking firms that carry hazardous materials differ from firms that do not carry such goods in terms of characteristics, such as size, and their accident experience. The paper contains two groups of analyses. The first makes comparisons between carriers of hazardous and non-hazardous materials to see if their physical characteristics, accident experience, and compliance with safety regulations differ. The second deals solely with the hazardous materials carriers. We report on an investigation that is designed to establish whether certain types of cargoes are more risky, and whether firm characteristics such as size, age and compliance with safety regulations influence accident rates.

The paper extends work that was reported at a Northwestern University conference in 1991 (Moses and Savage, 1991). As a result of the interest generated by that paper, the authors approached the Federal Highway Administration (FHWA) to obtain a larger database. The FHWA made available their entire record of motor carrier safety audits. Consequently in our present study we are able to expand the number of observations from 13,000 firms, of which 2,000 carried hazardous materials, to 75,500 firms, of which 13,500 carry hazardous materials. Some of the results obtained with the larger data set differ in non-trivial ways from those of the first paper.

2. DATA SOURCES

The data for the investigation are derived from the initial "Safety Review" audits of U.S. interstate motor carriers that are mandated by the 1984 Motor Carrier Safety Act. In the course of these audits, data are collected on firms' physical characteristics, goods carried, accident record, and compliance with federal motor carrier safety regulations. The data are kept in the FHWA's Motor Carrier Management Information System (MCMIS). We obtained the entire database for 92,529 firms that were audited between October 1986 and November 1991. We removed from the dataset Canadian and Mexican firms, bus companies, and firms that did not operate any trucks. We then cleaned the data by removing obvious data entry errors. As a result we have a usable dataset of 75,577 firms of which 13,498 (18%) indicated they carry hazardous materials.

3.   ANALYTICAL METHOD

In our analyses we adopt a type of regression technique based on the Poisson distribution in preference to the more common, ordinary least squares (OLS) approach. Professional opinion suggests that the Poisson distribution offers several distinct advantages when dealing with count data (Cameron and Trivedi, 1986; Hausman, Hall and Griliches, 1984). It has been applied to accident data for the airlines (Rose, 1990) and the shipping industry (McCullagh and Nelder, 1983).

In the Poisson formulation the number of accidents is the dependent variable, the explanatory variables are multiplicative and one takes the exponent of a coefficient in order to interpret it. Exposure to accidents, interpreted as truck miles in our study, is one of the explanatory variables. This contrasts with OLS style regressions which typically have accident rates (accidents per mile) as the dependent variable. The Poisson regression is by definition non-linear and fits an exponential curve to data.

As measures of goodness of fit, the percentage of variation in the dependent variable explained by the regression and the log-likelihood statistic are typically presented. The latter statistic is usually compared to the log-likelihood of a regression with only a constant. However, because accidents are heavily related to exposure we felt that the correct base would be a regression with a constant and the log of fleet miles.

A major advantage of the Poisson regression process is that it can deal effectively with datasets where a large proportion of the observations on the dependent variable, i.e. accidents, take the value zero. Previous work, using OLS analysis on truck accident rates, have experienced difficulty in this regard (Corsi et al., 1984, 1988, 1989). The underlying Poisson nature of accident occurrence has a specific assumption that the mean and the variance of the accident distribution are identical. This equivalence leads to heteroscedasticity, a serious econometric problem in any OLS regression albeit one that most statistical packages can correct for. Corsi et al. recognize this problem and try to cope with it by taking the natural logarithm of accidents, the dependent variable. However, this creates a second problem. Many firms have zero accidents in a given year and the logarithm of zero is undefined. Corsi et al. add an arbitrary constant to the accident rate of each firm in an effort to avoid this second problem.

4.   VARIABLES EMPLOYED

4.1   The Accident Experience of the Firm in the Previous 365 Days

Accident data are notoriously unreliable in the trucking industry. The widely used national truck accident database of the FHWA's Office of Motor Carriers is flawed because accidents are self-reported. It is generally believed that there are serious inconsistencies and under-reporting of damage-only accidents. Our data come from questions asked directly of managers by inspectors, and therefore should be more reliable. In previous analyses (Moses and Savage, 1991, 1992) the accident measures used were the total accident experience of firms, and the total number of fatalities and injuries. The FHWA ceased collecting these data in audits conducted after November 1, 1990 and turned instead to a measure called "reportable accidents". Reportable accidents are defined as accidents involving a fatality, an injury, or more than approximately US$5,000 in property damage.[1] We also use the total fatalities and injuries measure in our analyses, but must then limit ourselves to the audits conducted prior to November 1, 1990. We did this because we felt that carriers of particularly hazardous materials might be more scrupulous about keeping records of property-damage-only accidents, but that record keeping of fatality and injury data would be more consistent across carriers.

4.2   The Log of Total Fleet Miles of the Firm in the Past Year

We use these data to capture both the amount of exposure to accidents and any firm size effects on accident rates. Testing of the coefficient against 1 determines whether accidents increase more or less than proportionately with miles. Inclusion of this variable allows us to colloquially refer to "accident rates" when interpreting the coefficients on other explanatory variables.

4.3   The Percent of Drivers Employed on Trips over 100 Miles

We hypothesized that firms whose primary work involves short distances, typically in urban areas, would have a different accident experience from firms whose operations primarily involve long

---

[1] If a correlation is calculated between reportable accidents and total accidents, using those inspections prior to November 1, 1990, we find a correlation of 0.82. Analyses that we have conducted have revealed that by dealing with reportable accidents exclusively, we exclude a large number of minor, damage-only accidents that result from urban pick-up and delivery operations.

distance service on the Interstate Highway System, or rural highways. Urban firms may be involved in a higher number of accidents, but many of these will be minor property-damage-only accidents which are not included in the federal definition of a reportable accident. This variable cannot be expressed in logarithms because several firms report zero long distance drivers.

4.4 Private Carrier Status

We use a dummy (0-1) variable to indicate if the firm is a private carrier rather than a for-hire carrier.

4.5 The Type of Goods Hauled

goods and 21 sub-categories of hazardous goods from which the firm can specify what cargoes they carry. They can classify themselves in as many categories as they wish.

The data only permit the determination of whether hazardous materials are carried or not. We cannot tell what proportion of a firms business hazardous materials represent, although we can tell whether a firm is a general freight carrier. This allows us to differentiate between carriers, such as bulk tank-truck firms, which are likely to be exclusively hazardous materials haulers and other firms, especially the large less-than-truckload (LTL) operations, that may have a very small proportion of their ton-miles being hazardous materials.

As has already been stated, firms indicate which of 21 categories of hazardous materials they carry. In addition, they indicate whether the commodity is carried in tanks or packages or both. Packages seem to be the most common way of moving most hazardous materials. We decided to consolidate the potential 42 categories, 21 hazardous materials and 2 kinds of packaging, into nine categories. The primary motivation was to avoid collinearity. A correlation table was used in deciding on the consolidations. After the consolidations were carried out, we found that on average firms carried 1.6 of the 9 categories. The nine groups appear below. The amalgamations are shown in parentheses:

- Explosives (combination of categories explosives A, explosives B, explosives C, and blasting agents);

- Liquids in tanks (flammable liquids, corrosives, oxidizers and combustible liquids);

- Liquids in packages (flammable liquids, corrosives, oxidizers and combustible liquids);
- Gases in tanks (flammable gas, non-flammable gas);
- Gases in packages (flammable gas, non-flammable gas);
- Poisons (poison A, poison B);
- Radioactive materials;
- Hazardous wastes; and
- Other hazardous commodities (flammable solid, organic peroxide, irritating material, "other regulated materials", etiologic agent, "hazardous substances", and cryogenics).

4.6 The Log of the Years of Experience of the Firm

A difficulty in our investigation is that the data available to us are dates of incorporation rather than initial year of operation. This restricted the analysis of this variable to the 91% of the hazardous materials firms who are incorporated. It is, of course, true that many firms operated as a sole proprietorship or partnership prior to incorporation. Other authors have tried to base age on the date that carriers were issued their operating rights by the Interstate Commerce Commission (ICC). However, this procedure also has a weakness. Our data reveal that there are corporate firms that operated for years prior to ICC certification.

4.7 Performance in the Federal Safety Audit

Federal and State inspectors visit the operating bases of firms and make assessments of carriers' compliance with federal safety regulations and safety management policies such as those governing maintenance, driver hiring and training. The inspectors examine records and interview management officials, but do not actually inspect any equipment or test drivers. The inspectors have a standard list of 75 questions. They mark a pass or fail on each question, but can also append comments and supporting documentation. The carrier is then rated as satisfactory, conditional or unsatisfactory based on the answers to the questions, and a weighting scheme that is not known to the public. Firms that appear to have questionable safety practices, yet have not actually violated federal regulations are typically rated conditional pending further investigation.

The question of the effectiveness of these audits has been addressed by the authors elsewhere (Moses and Savage, 1992). For the purposes of this paper, we represent the audits by employing a dummy variable to indicate firms that are rated unsatisfactory. Summary statistics indicate that satisfactory and conditional firms have broadly similar accident rates, while those for unsatisfactory firms are much worse.

5.   COMPARISON OF CARRIERS OF HAZARDOUS AND NON-HAZARDOUS MATERIALS

5.1 Overall Means

Mean values of various leading characteristics were calculated for the 13,498 carriers of hazardous materials and 62,079 carriers of non-hazardous materials. Standard t-tests were conducted to see whether the two groups are significantly different. The results are shown in Table 1. The leading conclusions are:

- Hazardous materials carriers are five times larger than carriers of non-hazardous materials in terms of fleet mileage.

- Hazardous materials firms are 2 1/2 times more likely to be general commodity carriers. This, combined with the first conclusion, suggests that a large proportion of hazardous materials are carried by the very large LTL firms.

- About two thirds of the firms that haul hazardous materials are private carriers. The proportion is about the same for the non-hazardous commodity groups.

- Hazardous materials firms are 20% more likely to be incorporated.

- Based on group means, hazardous materials carriers have an accident rate 7% higher than that of carriers of non-hazardous materials, and a rate of fatalities and injuries that is 19% higher.

5.2 Multiple Regression Approach

One cannot draw strong conclusions on relative accident rates from simple comparisons of means. For example, hazardous materials firms are very large, and in earlier work we found that large firms have lower accident rates than small firms (Moses and Savage, 1992). A Poisson multiple regression approach is therefore used.

Table 1:   Comparison of carriers of hazardous and non-
hazardous commodities.

	Hazardous	Non-Hazardous	t-Statistic
Number of Firms	13,498 (18%)	62,079 (82%)	–
Average annual fleet miles	1,823,200	342,000	24.17
Percent private carriers	64%	63%	2.19
Percent general freight firms	22%	9%	34.70
Percentage of firms incorporated	91%	74%	56.15
Average years since incorporation	21.6	19.3	13.27
Average percent of drivers employed on trips over 100 miles	60%	67%	15.15
Percent rated satisfactory	66%	45%	46.26
Percent rated conditional	28%	46%	41.36
Percent rated unsatisfactory	5%	9%	18.19
Reportable accidents per million miles	0.67	0.63	2.61
Fatalities and injuries per million miles[1]	0.49	0.41	4.47

[1] For audits conducted prior to November 1, 1990.

The results for the regressions on reportable accidents, and total
fatalities and injuries are shown in Table 2. We do not concern
ourselves with the interpretation of variables other than the dummy
variables for hazardous materials carriers at this point. The other
variables are discussed at length in section 6 below, which deals
with the hazardous materials carriers.

Carriage of hazardous materials is represented by two dummy
variables. The first indicates firms who carry hazardous materials
as well as general freight. The second dummy variable identifies
firms that only transport hazardous materials, i.e. they do not also
carry general freight. There are approximately 3,000 firms in the
first category, and 10,500 in the second. Thus the regression allows
a comparison of the accident rates of three kinds of firms: 1) those
that do not carry any hazardous materials; 2) those firms that carry

hazardous materials as well as general freight; 3) those that carry hazardous materials exclusively. The coefficients on the hazardous materials dummy variables are interpreted below:

Effect of Dummy Variable on Accident Rates

	Reportable Accidents	Fatalities & Injuries
Hazardous Materials & General Freight	+18.2% (t=10.21)	+24.2% (t=10.61)
Hazardous Materials Only	+10.6% (t= 5.94)	+21.5% (t= 9.07)

Table 2: Multiple regression incorporating hazardous materials variables.

Dependent variable	Reportable Accidents	Fatalities & Injuries
Audit Dates	All Audits	Before 11/1/90
Observations	75,577	62,532
Proportion of Variation explained	0.80	0.66
Log-likelihood	- 37,469	- 27,374
Log-likelihood (log of miles and constant)	- 38,027	- 27,751
Explanatory variables (with t statistics in parentheses)		
Constant	- 13.719 (283.30)	- 14.330 (229.70)
Log of total fleet miles[+]	0.963 (12.10)	0.971 (7.48)
Percent of drivers employed on trips over 100 miles	0.017 (0.97)	0.150 (6.51)
Dummy variable - private carrier	- 0.261 (17.02)	- 0.317 (15.86)
Dummy variable - general freight and hazardous materials	0.167 (10.21)	0.217 (10.61)
Dummy variable - hazardous materials only	0.101 (5.94)	0.195 (9.07)
Dummy variable - rated unsatisfactory	0.686 (27.44)	0.546 (15.32)

[+]   The coefficient on miles is compared against 1 so as to determine the effect
      of fleet miles on accident rate per mile.

We find that firms that carry hazardous materials exclusively have lower accident rates than those that carry both general freight and hazardous materials. However, the accident rates of firms that carry hazardous materials, whether alone or in combination with general freight, are higher than the accident rates of those that do not carry hazardous materials.

It might be argued that hazardous materials firms may be more scrupulous in recording accidents than others. However, the fatality and injury regression supports the finding of a higher accident rate. It is difficult to accept the proposition of reporting bias for this category of accidents. Moreover, further support for this position comes from the government safety audits, which show that firms who haul hazardous materials are more likely to be deficient in reporting accidents than non-hazardous materials firms.[2]

6.   ANALYSIS OF THE HAZARDOUS MATERIAL CARRIERS

We now turn our attention to the 13,498 firms that indicated they carry hazardous materials. Initially we classify these firms by the specific hazardous cargoes they carry, and then compare the characteristics of these various sub-groups. We then use a multiple regression model to explain the accident rates of the sub-groups. In this latter analysis we comment on the effect of firm size, age and other characteristics, as well as the impact of performance in the safety audit on accident experience.

6.1 Characteristics of Firms that Haul Different Hazardous Cargoes

In Table 3 a comparison is made of the leading characteristics of the firms that haul the various hazardous materials. The leading conclusions are:

- Tank truck companies are about a third of the size of the average hazardous materials firm in terms of fleet miles. Also, tank

---

[2] Question 394-3 of the federal safety audit asks if the firm complies with the reporting of accidents. 9.4% of firms who haul hazardous materials, either exclusively or as part of a general freight business are assessed a "no" to this question, compared with 7.9% of non-hazardous materials firms. The difference is statistically significant with a t statistic of 5.57. Previous work has revealed that as the proportion of non-compliance rises accident rates increase (Moses and Savage, 1992).

truck firms are more likely to be private carriers.

• Tank truck companies, and haulers of hazardous wastes are more likely to be specialized carriers, and not carry general freight.

• Haulers of hazardous wastes have a higher proportion of drivers on long trips compared with the average hazardous materials firm.[3]

• Radioactive, explosive and poisonous materials are hauled by the very largest carriers, with an average size two or more times the overall mean.

Table 3: Comparison of firms hauling various hazardous cargoes (in descending order of average fleet mileage).

Commodity	% Long Distance	% General Freight	% Private	Inc. %	Fleet Miles
Radioactive	64*	54*	30*	96*	22,100,000*
Explosives	62*	44*	51*	95*	10,100,000*
Poisons	63*	45*	43*	96*	8,600,000*
Hazardous wastes	75*	13*	52*	93	5,100,000*
Other hazardous commodities	64	35*	54*	95*	4,100,000*
Gases in packages	54*	27*	70*	93	3,900,000*
Liquids in packages	63*	34*	57*	93	2,700,000*
Gases in tanks	51*	5*	75*	91*	1,400,000*
Liquids in tanks	54*	5*	73*	90*	1,100,000*
Average	60	27	61	93	3,600,000

* indicates significantly different from average at 5% level. The average used in determining the significance may be different from that shown in Table 1 as firms may carry several commodities.

---

[3] Officials at the FHWA offer an explanation for this finding. The hazardous wastes industry is made up of local and line-haul operators. The large hazardous wastes corporations perform the latter function, consolidating waste collected by small local independent contractors. The federal government has poor or non-existent records on these local operators.

6.2 Multiple Regression Approach

Poisson regression analyses of reportable accidents, and fatalities
and injuries were conducted for the 13,498 hazardous materials
carriers. Explanatory variables are firm size measured by annual
fleet miles; the percentage of drivers on long distance trips; a
dummy variable for private carriage; dummy variables for the
different categories of hazardous materials (excepting the "other"
category); and a dummy variable representing an unsatisfactory
audit rating. The results are shown in Table 4. The coefficients in
these regressions are reviewed below under two headings: firm
characteristics, and the commodity carried.

Conclusions on Firm Characteristics

• Accident rates decline with firm size.

The coefficient on total fleet miles has a value significantly less
than unity in both regressions, meaning that accidents increase
less than proportionately with miles, i.e. the rate of accidents
declines with size. In Figure 1, the results are shown as an index.
This index is the ratio of the predicted accident rate of a given size
group to that of the smallest group of firms in our study, namely
those with fleet miles of 5,000 per year. The results suggest that
accident rates decline by about 23% when comparing a firm with 1
1/2 million miles with a very small firm. Firms with annual
mileage of 5 million have accident rates 27% below the very
smallest firms. The rate of fatalities and injuries declines more
slowly with size. Firms with annual mileage of 5 million miles
have fatality and injury rates about 17% below those of the very
smallest firms.

• Long distance operations are associated with higher accident
  rates.

Long distance operators, defined as firms whose drivers are all
involved in trips that exceed 100 miles, have a total accident rate that
is 22% higher than that of firms that are exclusively involved in
short distance operations, and a rate of fatalities and injuries that is
53% higher. This result is not surprising. The measures of
accidents used in this analysis include only the most serious
incidents. While one would expect that congested urban areas would
have the highest number of accidents of all kinds, most of the
accidents that do occur are minor in nature. The accidents on long
distance trips tend to be more serious and result in a higher rate of
accidents that involve fatalities and serious injuries, as well as
more property damage.

Table 4: Multiple regression analysis of hazardous materials
carriers.

Dependent variable	Reportable Accidents	Fatalities & Injuries
Audit Dates	All Audits	Before 11/1/90
Observations	13,498	11,732
Proportion of variation explained	0.86	0.74
Log-likelihood	- 11,659	- 9,547
Log-likelihood (log of miles and constant)	- 11,904	- 9,790
Explanatory variables (with t statistics in parentheses)		
Constant	- 13,680 (169.65)	- 14,393 (142.97)
Log of total fleet miles[+]	0.954 (9.00)	0.973 (4.20)
Percent of drivers employed on trips over 100 miles	0.204 (7.37)	0.423 (11.71)
Dummy variable - private carrier	- 0.294 (12.07)	- 0.409 (13.16)
Dummy variable - explosives	0.015 (0.65)	0.011 (0.37)
Dummy variable - liquids in tanks	0.071 (2.94)	0.086 (2.92)
Dummy variable - liquids in packages	- 0.019 (0.77)	- 0.015 (0.48)
Dummy variable - gases in tanks	- 0.169 (4.23)	- 0.038 (0.81)
Dummy variable - gases in packages	0.115 (5.54)	0.099 (3.95)
Dummy variable - poisons	0.025 (1.07)	- 0.002 (0.08)
Dummy variable - radioactive	0.205 (7.60)	0.033 (1.00)
Dummy variable - hazardous wastes	- 0.133 (5.34)	- 0.178 (5.70)
Dummy variable - rated unsatisfactory	0.425 (8.29)	0.002 (0.03)

[+] The coefficient on miles is compared against 1 so as to determine the effect of
fleet miles on accident rate per mile.

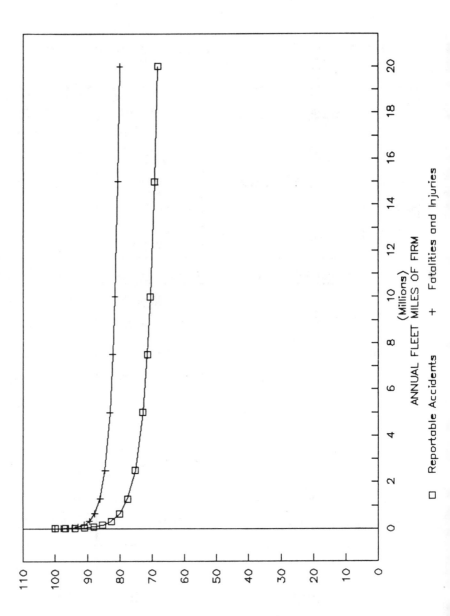

Figure 1: Index of accident rate by firm size.

- Private carriers appear to have accident rates that are about 30% lower than those of comparable for-hire carriers. This is true for both reportable accidents, and fatalities and injuries.
- Reportable accident rates do not vary with firm age.

We also decided to investigate the effect of firm age on accident performance, using as our variable the number of years since incorporation. Unfortunately, 9% of the firms are not incorporated, and hence regressions including this variable are restricted to corporate firms. To save space, the full regressions are not shown. In the reportable accidents equation, the coefficient on the log of years of experience variable was insignificant taking the value -0.001 with a t-statistic of 0.13. In the equation for fatalities and injuries, the coefficient was positive, 0.06, with a t-statistic of 4.45. The coefficient implies that a firm that has been incorporated for thirty years has an accident rate that is 22% greater than a newly incorporated firm. What should be borne in mind in interpreting this counter-intuitive result is that the inclusion of the experience variable reduces the size of the log of miles coefficient from 0.973 to 0.961. Apparently, the negative effect of size on the rate of fatalities and serious injuries is smaller in the equation that does not include the experience variable. The positive age effect of a firm that has been incorporated for many years is balanced by a more pronounced negative size effect.

- Firms who are rated unsatisfactory in government safety audits have reportable accident rates 53% higher than firms rated conditional or satisfactory, though these additional accidents do not result in a higher incidence of fatalities and injuries.

This result may appear to be somewhat surprising because the regression results for the full sample of 75,577 firms, shown in Table 2, suggest that unsatisfactory firms have both a 90% higher rate of reportable accidents and a 75% higher rate of fatalities and injuries compared with firms rated conditional or unsatisfactory. The implication of these results is that, *ceteris paribus*, the unsatisfactory firms who have an inordinately high rate of fatalities and injuries are primarily those who do not carry hazardous materials.

Conclusions on Hazardous Commodity Hauled

We report the percentage effect on accident, and fatality and injury, rates that are derived from our multiple regression in Table 5. Our major conclusions are:

- Carriers of gases in packages and liquids in tanks have the highest rates of accidents and fatalities and injuries. They are 10% above that of carriers who do not haul commodities in these groups.

- Carriers of gases in tanks have a rate of fatalities and injuries that is no different from carriers that do not haul this commodity. However, the rate of reportable accidents is 15% less. The difference in the two percentages suggests that the carriers of this product type tend to have accidents that involve high rates of personal injury and death.

- Carriers of radioactive materials appear to have a high reportable accident rate, one that is 23% above that of firms that do not carry commodities in this group. However, it is likely that this finding is due to scrupulous record keeping by this type of carrier, because the rate of fatalities and injuries is not significantly different from that of other carriers.

- Haulers of hazardous wastes have considerably lower accident, and fatality and injury, rates than do other hazardous materials carriers. The difference is about 15%. As discussed earlier, the database only covers line-haul operations of wastes. The data do not permit implications to be drawn about the safety of operators who provide local collection of hazardous wastes.

7.   CONCLUDING REMARKS

7.1 Comparisons with Non-hazardous Materials Carriers

- Hazardous materials carriers are five times larger than carriers of non-hazardous materials in terms of annual fleet miles. They are much more likely to be incorporated, and also to be general commodity carriers.

- Firms that carry hazardous materials exclusively have an accident rate 11% higher than comparable firms that do not carry these commodities, and a rate of fatalities and serious injuries that is 22% higher. Firms that carry hazardous materials in combination with general freight, have an accident rate that is 18% higher and a fatality and injury rate that is 24% higher.

Table 5:   Effect on accident rates of hauling specific hazardous
commodities (in descending order of effect on fatalities
and injuries).

	Predicted Percent Effect on	
	Reportable Accidents	Fatalities & Injuries[1]
Gases in packages	+12.2%*	+10.4%*
Liquids in tanks	+7.4%*	+9.0%*
Radioactive	+22.8%*	+3.4%
Explosives	+1.5%	+1.1%
Poisons	+2.5%	-0.2%
Liquids in Packages	-1.9%	-1.5%
Gases in Tanks	-15.5%*	-3.7%
Hazardous Wastes	-12.5%*	-16.3%*

* Indicates statistically significant, from firms who do not carry this
commodity, at the 5% level.
[1] For audits conducted prior to November 1, 1990.

7.2 *Comparisons between Carriers of Different Hazardous Materials*

- Radioactive, explosive and poisonous materials are hauled by the very largest carriers of hazardous materials.

- Tank truck companies are about a third the size of the average hazardous commodity carrier in terms of fleet miles.

- Gases in packages and liquids in tanks are the most dangerous commodity groups in terms of accident rates.

- Accidents involving gases in tanks result in a higher than average incidence of personal injury and death.

- Line-haul carriage of hazardous wastes has a significantly lower accident rate than carriers of other hazardous materials.

- Among hazardous materials carriers, accident rates decline with firm size. Private carriers are 30% safer than for-hire carriers.

- Carriers rated unsatisfactory in federal safety audits have an average reportable accident rate that is 50% higher than that of firms rated conditional or satisfactory, though these accidents do not result in a higher incidence of fatalities and injuries.

Acknowledgement. The authors acknowledge the financial support of the U.S. Department of Transportation through its University Transportation Centers grant to the Great Lakes Center for Truck Transportation Research, University of Michigan, Ann Arbor. We also acknowledge the excellent research assistance of Yajai "M" Yodin and Mona Shah. We also wish to express our gratitude to the Federal Highway Administration for the data employed in our research. All opinions and interpretations are, however, solely those of the authors.

8. REFERENCES

CAMERON, A.C., & TRIVEDI, P. (1986). Econometric models of count data: Comparisons and applications of some estimators and tests. Journal of Applied Econometrics, 1, 29-54.

CORSI, T.M., & FANARA, P. (1988). Driver management policies and motor carrier safety. Logistics and Transportation Review, 24, 153-163.

CORSI, T.M., & FANARA, P. (1989). Effects of new entrants on motor carrier safety. In L.N. Moses, and I. Savage (Eds.), Transportation Safety in an Age of Deregulation. New York: Oxford University Press.

CORSI, T.M., FANARA, P., & ROBERTS, M.J. (1984). Linkages between motor carrier accidents and safety regulation. Logistics and Transportation Review, 20, 149- 164.

HAUSMAN, J., HALL, B.H., & GRILICHES, Z. (1984). Econometric models for count data with an application to the patents-R&D relationship. Econometrica, 52, 909-938.

McCULLAGH P., & NELDER, J. (1983). Generalized Linear Models. London: Chapman and Hall.

MOSES, L.N., & SAVAGE, I. (1991). Motor carriers of hazardous materials. Who are they? How safe are they? Proceedings of Hazmat Transport `91. Evanston, IL: Northwestern University Transportation Center.

MOSES, L.N., & SAVAGE, I. (1992). The effectiveness of motor carrier safety audits. Accident Analysis and Prevention, 24, 479-496.

ROSE, N. (1990). Profitability and product quality: Economic determinants of airline safety performance. Journal of Political Economy, 98, 944-964.

Procedures for the Development of Estimated Truck Accident Rates and Release Probabilities for Use in Hazmat Routing Analysis

Eugene R. Russell, Sr.
Department of Civil Engineering
Kansas State University
Manhattan, KS 66506
USA

Douglas W. Harwood
Midwest Research Institute
425 Volker Boulevard
Kansas City, MO 64110
USA

ABSTRACT

This paper presents a procedure that can be used by highway agencies to develop estimated average truck accident rates and release probabilities for different highway and area types. The procedure is demonstrated using data from three states. Users are encouraged to develop truck accident rates and release probabilities from data for their own jurisdiction, using the procedure described in this paper. However, when data appropriate for the users locality are not available, the estimates presented here based on data from the three states can be used as default values. The development of these rates was part of a Federal Highway Administration sponsored study by the authors.

1. INTRODUCTION

The authors recommend that the Federal Highway Administration (FHWA) guidelines for hazmat routing studies be revised to incorporate an improved method for estimating accident probabilities. Specifically, the use of truck accident rates is recommended in preference to the all-vehicle accident rates presently used in the FHWA routing guide (Barber & Hildebrand, 1980). In addition, a new term representing the probability of a hazardous material (hazmat) release given an accident involving a hazmat carrying truck has been introduced. The revised equation for determining accident probability is:

$$P(R)_i = TAR_i \times P(R \mid A)_i \times L_i \tag{1}$$

where:

$P(R)_i$ = probability of an accident involving a hazmat release for route segment i

TAR_i = truck accident rate (accidents per veh-mi) for route segment i

$P(R \mid A)_i$ = probability of a hazmat release given an accident involving a hazmat carrying truck for route segment i

L_i = length (mi) of route segment i

The objective of the analyses performed in this paper is to determine values of TAR and P(R|A) in equation 1. Users are encouraged to determine expected values of TAR and P(R|A) from data for their own state. State-wide averages for specific highway and area types are generally much more reliable than estimates based on accident data for the specific highway segments being analyzed in a hazmat routing study, because the sample size of accidents for individual highway segments is often not large enough to allow statistically valid comparisons between alternative routes. Where the analysis segments are relatively short and the duration of the analysis period is limited to a few years, as is often the case, estimates based on actual accident histories will be unreliable.

There are cases, however, where accident rates may be substantially higher (or lower) than average, that warrant reliance on the accident history for a specific segment. It may be desirable to determine risk by highway segment. However, this should never be attempted unless a statistical test to determine when actual accident histories are preferable to system-wide averages. Appropriate statistical tests are discussed in detail in a report by Harwood and Russell (1990).

2.   PROCEDURE FOR DEVELOPING DEFAULT TRUCK ACCIDENT RATES AND RELEASE PROBABILITIES

The truck accident rate data used as default values in hazmat routing studies should reflect the influence of highway geometric and traffic variables that have a demonstrated relationship to truck safety. Two key variables whose strong relationship to truck accident rates has been demonstrated are highway type (two-lane highway, freeway, etc.) and area type (urban/rural). Several studies demonstrate such relationships. Freeways generally have lower accident rates than any other highway type, and should generally be preferred to other highway types for hazmat shipments. Rural highways also generally have lower truck accident rates than urban highways for the same highway type. Thus, routes that avoid urban areas are generally preferable, unless they are substantially longer or involve a less suitable type of highway. It would be desirable for the default accident rates used in hazmat routing studies to also reflect the influence of other geometric features of highways such as lane width, shoulder width, curves, grades, and intersections. Some of these relationships have been demonstrated for all-vehicle accident rates, but none of these features have been specifically related to truck accidents. The relationships developed in this paper quantify the effect of highway type and area type on truck accident rate, but not the effects of other geometric features.

In addition to truck accident rates, the distribution of truck accident types also varies with highway and area type. Rural highways and urban freeways tend to have a larger proportion of single-vehicle noncollision accidents, while lower-speed urban highways tend to have a higher proportion of multiple-vehicle collisions. Analyses of the FHWA Motor Carrier Accident reports show that the probability of a release given an accident involving a hazmat carrying truck is much higher in single-vehicle noncollision accident than in single- or multiple-vehicle collision accidents (Harwood & Russell, 1990). Thus, the probability of a release given an accident is also expected to vary for different highway and area types.

The following discussion presents the procedures that were used to develop default accident rates and release probabilities and can be used by highway agencies to develop default values from their own data. Site-specific accident data for the particular alternative routes being evaluated should only be used where a statistical analysis indicates a need. Estimates of truck accident rates and release probabilities based on an agency's own data are preferred to the use of default values.

2.1 Data Needs

Three types of data are needed to estimate truck accident rates and release probabilities in a form useful for hazmat routing analyses. These are:

- Highway geometric data.
- Truck volume data.
- Truck accident data.

In order for the analysis to be accomplished efficiently, it is desirable for these data to be available in computerized form using a common location identifier (e.g., mileposts) so that the three types of data can be linked together. Many state highway agencies have been computerizing and linking their data files and now, or soon will, have the capability to perform this type of analysis.

No state of which the authors are aware currently has the necessary data and linking capability to analyze accident rates for all public highways in the state. The best systems currently available include all highways under the jurisdiction of the state highway agency. To obtain unbiased estimates, it is desirable for the highway geometric, truck volume, and truck accident files to cover the entire state highway system. If only a subset of the state highway system is used, this subset should be selected through a statistical sampling process to maintain the unbiased nature of the estimates.

Highway geometric files are needed to define the characteristics of segments to which truck volume and accident data can be added. Highway geometric files typically consist of relatively short route segments (0.35 mi [0.56 km] or less in length) for which data on the geometric features of the segment are included. The minimum data that should be available for this analysis are:

- Number of lanes
- Divided/undivided
- Access control (freeway/nonfreeway)
- One-way/two-way
- Urban/rural

Traffic volume files typically include the Annual Average Daily Traffic (ADDT) and may also include either the average daily truck volume or the percent trucks in the traffic stream. In order to be useful, truck volume data needs to be given in the same location reference system as the highway geometric and accident data.

In selecting accidents for inclusion in the analysis, it is important to use the same definition of a truck that was used in obtaining the truck volume counts. Since nearly 89 percent of the accidents in which hazardous materials are released involve combination trucks (i.e., tractor-trailers), it would be desirable to limit the accident analysis to combination trucks only. Unfortunately, however, truck volume data for combination trucks are seldom available on a system-wide basis. Therefore, it is often necessary to use truck volume data and accident data for all trucks or for all commercial vehicles. Traffic counts of "all commercial vehicles" typically include both trucks and buses. Thus, if traffic volume counts for *all commercial vehicles* are used, it is important to include both bus and truck accidents in the analysis.

The recommended accident type categories into which the truck accidents should be classified using these data are those shown in Table 1. Each accident-involved vehicle should be treated as a separate observation (i.e., an accident involving two trucks should be counted as two accident involvements).

Table 1: Recommended accident type categories.

A. SINGLE VEHICLE ACCIDENTS	
1. Noncollision accidents	2. Collision accidents
• Ran-off-road	• Collision with fixed object
• Jackknife	• Collision with a parked vehicle
• Overturn	• Collision with a train
• Separation of units	• Collision with a nonmotorist
• Fire	• Other collision
• Cargo spillage	
• Cargo shifting	
• Other noncollision	
B. MULTIPLE-VEHICLE ACCIDENTS	
• Collision with a passenger car	
• Collision with a truck	
• Collision with other vehicle types	

2.2 Data Analysis

The average truck accident rate for each highway class can be computed as the ratio of total truck accidents to total vehicle-miles of truck travel for that highway class. In other words:

$$TAR_j = \frac{A_{ij}}{VMT_{ij}} \qquad (2)$$

where: TAR_j = Average truck accident rate for highway class j

A_{ij} = Number of truck accidents in one year on route segment i in highway class j

VMT_{ij} = Annual vehicle miles of truck travel on route segment i in highway class j

The values of TAR_j for each highway class from equation 2 can be used to replace default truck accident rates with values more suited to local conditions.

The probability of a hazmat release given an accident varies between highway types because it varies with accident type and because the distribution of accident types varies markedly between highway classes.

Harwood and Russell used FHWA motor carrier accident report data and developed a table that shows a probability of release given an accident by accident type (Harwood & Russell, 1990). The FHWA motor carrier accident reports show both whether the truck was carrying hazardous materials and whether the hazardous materials were released. It would be desirable for users to derive comparable values for their own state, but only three states currently have both data items in their accident records systems needed to make this determination .

If data is available, the probability of a release given an accident involving a hazmat-carrying vehicle can be computed as:

$$P(R \mid A)_j = \sum_k P(R \mid A)_k \times P(k)_j \qquad (3)$$

where: $P(R \mid A)_j$ = Probability of a hazmat release given an accident involving a hazmat-carrying vehicle for highway class j

$P(R \mid A)_k$ = Probability of a hazmat release given an accident involving a hazmat-carrying vehicle for accident type k (from state data)

$P(k)_j$ = Probability that an accident on highway class j will be of accident type k i.e., proportion of truck accidents for each accident type on highway class j (from state accident data)

The values of $P(R \mid A)_j$ from equation 3 should be used in place of the default values for the probability of release given an accident. Unfortunately, these values are not available in most states and default values must be used. Their development is illustrated in the remainder of this paper.

3.   DATA SOURCES

The development of system-wide estimates of truck accident rate for different highway and area types using the procedure presented above requires three types of data, preferably in computerized form. These data types are: highway geometrics, truck volumes, and truck accidents.

Only three state highway agencies that could provide the data needed to develop system-wide truck accident rates were identified. These agencies were the California, Illinois, and Michigan Departments of Transportation. The type of data available from each state is discussed below.

3.1   Highway Geometric Data

The highway geometric data available in all three states was quite extensive and only a portion of that data was used in the study. The geometric data file was used to define the highway type and area type for each highway segment. Nine highway classes (combinations of highway type and area type) were used in the study. These are:

• Rural two-lane highways.

• Rural multilane undivided highways.

• Rural multilane divided highways.

• Rural freeways.

- Urban two-lane streets.

- Urban multilane undivided streets.

- Urban multilane divided streets.

- Urban one-way streets.

- Urban freeways.

A few highway segments in each state could not be classified into one of these nine highway classes and were not considered.

3.2 Traffic Volume Data

Two forms of traffic volume data were obtained for each highway segment. These were: average daily traffic volumes and truck volumes. In all three states, average daily traffic volumes were available in the highway geometric file. However, in two of the three states, truck volumes had to be obtained from other sources.

The commercial vehicle volumes in the Illinois files were incomplete for many highway segments in Chicago and surrounding counties and for scattered segments elsewhere in the state. These missing data were estimated from the truck volumes for adjacent sections in the file or from the state's published commercial volume map.

3.3 Truck Accident Data

Accident data were obtained from existing files in all three states for their entire state highway system. Accident data were obtained for all commercial vehicles (combination trucks, single-unit trucks, and buses). Buses are not of direct interest to this study, but accident data for buses were included because buses are included in the commercial vehicle counts used as exposure data. The inclusion of the small percentage (5 percent) of buses in the truck volume and truck accident data is unlikely to have a major effect on the calculated truck accident rates.

The accident characteristics listed above in section 3.1 were used to classify accidents and to decide whether or not particular accidents met the criteria for inclusion in the study and should be counted.

3.3.1 Accident Classification

Accidents were classified by accident type using the categories listed in Table 1. Separation of Units, Fire, Cargo Spillage and Cargo Shifting were included in "Other Noncollision."

Accidents involving two or more trucks were treated as two or more accident involvements. The categories for multiple-vehicle accidents are based on the largest vehicle involved in the accident other than the vehicle under consideration.

The severity for each accident involvement was classified by the most severe injury in the accident as a whole, i.e., in a collision between a truck and a passenger car, injuries to the passenger car occupants are more likely than injuries to the truck occupants.

The accident data used in the analysis for California and Michigan covered the three-year period from 1985 through 1987, inclusive. Only two years of accident data, 1986 and 1987, could be used in Illinois.

4. DATA PROCESSING

All of the data files described above were obtained from the states on magnetic tape and were processed on an IBM-compatible mainframe computer using the Statistical Analysis System (SAS) following the step-by-step approach for merging the data from the available geometric, traffic volume, and accident files, as discussed in detail in a report by Harwood and Russell (1990).

5. ANALYSIS OF RESULTS

This section presents the results of the analysis of accident geometric and traffic volume data. First the accident rates, accident severity distributions, and accident type distributions for different highway and area type classes obtained in the analysis are presented. Next, a subsection on interpretation of results discusses the effects of accident reporting levels and the development of relationships between truck accident rate and variables other than highway and area class. Specific default values of truck accident rate and release probability are presented in the final section of the paper.

5.1 Truck Accident Rates

Table 2 shows a comparison of truck accident rates for all three states and includes an average[1] accident rate for all three states combined, weighted by veh-mi of truck travel. It is evident in Table 2 that there are substantial variations in accident rate among the three states. This, unfortunately, is the case in most accident studies.

Table 2: Truck accident rates by state and combined.

Highway class		Truck accident rate (accidents per million veh-mi)			
Area Type	Roadway type	California	Illinois	Michigan	Weighted average*
Rural	Two-lane	1.73	3.13	2.22	2.19
Rural	Multilane undivided	5.44	2.13	9.50	4.49
Rural	Multilane divided	1.23	4.80	5.66	2.15
Rural	Freeway	0.35	0.46	1.18	0.64
Urban	Two-lane	4.23	11.10	10.93	8.66
Urban	Multilane undivided	13.02	17.05	10.37	13.92
Urban	Multilane divided	3.50	14.80	10.60	12.47
Urban	One-way street	6.60	26.36	8.08	9.70
Urban	Freeway	1.59	5.82	2.80	2.18

* Weighted by veh-mi of truck travel.

A 1988 study has demonstrated from accident data (for all vehicle types, not just trucks) on two-lane highways in seven states that accident rates for seemingly identical conditions in different states can differ by a factor as large as 3 or 4 (Ng & Hauer, 1988). Other examples of large state-to-state differences in accident rate can be found in studies of two-lane highway safety and roadside clear recovery zones (Graham & Harwood, 1982; Smith et al., 1983).

The 1988 study mentioned above concludes that there are dangers in combining data from different states (Ng & Hauer, 1988). The authors agree and strongly encourage those performing hazmat risk analyses to develop default accident rates from data for their own state. However, it should also be recognized that the primary

---

[1] A weighted average by veh-mi of truck travel is equivalent to combining the accident rates for the three states by summing the numerators and denominators of the accident rate expressions.

objective in developing truck accident rates for hazmat routing analyses is to have accident rates that represent the relative differences in risk between highway classes and not to represent the absolute risk for any particular situation. The greatest state-to-state discrepancies in Table 2 tend to be those for highway classes with the smallest available sample sizes of truck accident involvements and truck travel. The weighted-average accident data in Table 2 minimize the influence of values based on small sample sizes and comes closer to representing the differences between highway classes than the data for any single state. Therefore, the three-state averages from Table 2 are appropriate for use as *default* values in hazmat routing studies when no better local estimates are available. *Locally generated data are always preferable when available.*

5.2 Accident Severity

Data was compiled to show the truck accident severity distributions by highway class in California, Illinois, and Michigan and were examined to compare the percentage of fatal and injury accidents in the three states and their combined data. The data in Table 3 suggest that it is likely that there are differences in accident reporting levels among the three states.

Table 3: Percentage of fatal and injury truck accidents by state and combined.

Highway class		Percent fatal plus injury accidents			
Area Type	Roadway type	California	Illinois	Michigan	Weighted average
Rural	Two-lane	40.8	32.2	27.4	34.2
Rural	Multilane undivided	40.2	36.6	25.8	35.2
Rural	Multilane divided	41.7	39.0	30.3	36.0
Rural	Freeway	40.1	31.9	28.3	34.9
Rural	All Types	40.6	32.6	28.1	34.7
Urban	Two-lane	34.5	23.8	25.2	26.3
Urban	Multilane undivided	30.3	22.7	23.4	24.2
Urban	Multilane divided	29.3	22.6	26.6	25.9
Urban	One-way street	35.4	21.4	20.8	22.1
Urban	Freeway	32.4	26.5	28.1	30.3
Urban	All Types	32.0	24.1	26.1	27.8
Rural and Urban	All Types	34.4	25.9	26.7	29.7

Experience indicates clearly that accident reporting levels increase
as accident severity increases, so reporting levels are likely to be
highest for fatal accidents and lowest for property-damage-only
accidents (American Association of State Highway and Transportation Officials, 1978; California Department of Transportation,
1974; Smith, 1973). However, reporting levels for less severe
accidents may vary widely between jurisdictions.

The three states differ in their reporting thresholds for property-
damage-only (PDO) accidents. Illinois uses a consistent $250
reporting threshold for PDO accidents and has for many years.
California has a state-wide $500 reporting threshold for PDO
accidents. Michigan has a $200 reporting threshold for PDO
accidents. However, PDO reporting levels in the various states
appear to be influenced as much by the characteristics of state-local
coordination as by the dollar threshold used for PDO accidents.

5.3 Accident Type Distribution

Data was compiled to show the percentage distribution of accident
types by highway class in California, Illinois, and Michigan,
respectively, using the accident type classifications presented
above. The accident data available from Michigan were not
sufficient to classify single-vehicle noncollision accidents into the
three subclasses shown in the table: run-off-road, overturned in
road, and other noncollision. Therefore, the relative proportions of
these subcategories of noncollision accidents were estimated from
the California and Illinois data. The data from the three states is
summarized in Table 4.

The data showed that the various highway classes have distinctly
different patterns of accident types. For example, the percentage of
single-vehicle noncollision accidents (which have the highest
probability of producing a hazmat release if an accident occurs)
shown in Table 4 is about two to three times as high on rural
highways as on urban highways.

On rural highways, the percentage of single-vehicle noncollision
accidents is generally higher for two-lane highways and freeways
than for multilane nonfreeways. In urban areas, two-lane
highways generally have a higher percentage of single-vehicle
noncollision accidents than other highway classes.

Table 4: Percentage of single-vehicle noncollision accidents by state and combined.

HIghway Class		Percent fatal plus injury accidents			
Area Type	Roadway type	California	Illinois	Michigan	Weighted average
Rural	Two-lane	15.5	9.7	6.7	11.1
Rural	Multilane undivided	15.0	8.8	3.1	10.2
Rural	Multilane divided	11.4	6.1	3.4	7.2
Rural	Freeway	10.6	20.5	13.7	12.9
Rural	All Types	13.1	14.3	8.3	11.2
Urban	Two-lane	7.5	3.3	4.1	4.3
Urban	Multilane undivided	3.4	2.3	1.4	1.7
Urban	Multilane divided	4.5	1.4	1.1	2.1
Urban	One-way street	3.1	1.4	1.6	1.6
Urban	Freeway	2.9	2.9	5.2	3.3
Urban	All Types	3.3	2.4	3.1	3.0
Rural and Urban	All Types	6.1	4.2	4.7	5.2

5.4 Probability of Release Given an Accident

An analysis of the FHWA Motor Carrier Accident Reports showed the probability of a hazmat release given an accident involving a hazmat carrying truck varies with accident type (Harwood & Russell, 1990). Table 5 summarizes the results. The distribution of accident types by highway class and the release probabilities for different accident types can be multiplied together to estimate the average release probability for accidents on each highway class. The release probability for a particular highway class is computed as the sum for all accident types of the proportion of each type of accident, times the probability of release given an accident for that accident type.

There was concern that the release probabilities in Table 6 might be substantially different if they were based on the accident type distribution for combination trucks only rather than the accident type distribution for all commercial vehicles. A supplementary analysis was performed and only minor variations in the values in Table 6 were found for combination trucks.

Table 5: Probability of release given that an accident has occurred as a function of accident type.

Accident Type	Probability of release
SINGLE-VEHICLE NONCOLLISON ACCIDENTS	
Run-off-road	0.331
Overturned (in road)	0.375
Other noncollision	0.169
SINGLE-VEHICLE COLLISION ACCIDENTS	
Collision with parked vehicle	0.031
Collision with train	0.455
Collision with nonmotorist	0.015
Collision with fixed object	0.012
Other Collision	0.059
MULTIPLE-VEHICLE COLLISION ACCIDENTS	
Collision with passenger car	0.035
Collision with truck	0.094
Collision with other vehicle	0.037

Table 6: Probability of hazmat release given that an accident has occurred.

Highway class		Probability of a release given an accident			
Area Type	Roadway type	California	Illinois	Michigan	Weighted average*
Rural	Two-lane	0.100	0.074	0.073	0.086
Rural	Multilane undivided	0.100	0.071	0.064	0.081
Rural	Multilane divided	0.087	0.064	0.062	0.082
Rural	Freeway	0.083	0.111	0.095	0.090
Urban	Two-lane	0.077	0.059	0.069	0.069
Urban	Multilane undivided	0.064	0.052	0.055	0.055
Urban	Multilane divided	0.068	0.048	0.058	0.062
Urban	One-way street	0.066	0.050	0.056	0.056
Urban	Freeway	0.062	0.055	0.067	0.062

* Weighted by veh-mi of truck travel.

6. FINAL VALUES FOR USE IN HAZMAT ROUTING ANALYSES

Table 7 presents the recommended default values for truck accident rate and probability of release given an accident by highway class. These final values are based on the combined three-state data given in Tables 2 and 6, respectively. The final values of truck accident rate and release probability can be used as default values for TAR and P(R I A) in equation 1 *when local estimates are not available.*

Table 7: Default truck accident rates and release probability for use in hazmat routing analyses.

Area Type	Roadway Type	Truck Accident Rate (accidents per million veh-mi)	Probability of Release Given an Accident	Releasing Accident Rate (releases per million veh-mi)
Rural	Two-lane	2.19	0.086	0.19
Rural	Multilane undivided	4.49	0.081	0.36
Rural	Multilane divided	2.15	0.082	0.18
Rural	Freeway	0.64	0.090	0.06
Urban	Two-lane	8.66	0.069	0.60
Urban	Multilane undivided	13.92	0.055	0.77
Urban	Multilane divided	12.47	0.062	0.77
Urban	One-way street	9.70	0.056	0.54
Urban	Freeway	2.18	0.062	0.14

Table 7 also shows the estimated releasing accident rate, in releases per million veh-mi, which is the product of truck accident rate and probability of release. Thus, the releasing accident rate is the product of the TAR and P(R I A) in equation 1 and represents the best available estimate of the relative risk of hazmat releases during transportation on different highway classes.

Acknowledgements The research on which this paper is based was funded by USDOT/FHWA. The data was compiled by Midwest Research Institute. The contents of this paper reflect the views of the authors and do not necessarily reflect the views, or any policy, of the U.S. Department of Transportation.

7. REFERENCES

AMERICAN ASSOCIATION OF STATE HIGHWAY AND TRANSPORTATION OFFICIALS (1978). A Manual on User Benefit Analysis of Highway and Bus-Transit Improvement - 1977, Washington, DC.

BARBER, E.J. & HILDEBRAND, L.K. (1980). Guidelines for Applying Criteria to Designate Routes for Transporting Hazardous Materials, Report No. FHWA-IP-80-15, Federal Highway Administration.

CALIFORNIA DEPARTMENT OF TRANSPORTATION (1974) (cited in Graham & Harwood, 1982). Accident Costs, internal memorandum.

GRAHAM, J.L. & HARWOOD, D.W. (1982). Effectiveness of Clear Recovery Zones, NCHRP Report 247.

HARWOOD, D.W. & RUSSELL, E.R., SR. (1990). Present Practices of Highway Transportation of Hazardous Materials, Publication No. FHWA-RD-89-013, Federal Highway Administration, McLean, VA.

NG, J.C.N. & HAUER, E. (1988). Accidents on Rural Two-Lane Roads: Differences Between Seven States, Presented at the 68th Annual Meeting of the Transportation Research Board.

SMITH, R.N. (1973). Predictive Parameters for Accident Rates, California Division of Highways.

SMITH, S.A., ET AL. (1983). Identification, Quantification, and Structuring of Two-Lane Rural Highway Safety Problems and Solutions, Report Nos. FHWA/RD-83/021 and FHWA/RD-83/022, Federal Highway Administration.

A Methodology for the Transfer of Probabilities Between Accident Severity Classification Schemes[*]

J. D. Whitlow
K. S. Neuhauser
Sandia National Laboratories[**]
Albuquerque, New Mexico,
United States of America

ABSTRACT

Accident-severity categories are used in many risk analyses for the classification and treatment of accidents involving vehicles transporting radioactive materials. Any number or definition of severity categories may be used in an analysis. A methodology which will allow accident probabilities associated with one severity category scheme to be transferred to another severity category scheme is described. The supporting data and information necessary to apply the methodology are also discussed. The ability to transfer accident probabilities between severity category schemes will allow some comparisons of different studies at the category level. The methodology can be employed to transfer any quantity between category schemes if the appropriate supporting information is available.

---

[*] This work performed at Sandia National Laboratories, Albuquerque, New Mexico, supported by the United States Department of Energy under Contract No. DE-AC04-76DP00789

[**] A United States Department of Energy Facility

1. INTRODUCTION

Evaluation of the radiological risks of accidents involving vehicles transporting radioactive materials requires consideration of both accident probability and consequences. The probability that an accident will occur may be estimated from historical accident data for the given mode of transport. In addition to an overall accident rate, distributions of accident severities and package environments across the range of all credible accidents are needed to determine the potential for damage to packaging shielding or release of material to the environment. These distributions are usually estimated with information from a variety of sources such as historical data, experimental data, analyses of accident and package environments, and expert opinion. The consequences of an accident depend on a number of factors including the type, quantity, and form of radioactive material being transported; the response of the package to accident environments; the fraction of material released to the environment; and the dispersion of any released material.

Several approaches have historically been taken for the classification and treatment of transportation accidents in risk analysis. One approach analyzes specific accident scenarios where each scenario describes an accident environment and the package response. Another approach uses event trees or fault trees in analysis of the risk. A third approach, and the main focus of this paper, involves the use of accident-severity categories.

With the accident-severity category approach, the complete range of critical accident environments (those environments with the principal radiological risk) resulting from all credible transportation accidents is divided into some number of severity categories. Each severity category represents a portion of all credible accidents, and the total of all severity categories covers the complete range of critical accident environments. This approach has been used in the risk assessment codes RADTRAN (Neuhauser, 1992) and INTERTRAN (Ericsson, 1983).

Accident-severity categories are ordinarily illustrated on a set of axes forming a grid, as shown in Figure 1 and Figure 2. The axes indicate critical accident environments and describe the range of parameters used to define the categories from zero (no accident) to the value for the most severe credible accident. The magnitude of the most severe credible accident depends on a number of factors including the mode of transport, regulations regarding transport, etc., and usually requires consideration of accidents that are plausible but have never actually occurred. Not all possible environments that may result from an accident are of equal concern because some environments have a much lower probability of occurrence or lesser capability to inflict damage on the package or both. The set of axes on which the severity categories are defined usually only represent the accident environments of primary concern (the critical accident environments) and the environments with relatively negligible risk are not included. Although Figures 1 and 2 show two critical accident environments, any number of environments may be considered in a severity category scheme.

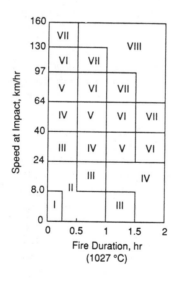

Figure 1: Eight-category accident-severity classification scheme (adapted from U.S. Nuclear Regulatory Commission, 1977).

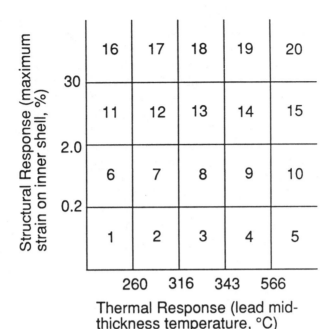

Figure 2: Twenty-category accident-severity classification scheme (adapted from Fischer, 1987).

No constraints are placed on the number or definition of severity categories by the accident-severity category approach, although the previously mentioned codes do place a maximum on the number of categories. As such, past studies have used different numbers and definitions of severity categories to represent the range of all credible accidents. In 1977, the U.S. Nuclear Regulatory Commission published a report entitled "Final Environmental Statement on the Transportation of Radioactive Material by Air and Other Modes" (U.S. Nuclear Regulatory Commission, 1977), which used several eight-category, accident-severity classification schemes similar to the scheme depicted in Figure 1. In 1987, the U. S. Nuclear Regulatory Commission published another study performed by the Lawrence Livermore National Laboratory to assess the level of safety provided to the public under severe highway and railway accident conditions during the transport of spent nuclear reactor fuel (Fischer, 1987). This 1987 study defined a scheme consisting of twenty response regions, comparable to severity categories, which represented the response of representative spent fuel casks to the complete range of critical accident environments and is similar to the scheme shown in Figure 2. The differences that can exist between severity category schemes are indicated by the differences between Figure 1 and Figure 2. Not only are the number of categories different, but the parameters used to define the categories are also different and no correspondence exists between any two categories in the different schemes. The differences in number and definition of severity categories used in different studies make direct comparisons of anything but total risk values extremely difficult. These differences may also lead to confusion and misinterpretation. An example of such misinterpretation is discussed in Luna, et al. (Luna, 1986).

To address this difficulty, a methodology has been developed which will allow accident probabilities associated with one severity category scheme to be transferred to another severity category scheme. This will permit meaningful comparisons of different studies at the category level. If a study used event trees, fault trees, or accident scenarios for the classification and treatment of transportation accidents, then the information from these trees or scenarios would first have to be transferred to a severity category scheme before this methodology could be utilized.

2. METHODOLOGY

A methodology for the transfer of probabilities between accident-severity category schemes was initially considered in a study performed at Sandia National Laboratories (Spanks, 1990). This earlier study developed a matrix to transfer accident probabilities from an eight-category scheme similar to the one shown in Figure 1 to a twenty-category scheme similar to the one shown in Figure 2. Spanks proposed mapping the two severity category schemes onto a common set of axes to form two overlying grids. In this case, correlations between the mechanical parameters of impact speed and cask structural response and the thermal parameters of fire duration and cask thermal response were needed to map the two schemes onto a common set of axes. The probabilities associated with the eight-category scheme were then transferred to the twenty-category scheme using an "equal area weighting" technique which assumes that the accident probability is constant within each severity category. The assumption that accident probability is constant across the range of accidents represented by each severity category is not representative of actual accident experience.

The methodology described in this paper maps the severity category schemes onto a common set of axes to form overlying grids, as was done by Spanks. However, the transfer of probabilities between accident-severity category schemes is based on the relationships between probability of occurrence and each parameter used to define the severity categories (the parameters along the axes of the overlying grids).

The first step in applying the methodology to transfer probabilities between accident-severity category schemes is to map the schemes onto a common set of axes. This step, depending on the parameters used with the original axes and available accident probability data, may require information or assumptions about numerous factors, such as cask characteristics, vehicle braking effects, target hardness, dispersion of flammable material, etc. Since this methodology transfers accident probabilities between schemes based on the relationships between probability of occurrence and each of the parameters defining the categories, the parameters for the common set of axes must be chosen such that these relationships can be obtained. The most commonly reported relationships of probability of occurrence to accident environments use simple accident parameters such as some form of impact velocity, fire duration, etc. Care should be taken not to misinterpret these or any other accident parameters. For example pre-accident speed, velocity change in an impact, and equivalent speed onto an unyielding target are different parameters but are all related to and might loosely be referred to as impact velocity.

After the severity category schemes are mapped onto an appropriate common set of axes, the two schemes that are to have probabilities transferred between them are overlaid. To illustrate this, consider the three-category scheme shown in Figure 3 and the four-category scheme shown in Figure 4. Both of these schemes are depicted on a common set of axes. For the purpose of illustration, consider that these simple accident-severity category schemes are for studies of spent nuclear fuel truck transport, have impact and fire as the critical accident environments, consider pre-accident speeds of zero to 160 km/hr to be credible, are only for accidents that involve fires, and consider a maximum credible fire duration to be two hours. Figure 5 shows the overlay of these two category schemes. The severity categories are not required to be graphically depicted and overlaid. The boundaries of every category in both schemes need only be accounted for mathematically; however, graphical depiction can provide a good physical awareness of the problem.

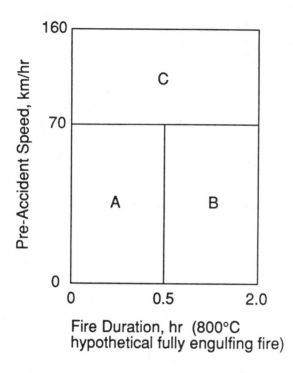

Figure 3: Three-category accident-severity classification scheme.

Relationships between probability of occurrence and both pre-accident speed and fire duration are needed to apply the methodology to the transfer of probabilities between the category schemes depicted in Figure 5. Information on the severities of transportation accidents provided in a study published by Sandia National Laboratories (Clarke, 1976) is used in this study to obtain the needed relationships. The cumulative probability distribution of pre-accident speed shown in Figure 6 and the cumulative probability distribution of fire duration shown in Figure 7 are both adapted from the aforementioned study (Clarke, 1976).

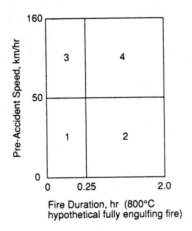

Figure 4: Four-category accident-severity classification scheme.

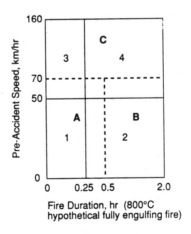

Figure 5: Overlay of three-category and four-category schemes.

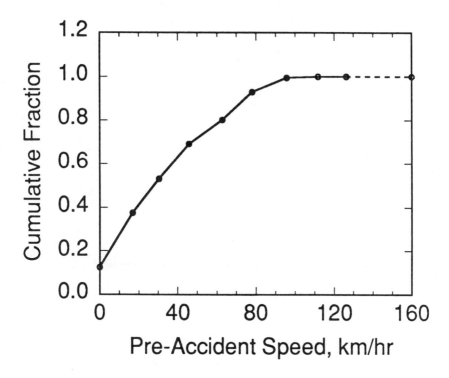

Figure 6: Cumulative probability distribution of pre-accident truck speeds (adapted from Clarke, 1976).

The cumulative probability distribution shown in Figure 7 was generated, because of the lack of historical accident data, by a Monte Carlo prediction scheme for a model of the expected duration of truck fires for trucks carrying only nonflammable cargo. The relationship shown in Figure 7 is assumed, for the purpose of illustrating the transfer methodology, to be equivalent to an 800 degree Celsius, hypothetical, fully engulfing fire that is used to define the severity categories in Figure 5. The lack of historical accident data and the assumptions and effort necessary to convert data or model results to the parameters used to define severity categories are often hindrances to applying the transfer methodology.

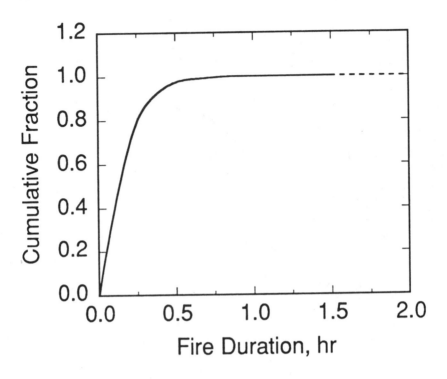

Figure 7: Cumulative probability distribution of fire duration in truck accidents involving fire (adapted from Clarke, 1976).

The relationships between probability of occurrence and each parameter used to define the severity categories should be as representative as possible of the information used to originally determine the accident probability associated with each category in the scheme that you are transferring from. The purpose of these relationships is to indicate how an accident probability associated with any severity category is distributed within that category so that it may be appropriately transferred to categories in another scheme. The relationships are not used to calculate probabilities directly, but since they are used to determine how accident probabilities are distributed to categories in a different category scheme, they should be consistent with actual accident experience.

To illustrate the transfer of probabilities between severity category schemes, consider that the three-category scheme shown in Figure 3 has a probability associated with each category and these probabilities are desired to be transferred to the four-category scheme shown in Figure 4. The overlay of these two schemes depicted in Figure 5 shows that the range of accident environments represented by category A of the three-category scheme encompasses all accident environments represented by category 1 and part of the environments represented by categories 2, 3, and 4 of the four-category scheme. The probability associated with category A, therefore, should be distributed to categories 1, 2, 3, and 4 of the four-category scheme.

The fraction of accident probability associated with category A to be distributed to each of the four categories in the four-category scheme is determined by use of the cumulative distributions shown in Figures 6 and 7. The joint probability of occurrence for category 1 and for the portions of categories 2, 3, and 4 encompassed by category A is calculated from these cumulative distributions. The value calculated by dividing each of these joint probabilities by the sum of the four joint probabilities gives the fraction of accident probability associated with category A to be distributed to each of the four categories (1, 2, 3, and 4) in the four-category scheme. The parameters of pre-accident speed and fire duration must be assumed to be independent of each other to use this procedure, which appears reasonable based on accident data (Clarke, 1976). To illustrate, note from Figure 6 that 86% of the truck accidents occur at pre-accident speeds less than 70 km/hr and 72% occur at pre-accident speeds less than 50 km/hr. Figure 7 shows, for truck accidents involving fires, that 97% have fire durations less than 0.5 hours and 79% have fire durations less than 0.25 hours. The joint probability of occurrence for category 1, which includes accident environments up to 50 km/hr and fire durations up to 0.25 hours, is calculated as

$$(0.72) \times (0.79) = 0.57.$$

Likewise, the joint probability for the portions of categories 2, 3, and 4 encompassed within category A is calculated;

for the portion of category 2 as $(0.72) \times (0.97 - 0.79) = 0.13$,

for the portion of category 3 as $(0.86 - 0.72) \times (0.79) = 0.11$,

and for the portion of category 4 as $(0.86 - 0.72) \times (0.97 - 0.79) = 0.03$.

The sum of these four probabilities equals 0.84.

The fraction of the accident probability associated with category A to be distributed to category 1 can now be calculated as

$$0.57 / 0.84 = 0.68.$$

Similarly, the fraction of the accident probability associated with category A to be distributed to categories 2, 3, and 4 is calculated;

for category 2 as 0.13 / 0.84 = 0.16,

for category 3 as 0.11 / 0.84 = 0.13,

and for category 4 as 0.025 / 0.84 = 0.03.

A convenient check on this step is that the fractions should add to one for each category that probabilities are being transferred from. The fractions calculated above for category A do add to one,

$$0.68 + 0.16 + 0.13 + 0.03 = 1.00.$$

Following a similar procedure shows that the range of accident environments represented by category B of the three-category scheme encompasses a portion of the accident environments represented by categories 2 and 4 of the four-category scheme. The accident probability associated with category B is distributed to categories 2 and 4 in the following fractions: 0.84 to category 2, and 0.16 to category 4.

Likewise, the range of accident environments represented by category C of the three-category scheme encompasses a portion of the accident environments represented by categories 3 and 4 of the four-category scheme. The fraction of the accident probability associated with category C to be distributed to category 3 was calculated to be 0.79 and the fraction to be distributed to category 4 was calculated to be 0.21.

The fractions calculated above that indicate how the accident probabilities associated with each of the three categories (A, B, and C) in the three-category scheme are distributed to each of the four categories (1, 2, 3, and 4) in the four-category scheme are displayed

in matrix form in Table 1. The total accident probability associated
with each of categories 1, 2, 3, and 4 is calculated by summing the
probabilities distributed to each of these categories from categories
A, B, and C. As Table 1 shows, the accident probability of category 1
is calculated by multiplying 0.68 by the accident probability of
category A. Likewise, the accident probability of category 2 is the
sum of the products of 0.16 multiplied by the accident probability of
category A and 0.84 times the accident probability of category B. The
accident probabilities of categories 3 and 4 are calculated in a
similar manner.

Table 1: Transfer fractions from example in text.

		Categories that accident probabilities are being transferred from.		
		A	B	C
Categories that accident probabilities are being transferred to.	1	.68	---	---
	2	.16	.84	---
	3	.13	---	.79
	4	.03	.16	.21

3. SUMMARY AND CONCLUSIONS

A methodology has been developed which allows the accident probabilities associated with one accident-severity category scheme to be transferred to another severity category scheme. The methodology requires that the schemes use a common set of parameters to define the categories and assumes these parameters are independent of one another. The transfer of accident probabilities is based on the relationships between probability of occurrence and each of the parameters used to define the categories. Because of the lack of historical data describing accident environments in engineering terms, these relationships may be difficult to obtain directly for some parameters. Assumptions or experienced judgement are often needed to obtain the relationships. The most commonly reported relationships of probability of occurrence to the severity of an accident environment are for simple accident parameters. The relationships used in this paper were adapted from a Sandia National Laboratories report (Clarke, 1976). Even though this report is more than 15 years old, it still provides the most complete information for the United States that the authors are aware of. The use of appropriate relationships, even if they are not exact, allows the accident probability associated with any severity category to be distributed within that category in a manner consistent with accident experience, which in turn will allow the accident probability to be appropriately transferred to a different category scheme.

The ability to transfer accident probabilities between severity category schemes will allow some comparisons at the category level of studies which used different category schemes. This may be useful when comparing, for a similar transport situation, older studies with more recent studies or studies done at different institutions or by different countries. The methodology will allow accident probability information from past studies to be used in current severity category schemes for comparison purposes. An example of the schemes that this methodology has been applied to are the eight-category and twenty-category schemes shown in Figures 1 and 2, respectively. Through a better understanding of how severity categories in different category schemes relate to one another, hopefully the methodology presented in this paper will reduce some of the confusion and misinterpretation associated with comparing different studies.

The ability to transfer accident probabilities between severity category schemes will not directly allow all quantities commonly associated with severity categories to be transferred between the schemes. Risk, for example, is a function of both accident probability and consequence. If this methodology were to be used for the transfer of risk between category schemes, one could transfer the accident probabilities as outlined above and then perform a consequence analysis on the new category scheme to obtain risk values in the new scheme. The basic methodology described in this paper can, however, be used for any quantity, not just accident probability, if the relationships between that quantity and the parameters used to define the categories are obtainable. Risk, therefore, could be transferred directly between category schemes by following the steps of the methodology and substituting the relationships between risk and each of the category-defining parameters for the relationships between probability of occurrence and each of the category-defining parameters. These risk relationships, however, could be difficult to obtain because of the many factors upon which risk depends and would apply only to the particular case for which it was developed.

4. REFERENCES

CLARKE, R. K., FOLEY, J. T., HARTMAN, W. F., & LARSON, D. W. (1976). Severities of Transportation Accidents. SLA-74-0001. Sandia National Laboratories. Albuquerque, NM, U.S.A.

ERICSSON, A., & ELERT, M. (1983). INTERTRAN: A System for Assessing the Impact from Transporting Radioactive Material. IAEA-TECDOC-287. International Atomic Energy Agency. Vienna, Austria.

FISCHER, L. E., et al. (1987). Shipping Container Response to Severe Highway and Railway Accident Conditions. NUREG/CR-4829, Volumes 1 & 2. U. S. Nuclear Regulatory Commission. Washington, DC.

LUNA, R. E., et al. (1986). Response to the Report Entitled "Transportation Risks: Appendix A, DOE Environmental Assessment-Analysis of RADTRAN II Model and Assumptions." SAND86-1312. Sandia National Laboratories. Albuquerque, NM, U.S.A.

NEUHAUSER, K. S., & KANIPE, F. L. (1992). RADTRAN 4: Volume 3 Users Guide. SAND89-2370. Sandia National Laboratories. Albuquerque, NM, U.S.A.

SPANKS, L. (1990). A Method To Expand Existing Severity Category Matrices for RADTRAN to 20 Categories. TTC-0845. Sandia National Laboratories. Albuquerque, NM, U.S.A.

U. S. Nuclear Regulatory Commission. (1977). Final Environmental Statement on the Transportation of Radioactive Material by Air and Other Modes. NUREG-0170. Volume 1.

A Regression Model for Estimating Probability of Vessel Casualties

Tai-Kuo Liu
U.S. Department of Transportation
Research and Special Programs Administration
Volpe National Transportation Systems Center
Kendall Square
Cambridge, MA
U.S.A. 02142

ABSTRACT

A waterway vessel casualty probability model was developed based on 10-year waterway specific historical vessel casualty rates of the self-propelled, deep draft, dry cargo and tanker vessels and waterway characteristics across the U.S. waterways. The model was used as the basis for developing waterway specific risk adjustment factors and estimating waterway specific vessel casualty probabilities for all vessel groups and casualty types. The estimated probabilities were applied over a 15-year period to project the number of vessel incidents and the consequences that could be resulted from the projected incidents. The results were used to assess the benefits attributed to reduced navigational risks in comparison to the costs of proposed Vessel Traffic Services (VTS) systems as an objective in the U.S. Port Needs Study.

1. INTRODUCTION

Estimating the probability of vessel casualties[1] by vessel type (characterized as collision, ramming, or grounding) in the

---

[1]"Vessel Casualty" is standard naming for vessel incidents in the U.S. Coast Guard Casualty Maintenance Data Base (CASMAIN). VTS-addressable vessel

waterways of 23 study zones encompassing 82 deep draft ports was one of the key tasks conducted for the U.S. Port Needs Study (USDOT, 1991). Vessel casualty probabilities were applied to estimates of future waterway vessel movements over a 15-year period to project the number of vessel incidents and the associated consequences (i.e., physical damages, loss of life, resulting spills, and environmental damages). The objective of the study was to compare the costs of proposed Vessel Traffic Services (VTS) systems with the benefits attributed to reduced navigational risks in order to establish program priorities across the nation. This paper is emphasized on the methodologies developed for estimating waterway-specific vessel casualty probabilities whereas the associated consequences are discussed in more detail in the U.S. Port Needs Study.

Estimates of the probability of future vessel casualties in a specific waterway depend on an accurate profile of the historical casualty incidents and the associated vessel traffic patterns. Casualty rates, measured as the number of casualties per 100,000 units of vessel movements, can be a good indicator of waterway-specific casualty probability if the event population is sufficient to reflect the local trend of casualty occurrences. However, if vessel incidents in the period of study are very few, as was the case in many of the waterways in this study, the likelihood of a future casualty probability would have to be inferred from a broader data base.

National average aggregate casualty rates were calculated to include all the waterways and casualties defined in the study, thus representing the aggregate level of casualty rates by casualty type, vessel type, and vessel size. These national averages were, however, not adequate to represent the more diverse casualty rate pattern of each waterway. Large variations in historical casualty incidents and vessel movements were found among the individual waterways.

The premise in the study was that the national average casualty rates could be adjusted to reflect the unique navigational characteristics of each waterway defined by a specific set of parameters (e.g., waterway configuration, meteorological and hydrologic conditions,

---

casualties are defined as the most dynamic cases which are potentially avoid-able with an advanced VTS system operational at the time of incidents. They generally involve participating vessels greater than 20 meters. VTS address-ability is determined by a set of guidelines for the screening of original CASMAIN file. The screening results in a smaller casualty data base which is more reflective of addressable navigational situations.

traffic volume or density, the level of traffic management, and the existence of VTS systems). Using a multiple regression approach, the probability of vessel casualties by subzone, as the dependent variable, was regressed against a set of predetermined variables characterizing each waterway. The most significant variables from a large pool of variables were selected as the independent, or explanatory variables to predict the dependent variable, or the probability of vessel casualties.

2.   ANALYSIS FRAMEWORK AND DATA ·

Development of the model depended on extensive cross-subzone and multivariate data analyses, therefore, data consistency was explicitly addressed in the data gathering and analysis process. Data sufficiency was also an important issue because the distribution of casualties could not satisfy all the dimensions (i.e., casualty type, vessel type, and vessel size) in each unique waterway if separate models were to be calibrated.

The best model resulting from the calibration process was estimated from the self-propelled, deep draft, dry cargo and tanker vessels and their respective historical casualties. They represented not only the most sufficient sample of observations, but also the most consistently estimated traffic movements reflecting the navigational risk conditions of the primary traffic routes that characterize each waterway.

The predicted casualty probabilities of this vessel group, weighted with historical casualty rates, were subsequently developed into waterway-specific risk adjustment values, thus representing the relative risk levels of each waterway. These waterway-specific risk values were then applied to the national average casualty rates of all vessel types to derive waterway-specific casualty probabilities by casualty type, vessel type, and vessel size. Figure 1 describes the key components of the analytical process. This process and sources of data are summarized in the following five sections.

2.1 Waterway Types

The waterways in the 23 study zones were subdivided into 99 sub-zones defined by six waterway types: open approach, convergence, open bay or harbor, enclosed harbor, constricted waterway, and river. The waterway type was typically defined according to the generic waterway configuration, channel characteristics, and route

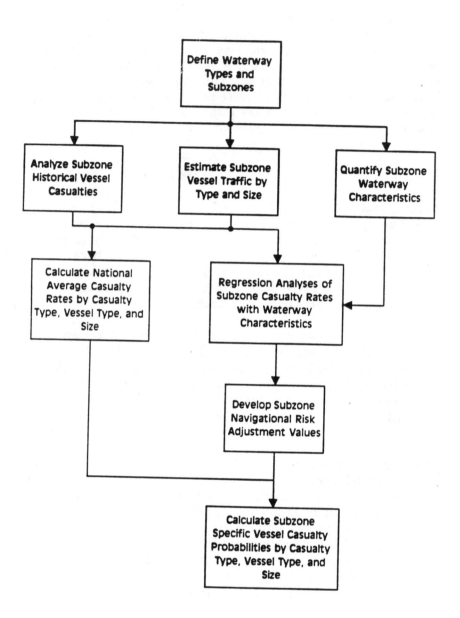

Figure 1: Analytical process of estimating waterway vessel casualty probabilities.

alignment. A typical study zone usually included several subzones, each representing one of the generic waterway types within the study zone boundary (e.g., Boston, Figure 2). Other study zones contained just one uniform subzone in the entire waterbody, such as Santa Barbara with a straight shore line and wide open water. A large study zone, such as Puget Sound, had multiple waterways of similar types throughout the study zone.

A dummy variable was assigned to the waterway type for each subzone. In the regression analyses, each subzone represented a unique and independent observation in which the historical vessel casualties and vessel traffic movements were analyzed along with the waterway characteristics.

2.2 Vessel Transit Casualties

The Coast Guard's Casualty Maintenance (CASMAIN) Data Base is the central source of vessel casualty data. For the 23 study zones, the historical vessel casualty records were examined for the period from 1980 to 1989. The original data base contains about 36,000 records for the most recent 10-year period and was reduced to 2,337 records after a stringent screening process to eliminate the records that were not qualified as VTS-addressable (e.g., fires, explosions, docking errors, equipment failures, etc.). Also, a large portion of the vessels under 20 meters[2] were excluded from the scope of the analysis unless they were passenger ferries or were involved in incidents with other VTS-addressable vessels. Three dynamic casualty types that are potentially addressable by VTS were retained and defined as collision, grounding, and ramming.

The VTS-addressable casualty file was further reconstructed to combine multiple barge records with tugboat/towboats into barge-tow transits. In other words, each barge transit was comprised of a tugboat or towboat and any number of barge vessels in a transit movement record, rather than being treated as separate records as in the CASMAIN data base. As a result of this barge to barge tow conversion, the total number of vessel casualties were reduced from 2,337 individual vessels to 1,492 transit movements and a total of

---

[2]The 20-meters threshold was used to conform to a Coast Guard regulation governing mandatory participation of vessels in a VTS system which was proposed at the beginning of this study. In the latter stages of the study when the data had already been taken, the Coast Guard made the decision to remain with 300 gross tons as the mandatory participation criteria.

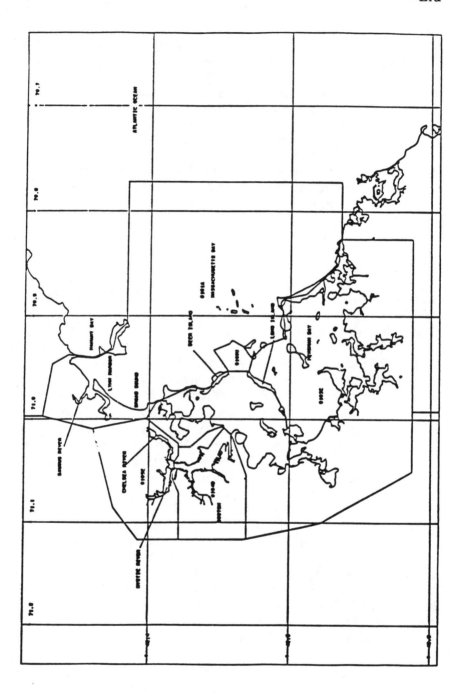

Figure 2: Zone 1 - Boston, MA - Zone and subzone boundaries.

1,084 casualty events. There were usually two transits involved in a case of collisions. Those vessel transit casualties were then assigned to specific subzones as determined by the subzone boundaries and the latitudes and longitudes of the incidents.

2.3 Vessel Traffic

The main data source of vessel types and commodity movement was based on the Army's Corps of Engineer's (COE) Waterborne Statistics. This series of statistics documents the annual departures and arrivals of five broadly classified vessel types and over 20 commodity types at each of the Corps of Engineers' waterway codes. The data, however, lacks specific origin, destination, and routing information except for the general direction of the traffic in and out of study zones.

The COE vessel traffic data are generally found more consistently and thoroughly compiled for the deeper draft vessels with drafts over 18 feet, but are less reliably reported for the shallower draft vessels (i.e., 18 feet or less). Local data sources including port authorities, Coast Guard district offices, cruise and ferry services, barge companies, and state registration of fishing and recreation fleets were therefore sought to supplement the missing data links for passenger ferries, cruise vessels, recreation, and small fishing fleets.

The vessel traffic volumes were analyzed in a traffic assignment process over the vessel route network for each study zone developed in the Port Needs Study. The traffic volumes allocated to each subzone were then divided by the subzone vessel transit casualties to calculate historical vessel casualty rates. Six vessel types and three vessel sizes were classified as indicated in Table 1 based on the 10-year traffic estimates for each subzone.

2.4 Vessel Traffic Routes and Network

The traffic routes and network for each study zone were developed in a PC-based relational data base. Each route is comprised of a number of route segments with a starting node and ending node. Each route segment, represented in a separate record file, contains homogeneous information such as length of the segment, minimum width, channel depth, number of hazards, starting node and ending node, and whether the segment intersects with a subzone boundary. The node file contains only the longitude and latitude of the node in addition to the study zone ID code.

Table 1: National average casualty rates by vessel type, vessel size, and casualty type - 1980 to 1989. 23 study zones (Number of casualties per 100,000 transits).

		Collision	Ramming	Grounding	Total
	Small	0.218	0.056	0.343	0.617
Passenger	Medium	8.425	0.000	16.764	25.189
	Large	—	—	—	—
	Small	0.582	0.114	0.162	0.858
Dry Cargo	Medium	1.552	0.507	1.123	3.182
	Large	3.872	1.336	8.717	13.925
	Small	0.462	0.000	0.578	1.040
Tanker	Medium	0.960	0.183	1.069	2.212
	Large	7.718	3.634	19.373	30.725
	Small	2.986	1.551	1.907	6.444
Dry Cargo Barge	Medium	—	—	—	—
	Large	18.901	0.000	29.270	48.171
	Small	3.221	0.966	3.455	7.642
Tanker Barge	Medium	—	—	—	—
	Large	2.277	2.167	2.708	7.152
	Small	0.388	0.226	0.454	1.068
Tug/Tow Boat	Medium	—	—	—	—
	Large	—	—	—	—

The traffic assignment was largely done manually at the route level with the consideration of the study zone network configuration and in consultation with local authorities. The vessel transits were assigned to routes and aggregated at the segment level. The subzone level vessel traffic was determined by the highest vessel traffic volume in any route segment falling within the subzone boundary, and the dominant traffic route connecting these segments was selected to represent the traffic in the subzone. The segment information was the basis for further transformation of the route and waterway characteristics.

2.5 Waterway Characteristics

The waterway specific characteristics may be described as waterway physical measurements as well as meteorological and hydrologic conditions. Waterway physical measurements include channel width, depth, route alignment in terms of number of turns, average or sum of total degrees of course changes, number of obstructions (e.g., bridges, anchorages, crossing lanes, etc.), total length along the primary traffic route, and total area of water surface in the subzone. Those attributes were mostly measured at each route segment. They may be enumerated in many different ways to describe the physical characteristics or the waterway configuration for each subzone.

Meteorological and hydrologic factors included are wind speed, visibility and current velocity. Although the meteorological data are available on an hourly basis, they are often not differentiated at the subzone level since the weather information is usually surveyed at only one weather station located within the study zone. The current velocity is marked on NOAA charts in very small segments. And translating these small segments into an aggregate measure is also very difficult. The average conditions or the percentage times of the most critical conditions (i.e., visibility less than one nautical mile or wind velocity greater than 20 knots) have been used to represent the meteorological or hydrologic variables in the calibration of the model. They were, however, not captured as significant explanatory variables in the regression model as compared to other waterway-specific variables.

3. WATERWAY EXPLANATORY VARIABLES

The waterway physical measurements, meteorological and hydrologic conditions, traffic volumes and densities, as well as subzone type variables comprised a pool of the potential variables for the regression analyses (Table 2). The selection of the independent, or explanatory variables for the model was conducted in a calibration process in which various combinations of the variables were systematically tested and analyzed to fit a model. The best model was determined in the stepwise process by its overall statistical fit, the significance of the parameters, and the signs of the parameters.

4. CALIBRATION OF THE MODEL

A linear, multiple regression model was selected as the functional form for the model. Both linear as well as generalized logit regressions were tested in the step-wise calibration procedures. They resulted in similar combinations of the independent variables and similar effects to the probability of vessel casualties. The most significant variables selected in the stepwise linear regression were: the waterway/channel characteristics, a local traffic density factor, and two subzone type dummy variables. Meteorological and hydrologic variables show little effect or incorrect signs.

Although meteorological and hydrologic conditions (i.e., visibility, wind speed, and current velocity) are considered highly correlated with navigational risk, the relationships of these variables with casualty probabilities were, however, not revealed in this analysis. As subzone level meteorological and hydrologic data were not available, it is unclear to what extent improved data would have better explained these relationships, given the form of the model. Also, there is an argument that the average meteorological or hydrologic conditions may not be as indicative as the actual, or the variance of these conditions, to casualty occurrences. It is therefore suggested that a disaggregate approach with more refined data may be more appropriate and revealing of the effects of meteorological and hydrologic conditions on casualty occurrences.

Traffic variables, whether presented as volumes (e.g., annual deep draft vessel transits or total annual vessel transits) or as densities (e.g., number of transits per route mile) were not proved to be significant independent variables that explain the casualty probabilities in the model. Vessel traffic volume (i.e., the self-propelled, deep draft vessels) was, instead, more effectively applied as a weight variable in the calibration of the parameters. As the estimated subzone vessel casualty probability was subsequently multiplied by the forecast vessel transits to predict the number of casualties, the weighted least-square procedure yielded the most efficient estimators for prediction of future vessel casualties.

Besides the fact that vessel casualty rate was defined as the number of casualties per 100,000 transit movements per subzone, an alternative casualty rate measurement (i.e., number of casualties per 100,000 transit-miles) was also tested during model calibration. Although the casualty rate per transit-mile implies an additional dimension of traffic exposure measure (i.e., distance travelled), it was, nevertheless, not predicted well by either the subzone variables,

Table 2: Waterway explanatory variables.

Subzone Type Constant Variables

- Open Approach
- Convergence
- Open Harbor or Bay
- Enclosed Harbor
- Constricted Waterway
- River

Meteorological and Hydrographic Conditions

- Average maximum current
- Visibility
- Wind speed

Route Characteristics

- Route length of the primary traffic route
- Minimum width along the primary traffic route
- Average channel width
- Minimum channel depth
- Number of course changes
- Sum of total degrees of course changes
- Average degrees of a course change
- Number of Obstructions (e.g. bridges, anchorages, crossing lanes, etc)

Vessel Traffic

- Medium and large dry cargo and tanker vessel transits
- All vessel transits
- Ferry miles
- Local registered vessels

Traffic or Route Characteristics Per Route Mile

- M&L dry cargo and tanker transits per route mile
- All transits per route mile
- Ferry miles per square mile
- Other vessels per route mile
- Number of course changes per route mile
- Sum of total degrees of course changes per route mile
- Number of obstructions per route mile

or the traffic density and route characteristics divided by route length. The distance factor (i.e., the route length of the primary route within subzone) was, instead, represented as an independent variable and explained well the subzone casualty probability along with other subzone variables.

5. THE MODEL

The results of the final model estimated from the weighted least-square procedure is shown in Table 3. The explanatory variables in combination explain 75% of the total variations of the historical casualty rates among the subzones with a highly significant F-value. The t-statistics for each of the parameters are all significant at or greater than 92.5% level.

Table 3: Estimates of Variables

Explanatory Variable	Parameter Estimate	t-statistic	Prob > |t|
Constant	-0.372321	-0.235	0.8145
Open	-3.529773	-1.862	0.0662
Narrow (Constricted Waterway)	16.327722	7.187	0.0001
Route Length	0.228527	4.802	0.0001
Average Width	-0.000407	-1.879	0.0638
Sum Degrees of Course Changes	0.012121	4.251	0.0001
Other Vessels Per Mile	0.000392	1.809	0.0741
R-square	0.7476		
F Value	39.488		
Prob > F	0.0001		

Among the explanatory variables in the model, Open (i.e., Open Approach) and Narrow (i.e., Constricted Waterway), are the two subzone type variables that indicate a constant effect on the risk level in addition to the effects of the subzone-specific variables. The estimators for the other four subzone type constants (i.e., convergence, open bay or harbor, enclosed harbor, and river) are not as significant as the other variables in the model. Of the four subzone-specific variables, Route Length, Average Width, and Sum of Degrees of Course Changes jointly describe the physical characteristics of route alignment and waterway/channel width. The fourth variable, Other Vessels Per Mile, represents a proxy of the density of other local vessel activities and contributes a marginal effect to the predicted vessel casualty probability. The four continuous variables plus two subzone type variables and the constant factor explain the subzone vessel casualty probability sufficiently well.

6. DEVELOPMENT OF SUBZONE VESSEL CASUALTY PROBABILITIES

While the estimated subzone vessel casualty probabilities could have been used to project vessel casualties for self-propelled, deep-draft dry cargo, and tanker vessels, a normalized, subzone-specific risk value for each subzone was also developed for estimating vessel casualty probabilities and vessel casualties among all other vessel groups. In other words, the relative risk levels among subzones, estimated from the deep-draft vessel group and subzone waterway characteristics, were assumed to be independent of all other vessel types, vessel sizes, and casualty types. The normalized subzone risk values could then be applied to the national average vessel casualty rates to derive a matrix table of subzone vessel casualty probabilities by casualty type, vessel type, and vessel size.

6.1 Weighted Subzone Vessel Casualty Probabilities

The historical casualty rate and predicted casualty rate of a subzone are described in equations 1 and 2 below:

$$HR = \frac{\text{subzone historical casualties}}{\text{subzone vessel movements}} * 100{,}000 \qquad (1)$$

Predicted Casualty Rate:

$$
\begin{aligned}
PR = \ & \text{-}0.372321 - 3.529773*OPEN \\
& +16.327722*NARROW + 0.228527*RTLENGTH \\
& \text{-}0.000407*AVGWIDTH + 0.012121*SUMHEAD \\
& +0.000392*OTHER\_ML
\end{aligned}
\qquad (2)
$$

Overall, the model predicted the subzone vessel casualty probabilities sufficiently well with an R-square of .75. But when the predicted subzone casualty rates and casualties were compared with the historical casualty rates and casualties, more than two-thirds of the observations were predicted well, whereas less than one-third of the subzones showed larger discrepancies between the predicted and observed casualties. It was apparent that some of the variations of the casualty rates could not be predicted by the systematic measures specified in the model, although there were no systematic errors other than a few outliers found to cause the largest discrepancies. In order to reduce the gap between the predicted and observed casualty probabilities for any unknown factors, the weighted average of the subzone historical casualty rate and the predicted casualty probability, as shown in equation 3, was taken to represent the vessel casualty probability for each subzone.

Weighted Casualty Rate:

$$WR = (HR + PR) / 2 \tag{3}$$

6.2 Subzone Risk Adjustment Factors

The weighted subzone vessel casualty probabilities were then normalized by the national average of this particular vessel group. More specifically, each weighted subzone vessel casualty probability was divided by the mean casualty rate (9.62141 casualties per 100,000 movements) of all the medium and large self-propelled vessel transits of all subzones over the 10-year period (equation 4). The results were the subzone risk values as shown in Table 4. These subzone risk values were then applied, as subzone adjustment factors, to the national average vessel casualty rates (see Table 1) to derive the subzone vessel casualty probabilities.

Subzone Risk Adjustment Factor:

$$RF = WR / 9.62141 \tag{4}$$

6.3 Subzone Vessel Casualty Probabilities

The subzone risk adjustment factors were applied to the national average casualty rates by vessel type, vessel size, and casualty type as shown in equation 5. As the subzone risk adjustment factors were the estimated subzone risk probability relative to the national average, multiplying the national average of any vessel casualty category by the corresponding subzone risk value resulted in a vessel casualty probability of that vessel category and casualty type for that subzone (equation 6). This process resulted in a matrix of 5,130 values for vessel casualty probabilities.

National Casualty Rate (by vessel type, vessel size and casualty type):

$$NR = \frac{\text{sum of all subzones casualties}}{\text{sum of all subzones vessel movements}} \tag{5}$$

Table 4: Subzone risk adjustment values.

Subzone	Type	Risk Value		Subzone	Type	Risk Value
1 BOSTON, MA				12 LONG ISLAND SOUND, NY		
1	A	0.37508		1	A	0.02232
2	B	0.03127		2	B	0.07547
3	C	0.75154		3	C	1.01728
4	D	0.46461		4	D	0.04759
5	E	2.81203		5	D	0.05255
2 PUGET SOUND, WA				6	E	1.04856
1	A	0.91939		13 PHILADELPHIA, PA		
2	B	0.30525		1	A	0.50696
3	C	0.64297		2	B	0.33529
4	E	1.04813		3	C	1.08857
5	C	0.01971		4	F	1.91007
6	D	0.09593		14 SAN FRANCISCO, CA		
7	D	0.78129		1	A	0.14195
9	E	2.90479		2	B	0.45094
10	D	0.03905		3	C	0.84060
3 LA/LONG BEACH, CA				4	D	0.46885
1	A	0.02371		5	F	2.53756
2	B	0.44709		15 PORTLAND, OR		
3	C	0.23691		1	A	0.17350
4	D	0.60029		2	C	1.96100
4 SANTA BARBARA, CA				3	F	3.36973
1	A	0.26169		16 ANCHORAGE, AK		
5 PORT ARTHUR, TX				1	A	0.43966
1	A	0.53874		2	C	5.84886
2	E	2.38349		3	D	1.36514
3	E	4.38490		17 PORTLAND, ME		
4	F	1.07481		1	A	0.00920
6 NEW ORLEANS, LA				2	C	0.13546
1	A	0.85570		3	D	0.18200
2	E	1.94588		18 PORTSMOUTH, NH		
3	F	3.02567		1	A	0.02258
4	E	4.51479		2	B	0.04338
5	F	1.63881		3	D	0.11890
6	F	5.92739		19 PROVIDENCE, RI		
7 HOUSTON, TX				1	A	1.43090
1	A	0.03408		2	C	1.76036
2	E	2.91751		3	D	1.34687
3	D	0.18659		20 WILMINGTON, NC		
8 CHESAPEAKE SOUTH, VA				1	A	0.00840
1	A	0.04265		2	E	0.85509
2	B	0.44280		3	F	1.67455
3	C	0.30003		21 JACKSONVILLE, FL		
4	D	0.37894		1	A	0.22962
5	E	1.25083		2	E	3.04065
6	C	0.35879				
9 BALTIMORE, MD				22 TAMPA, FL		
1	C	1.91003		1	A	0.79077
2	D	0.32546		2	C	5.12433
3	F	1.73354		3	D	0.51059
10 CORPUS CHRISTI, TX				23 MOBILE, AL		
1	A	0.06922		1	A	0.04222
2	B	0.50529		2	E	2.10450
3	E	1.72868		3	C	0.44332
4	F	0.83066		4	E	4.19989
11 NEW YORK CITY, NY				5	F	0.44424
1	A	0.10112				
2	B	0.21879				
3	C	0.14023				
4	D	0.15273				
5	E	1.68713				
6	C	0.42998				
7	E	1.26651				

Note: Subzones No. 2-8, 13-5, 17-4 and 18-4 are not included because they have no dominant vessel route nor VTS addressable transits.

Subzone Vessel Casualty Probability (by vessel type, vessel size and casualty type):

$$P = NR * RF \qquad (6)$$

7. FINAL COMMENT AND RECOMMENDATION

The steps involved in estimating waterway-specific vessel casualty probabilities are summarized as:

1) estimate the national average vessel casualty rates by casualty type, vessel type, and vessel size;

2) develop subzone risk adjustment values based on subzone characteristics in a multivariate regression analysis; and

3) apply the subzone risk adjustment values to the national average casualty rates to estimate subzone vessel casualty probabilities by casualty type, vessel type, and vessel size.

While estimating separate models for each casualty type and vessel category requires more sufficient sample representation, this approach does offset the problem of data deficiency and is considered analytically sound and efficient for the purpose of the study. For the calibration of the vessel casualty probability model and development of subzone specific risk adjustment values, the model-based, self-propelled, deep draft dry cargo and tanker vessels represent the most crucial component of the VTS-addressable vessel population and provide the most sufficient data base for model development.

The regression model fits the 10-year historical vessel casualty rates reasonably well with the predominant waterway physical characteristics (i.e., route length, average channel width, and sum degrees of course changes) and waterway type variables (i.e., open or narrow waterway). Local traffic density factor, represented by a proxy of local registration of small vessel fleets, also contributes a marginal effect on projecting the probability of vessel casualties. The model might have been further improved had more refined subzone level data (i.e., subzone meteorological and hydrologic measurements and local vessel traffic densities) been available. Although repeating a comprehensive, large scale study with improved data is not planned in the near future, validation and further model refinement focusing on a few areas such as geographically diversified Puget Sound or a high risk area like New Orleans is worth consideration. It is suggested that more refined data at the subzone level, over a longer time period, is required to improve the existing model.

Variations, between the predicted casualty probabilities and historical casualty rates, are found expected in some study zones. Although the largest discrepancies among the outliers can not be explained by the model, these discrepancies are corrected by using the weighted average of historical casualty rate and predicted casualty probability. More rigorous weight factors between the historical casualty rates and predicted probabilities could be developed based on the level of variations of the historical occurrences and predicted casualties. This, however, also requires a richer data set over a longer period of time.

Overall, this approach suggests a means for generalizing location specific casualty probabilities of a particular transportation mode based on cross-sectional and time-series data in multiple regression analyses. The process involves analysis of historical casualty incidents, estimation of traffic exposure measures, analysis of location specific risk variables, and development of localized risk adjustment factors for the application to aggregate casualty rates. The experience learned from the process of estimating waterway specific vessel casualty probabilities might be applicable to other transportation modes such as a highway or rail incident study. The critical elements that need to be defined in an application are: corridor/section types, route characteristics, particular vehicle class and accident rates, traffic exposure measures, and perhaps meteorological measurements.

8. REFERENCES

BUREAU OF MANAGEMENT CONSULTING, CANADA (1988). Chapter 3.0 Navigational Risk. In VTS Benefit/Cost Update Study, prepared for the Canadian Coast Guard.

BUREAU OF MANAGEMENT CONSULTING, CANADA (1989). Risk Assessment and VTS Needs: A Study Approach and Activity Plan.

BUREAU OF MANAGEMENT CONSULTING, CANADA (1990). Analysis of VTS Risk Factors and Effectiveness Measures in the Tofino Vessel Traffic Zone, prepared for the Canadian Coast Guard.

CANADIAN COAST GUARD (1984). Vessel Traffic Services, Final Report. National Vessel Traffic Services Study. Document TP5965-1E. Ottawa.

COCKCROFT, A.N. (1978). The Incidence of Sea and Harbor Collisions. In Third International Symposium on Vessel Traffic Services: Proceedings of the Symposium. Liverpool.

DEGREE, THOMAS, & COLES, P. (1988). Assessment of the Effects of Vessel Traffic Services on Marine Traffic Safety by Mathematical Modelling. In Sixth International Symposium on Vessel Traffic Services: Proceedings of the Symposium. Gothenburg, Sweden.

DRI/MCGRAW-HILL, et al. (1990). Fleet Forecasts for the United States to 2020, Draft, prepared for U.S. Army Corps of Engineers. Fort Belvoir, Virginia.

HARWOOD, D.W., VINER, J.G., & RUSSELL, E.R. (1990). Truck Accident Rate Model for Hazardous Materials Routing. Transportation Research Record 1264.

HAUER, E. (1986). On The Estimation of The Expected Number of Accidents. Accident Analysis & Preview, Vol. 18, No. 1, pp. 1-12.

U.S. ARMY CORPS OF ENGINEERS (1990). Waterborne Commerce Statistics Center. (1990). Waterborne Commerce of the United States, Calendar Year 1988, Parts 1-4. U.S. Army Engineer District, New Orleans, LA.

U.S. COAST GUARD (1973). United States Coast Guard Study Report: Vessel Traffic Systems Analysis of Port Needs.

U.S. DEPARTMENT OF TRANSPORTATION, RESEARCH AND SPECIAL PROGRAMS ADMINISTRATION. (1991). John A. Volpe National Transportation Systems Center. Port Needs Study - Vessel Traffic Services Benefits, prepared for U.S. Coast Guard, Office of Navigation Safety and Waterways Services.

U.S. DEPARTMENT OF TRANSPORTATION, MARITIME ADMINISTRATION (1988 and 1989). Merchant Fleets of the World as of December 31, 1987, and December 31, 1988, Washington, D.C.

ZECKHAUSER, R.J., & VISCUSI, W.K. (1990). Risk Within Reason. Science, Vol. 248.

Site Specific Transportation Risk Analysis for the Manitoba Hazardous Waste Management Corporation

Edwin J. Yee
Manager, System Development
Manitoba Hazardous Waste Management
Corporation
Winnipeg, Manitoba
Canada R3H 0Y4

ABSTRACT

A site specific transportation risk analysis of two potential sites for a hazardous waste management facility is examined with respect to the overall dangerous goods transportation risk currently existing in Manitoba. Background dangerous goods transportation risk is established and compared with the incremental risks associated with optimum transport routing networks developed for the proposed siting options. Emphasis is placed on the methodology for the establishment of the waste transport networks. Transportation risk mitigative measures are also examined.

1. INTRODUCTION

In 1982, the Manitoba government initiated a long term, multi-phased hazardous waste management program. From the outset, the program contemplated the development of an integrated "system" that provided for the minimization of hazardous waste through reduction, reuse, recycling and recovery at source, as well as addressing treatment and disposal of material not managed at source. In November of 1986, the Manitoba Hazardous Waste Management Corporation was established as a commercial Crown Corporation, reporting through its Board of Directors to a responsible minister with its equity held by the Minister of Finance. The Corporation is mandated to establish and operate a hazardous waste management system which will safely treat and dispose of hazardous waste in accordance with regulations and standards prescribed by the Minister of Environment.

Historically, the siting of hazardous waste treatment and disposal facilities has been a difficult undertaking, raising public fear and concern respecting a wide range of issues. Early in the development process for an overall system to manage regulated hazardous waste, the Corporation identified the need to address issues related to the transportation of waste within the system. This includes the transportation of these materials from their source to the proposed central and out-of-province facilities. To accomplish this evaluation, consideration was given to the characteristics of the waste, the nature of potential receptors along the transportation network, historical traffic and accident statistics along potential routes, and the range of accident scenarios that could potentially occur. It was also important to consider the impact of hazardous waste transportation in the context of the existing risk associated with the transport of all dangerous goods. Hazardous waste movements are a small component of the total amount of dangerous goods transported on the roadways today. The proposed provincial hazardous waste management system intends to collect and properly manage a greater volume of these materials than is currently being done. As a result, some increase in the total number of dangerous goods shipments is expected, however, the increase in the percentage of total dangerous goods shipments will be small.

2.   ASSESSMENT APPROACH

The assessment of transportation risk was limited to the estimation of accident frequency and spill frequency as a result of these accidents. The consequences of such spills with respect to public health, including other road users, is documented in a report entitled "Site Specific Risk Assessment Montcalm Site Volume 1 - Main Report" (Reid Crowther & Partners, 1992).

In 1989, the Corporation initiated research based on the rationale that hazardous waste represents a small component of the existing dangerous goods traffic, and completed Part 1 of a multi-phased study on Transportation Risk Assessment entitled "Transportation Risk Assessment Part 1 - Data Collection and Model Development" (M.M. Dillon, 1989). This work included a comprehensive data collection program of roadway and rail travel information, geo-referencing of provincial waste generators, the development of a transportation model, and the development of an initial risk assessment methodology. This initial effort was generic in nature, as sites and system specifics had not been identified at the time and excluded Winnipeg traffic data.

The Corporation initiated Part 2 of the study in 1991 and expanded the scope of work to include travel on the Winnipeg truck route system. The purpose of the study was to provide Transportation Risk input to the overall environmental impact assessment of the Hazardous Waste Management System as a whole and the facility development in particular. Two candidate communities, Winnipeg and Montcalm, were identified to host the facility and specific sites were identified, allowing site-specific work to be undertaken. The Winnipeg site is approximately 3 kilometers south of Highway 100 and 2.5 kilometers west of Waverley Street. The Montcalm site is located approximately 1.5 kilometers west of Highway 75 on the south side of Highway 14, approximately 70 kilometers south of Winnipeg. The goal of the study was to develop and implement a risk assessment methodology that achieved the following objectives:

- Identify and optimize routing of waste from various generators to the facility

- Quantify baseline transportation risk associated with general dangerous goods

- Determine incremental risk associated with hazardous waste transported on routes previously identified

- Differentiate between the risk of an accident and the risk of a spill

- Evaluate risk mitigation measures and the overall system sensitivity to these measures

To accomplish this, the study was undertaken in two phases. Phase 1, "Data Collection and Methodology" (M.M. Dillon, 1991), provides updated data presented in the previous study, expands the transportation network to address waste and dangerous goods movements in the city of Winnipeg and refines the initial risk assessment methodology. Phase 2, "Site Specific Examinations" (M.M. Dillon, 1991), provides the actual Site-specific Transportation Risk Assessment.

Although mitigation of transportation risk was addressed as part of the site specific transportation risk assessment work, the

Corporation further examined risk mitigation issues respecting legislation and regulation, vehicle and equipment design, personnel training and operating procedures, detailed in the "Transportation of Hazardous Waste" report (M.M. Dillon, 1991).

2.2 Waste Quantity Estimates

The transportation assessment work is based on the latest available waste market research provided by the Corporation. In 1989, the initial work was based on an estimated 23,000 tonnes per year of waste, projected to be available for management by a central treatment facility. For the site specific transportation risk assessment work, this estimate was refined to 18,408 tonnes per year, based on further studies conducted by the Corporation to define the hazardous waste market.

2.3 Methodology

In order to examine transportation risk incrementally from the baseline condition, general dangerous goods movements and hazardous waste transport traffic data, accident data, hazardous waste generation nodes, unit generation areas, Provincial Trunk Highways and the city of Winnipeg Truck Routes were computer modelled. The assessment was divided into provincial and Winnipeg components to account for the difference between the urban and rural transportation and environmental settings.

Two types of modelling conventions were used to establish optimum routing selection:

*   Minimum Time Travel (M.T.T.), where the trucks were routed on the basis of least time from generator to receiver. This inherently minimizes risk as it reduces the time exposure of a vehicle on the roadway.

*   Minimum Accident Probability (M.A.P.), where the trucks were routed on the basis of least accident probability from generator to receiver.

The following environmental and public safety issues were identified for both the provincial and Winnipeg transportation components and applied to the modelling conventions:

Provincial Component

*   number of watercrossings;

- federal and provincial parks;
- ecological reserves and wildlife lands;
- provincial forests;
- number of at grade road/rail crossings;
- number of large residential groupings within a close proximity to route;
- ground water pollution hazard areas; and
- communities having emergency-response capability.

Winnipeg Component

- nature of zoning adjacent to truck routes;
- number of water crossings and other water bodies adjacent to routes (rivers, streams, storm drains, ponds reservoirs);
- park areas;
- number of at-grade railroad crossings;
- ground water pollution hazard areas;
- emergency-response capabilities;
- intersections on routes that converge on the facility; and
- wildlife areas.

Traffic volumes on the highway network and urban truck routes were collected from the Manitoba Department of Highways and Transportation and city of Winnipeg Streets and Transportation Department respectively. Based on a review of the data and available information from the Manitoba Department of Highways and Transportation, city of Winnipeg Streets and Transportation Department, and the Manitoba Department of Environment, estimates for commercial and dangerous goods traffic were derived from the total traffic volumes. Total, commercial and dangerous goods and historical accident information were modelled to adequately quantify existing baseline conditions and to provide the necessary interpolative tools to forecast risk associated with the incremental hazardous waste hauling activity.

The waste network assumed for the study was derived from the Corporation's research in waste characterization. The waste generators in the provincial component were grouped into contributing communities. Smaller towns with lower production were lumped together and the production location was assumed to be that which was farthest from the management facility. The urban component (Winnipeg) was defined as unit generation areas (UGA's). The production location was centralized within the UGA and was connected into the transportation network at one or more locations. The waste network also includes materials handled by the transfer operation and subsequent routing to out-of-province facilities.

3.   TRANSPORTATION NETWORK

Currently in Manitoba, all Provincial Trunk Highways and city of Winnipeg truck routes are considered dangerous goods routes. The Manitoba Dangerous Goods Handling and Transportation Act has provision for municipalities to specifically designate dangerous goods routes. At this time, the city of Winnipeg is examining the feasibility of designating specific dangerous goods routes. In order to quantify baseline risk, incremental risk and to facilitate the mitigation of transportation risks, an optimum transportation network was established for waste transport from the points of generation to the facility, using the established methodology.

3.1 Baseline Risk

As transport to the treatment facility will become part of the transportation risk associated with general dangerous goods hauling activity, it was necessary to make the evaluation relative to the baseline conditions. In assessing incremental risk associated with transporting hazardous waste to and from the treatment facility, identification of the characteristics of existing dangerous goods transport activity was undertaken. The data base, discussed in Section 2.3, is explicit regarding volume, accidents and network specifics, thus providing the following baseline observations:

- Commercial truck traffic on highways ranges from 10 to 20 percent of the total traffic, while it was assumed that 6 percent of total traffic in Winnipeg comprise commercial trucks. For this study, the 10 percent estimate was used to determine commercial truck traffic on the highways.

- Studies undertaken by Dillon (1989) and data collected by Manitoba Environment (ongoing) suggest that 10 percent of all commercial truck activity involves transport of dangerous goods. Furthermore, it was estimated that transport of dangerous goods by truck within the province could make up 730,000 to 910,000 truck movements annually.

- Based on the traffic volume of dangerous goods shipments over the length of roadway within the study scope, about 65 to 85 million vehicle-kilometres are logged in Manitoba each year, transporting dangerous goods.

- The frequency of dangerous goods truck accidents was derived from the accident frequency data for the overall traffic stream. For the provincial highway network, data suggest that the ratio of commercial truck accidents to total traffic accidents is similar to the ratio of commercial truck traffic to total traffic. In Winnipeg, other data suggest that the ratio of commercial truck accidents to total accidents is approximately one-half the ratio of commercial truck traffic to total traffic on the urban truck route network.

- The probability that an accident involving dangerous goods would result in a spill was determined to be 0.20 (1 in 5).

- The existing frequency of accident-related spills involving dangerous goods in Manitoba is about 20 per year.

- Review of environmental accident reports suggest that the average spill involving dangerous goods would be about 1000 L of material.

3.2 Optimum Routes

Assignment of background transportation network activity with respect to total traffic, commercial traffic and estimated existing dangerous goods traffic for highways and Winnipeg truck routes was undertaken. Estimated accident frequency was established and the links in the network were categorized into low, medium and higher accident probability zones. Historical frequencies of dangerous goods spills in the province of Manitoba were examined. Wastes generated by each node were assigned to the highway network using both modelling conventions.

The routes identified from both assignments were essentially the same for wastes on the provincial network. Assignment of waste to the urban truck route network in Winnipeg based on the two conventions yielded significantly different results. Using the M.A.P. convention resulted in the routing of vehicles away from central Winnipeg, providing positive implications with respect to public acceptance and public health and safety issues. The M.A.P convention was selected as the recommended method for the identification of optimum routes.

The proposed routing network established for both the Winnipeg and Montcalm sites is illustrated in Figures 1, 2 and 3. It should be noted that the Winnipeg routing component illustrated in Figure 3, is the same for both siting options as it is common to the overall waste transport network and both sites are located to the south of Winnipeg. A comparison of the resultant routing schemes with the Winnipeg system recommended routing plan indicates that much of the hazardous waste traffic is assigned to the same links in the highway network. Again, this is because many of the main routes are directed to Winnipeg and the Montcalm site has excellent direct access to Winnipeg along Highway 75. Redistribution of some traffic is evident, as illustrated in Figure 2, mainly from the south and west regions of the province. Much of the traffic eastbound and westbound on the Trans Canada Highway diverts to more southerly east/west routes.

3.3 Incremental Risk Comparison

Based on the proposed routing, an estimate of total dangerous goods transport (dangerous goods and hazardous waste) and accumulated accident and spill probability was obtained with the following results:

- Total dangerous goods transport on the optimum Winnipeg and Montcalm systems is estimated to be 180,000 and 260,000 vehicle-kilometers respectively.

- Accident probability is estimated to be 1 every 6 years for the Winnipeg system and 1 every 4 years for the Montcalm system.

- Accident related spill probability is estimated to be 1 every 29 years for the Winnipeg system and 1 every 22 years for the Montcalm system.

Table 1 provides a comparison of the transportation finds for the system option versus the baseline condition.

Table 1: Comparison of the baseline condition vs the Winnipeg and Montcalm transportation systems.

	Baseline Condition	Winnipeg System	Montcalm System
Total Dangerous Goods Transport (Vehicle-Kilometers)	65-85,000,000	180,000	260,000
Accident Probability	100 / Year	1 / 6 Years	1 / 4 Years
Accident Related Spill Probability	20 / Year	1 / 29 Years	1 / 22 Years

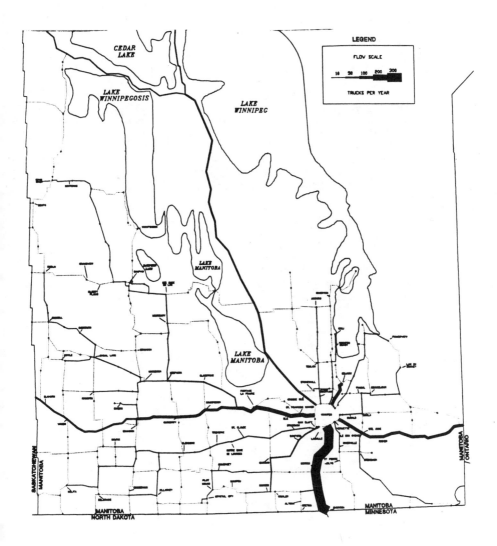

Figure 1: Optimum waste routing network for the provincial
roadway component of the Winnipeg site option.

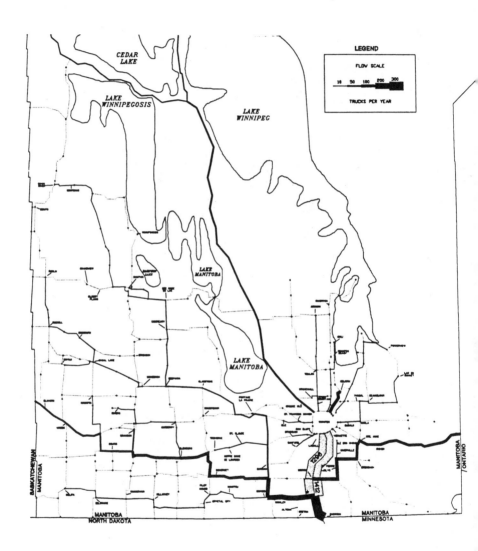

Figure 2:   Optimum waste routing network for the provincial
               roadway component of the Montcalm site option.

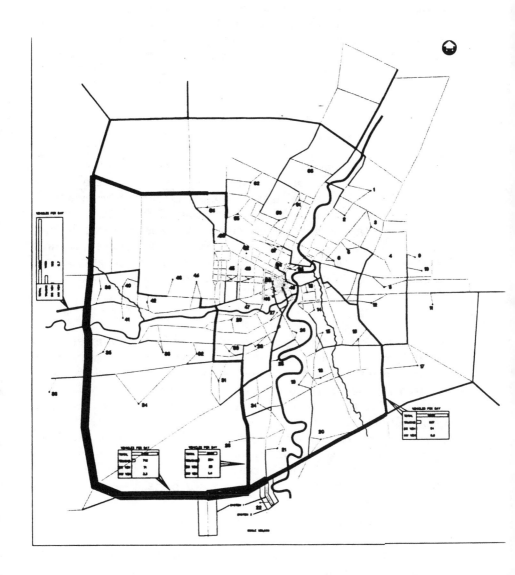

Figure 3: Optimum waste routing network for the Winnipeg roadway component.

The incremental increase in dangerous goods transport activity associated with hauling hazardous waste to the facility is very small (0.2 to 0.4 percent) as the roadway networks identified for hauling hazardous waste are merely a subset of the overall roadway network servicing dangerous goods transport. Similarly, the incremental increase in transportation risk associated with hazardous waste transport to the treatment/storage/transfer facility is very small (0.2 percent) and essentially equal between the two system options.

4. RISK MITIGATION

Although the incremental risk resulting from the addition of the proposed hazardous waste transport activity represents a minute incremental risk, risk mitigation measures should be examined and implemented. Through the application of the M.A.P. convention for the establishment the optimum or proposed routing, the first level of risk mitigation was achieved. Further mitigative measures including re-routing, roadway improvements, equipment design, training and operating procedures were examined.

4.1 Re-routing

The recommended routing strategy adopts a minimum accident probability convention. In conjunction with the environmental setting and public safety criteria, reduction in transportation risk was accomplished at the initial level. At the provincial level, traffic is attracted to the four-lane divided roadways for their speed and safety-related characteristics. At the Winnipeg level, traffic was directed away from central Winnipeg, where accident rates are higher than average, to the perimeter, where accident rates are lower than average. Further re-routing on this basis may be adopted as required on a perceived problem-specific basis.

4.2 Roadway Improvements

The recommended routing scheme was inspected with respect to link-specific accident rates. Because of the nature of routing assignment, unusually high accident rates were not observed on the chosen links. However, the transportation authorities for the province and Winnipeg are continuously evaluating plans for improving the transportation network. This includes measures for improving overall safety and other service-related characteristics of public roadways. Some programs planned by these agencies have some relevance to the recommended routing scheme. These include the following:

- Twinning of Highway 75 from Winnipeg to Emerson including the Morris bypass and an interchange at Highway 1

- Twinning of Highway 14 from Morden to Winkler

- Interchanges with Highway 100 (Winnipeg South Perimeter) at St. Mary's Road, St. Anne's Road, Waverley Street, Highway 3 and Highway 2

- Twinning of Highway 59 from Winnipeg to Ilse des Chenes

- Completion of the north bypass Highway 101 (Winnipeg North Perimeter)

- Twinning of roadway, Pembina Highway - St. Norbert

- Railway underpass, Route 90 (Kenaston Boulevard) near Wilkes Avenue

- Twinning of roadway, St. Mary's Road near Perimeter Highway

4.3 Equipment Design

Under current regulatory requirements, there are no specifically designed equipment or safety features that must be incorporated for the transport of dangerous goods or hazardous waste compared to those for other commodities. Vehicles that transport dangerous goods and hazardous waste meet the same requirements as for the transport of other commodities. Vehicle design features are intended to protect against leakages or spills. Vehicle types and specialization are a function of the commodity they are intended to haul. Transportation of dangerous goods comprises about 10 percent of all commercial hauling activities. Vehicle technology has been developed to service this significant market share of the transport industry. This includes design features intended to protect against leakage or spills. Based on the physical properties of hazardous waste and the categories of pumpable, non-pumpable, bulk or drummed, the vehicle requirements for the Corporation include the small load tanker (12 tonnes), large load tanker (22 tonnes), lugger box (12 tonnes), dump truck (12 tonnes) and van (18 tonnes).

Equipment used for the hauling of dangerous goods, including hazardous waste is regulated under the federal and provincial Transport of Dangerous Goods Regulations. Additionally, the

Canadian Standards Association (CSA) Code pertaining to structural integrity of transport containers has been drafted with input from the transportation industry. Most of the active truck fleets already meet or exceed these standards. Safety features contained in this code and in the nature of vehicles currently hauling dangerous goods include the following:

- Compliance with regulatory standards respecting, vehicles, containers, labels, placards, documentation and safety

- Rear-end protection to absorb the shock of rear-end collisions and avoid damage to the container

- Reflective devices or markings for trailers

- Container appurtenances (valves, fittings, gauges) protected to guard against damage due to collision and overturning

- Tanks and vessels designed and constructed in accordance with the best known and available practices

Additional safety features that could be considered include double-walled tank containers, overflow reservoirs built into van trailers and tank baffles to limit volume of release should a spill occur.

The Corporation will develop emergency response capability with appropriate equipment to promptly and comprehensively respond to all emergency situations associated with the overall management system.

4.4 Training

Professional truck drivers must complete extensive and more detailed training than regular passenger-vehicle car drivers are required to take. In addition to the basic vehicle-oriented training, the drivers of dangerous goods vehicles are required to be trained in health, safety and environmental concerns for the goods to be handled and emergency management procedures. Drivers of hazardous waste transporting vehicles are required to be familiar with relevant waste management and emergency response equipment, legislation, regulations and guidelines.

Extensive internal training programs will be mandatory for all of the Corporation's staff and any carriers contracted for the handling of materials in the proposed system. These programs will address the following areas:

- Workplace Safety and Health

- Workplace Hazardous Materials Information System (WHMIS)
- Transport of Dangerous Goods
- Specialized Waste Management Equipment
- Emergency Response
- First Aid

4.5 Operating Procedures

Despite the recent deregulation of the trucking industry, highway truck drivers are tightly controlled by government and individual company policies. Hours of driving are regulated and must be recorded in log books. Manifests, detailing classifications, concentrations, quantities, packing information, emergency handling procedures, schedules, origin and destination of waste must accompany all shipments of hazardous waste.

In addition to the above, the Corporation will develop specific operating procedures particular to its waste management system operation. The procedures specific to the transport operation will address the following:

- Routing
- Speed restrictions
- Frequency of on-route vehicle inspections
- Operating hours and conditions
- Pre-transit and post-transit inspections
- Communications
- Emergency Response
- Extended vehicle maintenance standards
- Record keeping and reporting

4.6 Residual Impacts

The potential transportation impacts, accidents and spills resulting from accidents, associated with the proposed transportation network will be mitigated through the route selection process establishing the network and the additional measures discussed in the previous section. The probability of their occurrence has been demonstrated to be a very small incremental increase over the existing dangerous goods traffic. However, total avoidance of accidents and spills resulting from accidents over the life of the management system is not possible. It is recognized that residual impacts as a result of a spill will be those posed by residual contaminants remaining after spill clean-up.

As part of the Corporation's overall mandate and environmental policy statement, "Ensure that all activities involved with the handling, storage, treatment and disposal of waste and residue products are consistent with the protection of public health and environmental quality.", these residual impacts will be mitigated through a prompt and comprehensive spill clean-up program. This program will include monitoring of clean-up operations to ensure all impacts are addressed and minimized, and spill sites are restored to the maximum degree possible.

5.   CONCLUSIONS

The fundamental principles underlying the approach to this analysis relate to the fact that transportation of dangerous goods is an ongoing condition that contributes to an existing level of risk on the transportation network. Hazardous wastes currently make up a very small part of this activity. The proposed development of a management system to properly treat hazardous waste will require transporting these materials to a treatment/storage/transfer facility; thus resulting in an increase in this activity.

The route optimization exercise focused on assignment of hazardous waste traffic by minimum travel time (M.T.T.) and minimum accident probability (M.A.P.). In the provincial network, route selection was insensitive to these approaches. In Winnipeg, M.A.P. routing showed favourable characteristics and was selected as the recommended convention for the establishment of the waste transport networks.

Baseline transportation risk was established from existing data on dangerous goods traffic activity. This risk exists throughout the transportation network and accompanying environmental setting.

In assessing incremental risk of the proposed activity, the analysis generally found the resulting transportation network servicing either system, to be very similar for two main reasons:

- The Winnipeg area is the largest contributor to the hazardous waste stream and the routes accessing the Winnipeg system site and the highway accessing the Montcalm system site are the same; and

- The provincial network is a radially based system. Therefore, much travel from remote areas in Manitoba is attracted to Winnipeg and redirected to its eventual destination.

In the context of baseline transportation risk associated with the overall dangerous goods activities, the incremental risk associated with hazardous waste under either system option is very small (0.2 percent). This is directly related to the very low traffic flow implication resulting from hauling hazardous wastes generated in Manitoba.

Comparing the Winnipeg system with the Montcalm system, produces a more measurable difference. Travel requirements of the Montcalm system are about 40 to 45 percent greater than the Winnipeg system. This translates to about a 30 to 35 percent higher risk of accident or spill potential. This increase is predominantly a result of the difference in traffic between Winnipeg and the Rural Municipality of Montcalm on Highway 75, between the two system options. This shift in traffic is also responsible for any change in environmental setting comparisons between the two system options, which primarily effect:

- The north/south corridor, south of Winnipeg: and

- The east/west corridors close to Winnipeg, which shift southerly to service the 8 Montcalm site.

The analysis also found that the chances of a spill resulting from an accident are about 1 in 5 (0.20). It should be noted that this spill probability would be considered a conservative assumption because other factors exist affecting whether a spill would result in any measurable consequences (i.e., volume and nature of release, surrounding environment, emergency response availability, etc.).

Evaluation of further risk mitigation measures suggest safety factors exist on the predictions of transportation risk outlined in this analysis. However, routing to optimize risk probability mitigation, as was completed in this work, is the most significant contributing mechanism to system safety.

6. REFERENCES

M.M. DILLON LIMITED (1989). Manitoba Hazardous Waste Management Corporation Transport Risk Assessment Part 1 - Data Collection and Model Development. Winnipeg: M.M. Dillon.

M.M. DILLON LIMITED (1991). Manitoba Hazardous Waste Management Corporation Site-Specific Transportation Risk Assessment Phase 1 - Data Collection and Methodology Development. Winnipeg: M.M. Dillon.

M.M. DILLON LIMITED (1991). Manitoba Hazardous Waste Management Corporation Site-Specific Transportation Risk Assessment Phase 2 - Site-Specific Examinations. Winnipeg: M.M. Dillon.

M.M. DILLON LIMITED (1991). Manitoba Hazardous Waste Management Corporation Transportation of Hazardous Waste. Winnipeg: M.M. Dillon.

REID CROWTHER & PARTNERS LIMITED (1992). Manitoba Hazardous Waste Management Corporation Site Specific Risk Assessment Montcalm Site Volume 1 - Main Report. Winnipeg: Reid Crowther & Partners.

The Need and Value of Routing of Dangerous Goods Transports

Göran Stjernman
Department of Transportation and Logistics
Chalmers University of Technology
S-412 96 Göteborg
Sweden

ABSTRACT

The paper starts out from a systems point of view and from a brief analysis of dangerous goods transports, three lines of investigation are suggested. One of these lines is developed here. The main objective is to assess the need and value of routing of dangerous goods transports. The quantifying of the risk of dangerous goods transports is a prerequisite for an objective assessment of the need and value of routing. It is not possible to assess the risk of dangerous goods transports directly from statistics on dangerous goods accidents. However, it is possible to assess the effect of some road parameters on traffic safety, although with relatively low accuracy. Methods for determining the portion of dangerous goods vehicles in the estimated number of accidents, and the number of accidents that result in the dangerous goods affecting the environment do not exist. This means that the probability of a dangerous goods accident cannot be assessed, which consequently means that the societal risk cannot be expressed in quantitative terms. The practical use of the technological advances of the past few years in directing traffic, in order to better utilize the infrastructure, is near at hand. This offers an opportunity to implement dynamic and efficient routing of dangerous goods transports. With such an implementation of routing, the requirement for accuracy of the probability assessment may be much less than it is today.

1. INTRODUCTION

The requirements for transports of goods are changing due to the rationalization (capital rationalization) of industry and trading, as well as to the changed characteristics of goods. These requirements include precision in timing, flexibility, control/rearrangement, low damage frequency and information. The decreasing storage level in industry, affecting buffer storage as well as stock, makes it necessary for deliveries to arrive within very narrow time margins ("just in time"). Accordingly, with shortened lead times and increased sensitivity for disruptions, the flexibility of time tables and capacity is of greater importance as is the capacity for emergency deliveries. As goods increase in value and stock levels decrease, it becomes more critical to have low damage frequency. Regarding primary materials, for which margins in the material supply are diminishing, low damage frequency is of paramount importance.

During the past decade, environmental issues have been increasingly debated in society. The increased use of chemical products has been questioned, among other things, and in this context, the increased transportation of these products, as well as substances needed for their production, has been discussed. Transportation of the most common petroleum products, fuel oil and gasoline, which in case of an accident can cause both environmental harm and acute damage to the surroundings, did not attract very much attention - at least not until 1987. In that year several tank car accidents occurred in Sweden as well as the accident in Herborn, Germany, where a petrol truck turned over with the result that five people were killed and 24 were injured. The debate on transportation of dangerous substances consequently became quite animated and routing of dangerous goods transports was put forward as a way of increasing the safety of human beings and property.

A definition of dangerous goods is given in ADR (Accord européen relatif au transport des marchandises Dangereuses par Route), which is a European agreement on accepting international road transports provided they are performed according to the requirements in the appendices to the agreement. In the ADR certain substances are classified as dangerous goods partly depending on how dangerous the substance is and partly depending on the quantities of the substance that are carried on the specific transport.

2. SYSTEM VIEW

2.1 Identification of the Problem

Transports of dangerous goods, especially by road, are recognized as a problem by some members of communities and there are groups that consider it to be a very risky activity. Briefly, these groups demand that such transports be directed to "harmless" roads and not pass centres of population, hospitals and similar sites. Accepting the given description of the problem, the solution put forward could result in the symptoms being addressed rather than the real problem, which in the worst case could lead to an even more serious situation.

A more appropriate approach is to find out if transports of dangerous goods are a problem to the extent that this could be considered a societal problem and, if so, to search for the most suitable solutions or actions. The fact that transports of dangerous goods are experienced as a dangerous activity by at least some people in our society is chosen as a basis for this search.

In order to investigate whether dangerous goods transports should be considered as a problem and, if so, how to formulate it as well as what solutions, taking into account various circumstances, might be possible, the society is regarded as a system. This system is very complex and it encompasses many subsystems. However, it should be sufficient to study a much smaller number of these systems.

One system to be studied is, of course, the transport system and in the process further subsystems can be identified, for example those who are dependent on the transport system and, furthermore, those who affect or are affected by the transport process.

2.2 System Relations

The set of subsystems shown in Figure 1 were chosen to illustrate some important interrelations between them.

As the performance of the transport system is of great importance to its consumers and their ability to perform satisfactorily, reduced capability of the transport system produces a negative influence on society.

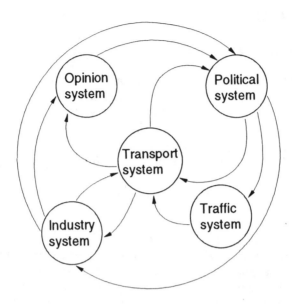

Figure 1: Some essential interrelations between some of the subsystems of society.

It should be noted that among the physical resources used by the transport system, the number and design of roads cannot be influenced by the transport system itself (regardless of its capital resources). However, the transport system can exert an influence on the traffic system and, through this and other systems, indirectly influence the development of the road network, as can be seen in Figure 1. Other physical resources such as vehicles, drivers etc. are available to the extent that the transport system can invest and pay for the operation of these resources.

The transport system is a consumer in relation to the traffic system which includes the road network, bridges, traffic rules, regulations concerning vehicles and drivers etc. The industry system is mainly the processing industry and pulp and paper industry, as well as the manufacturing sector.

The rest of the systems in Figure 1, the opinion system and the political system, produce a feedback between the subsystems of society. The opinion system consists of, in addition to the media, several groups, organized (e.g. lobby groups) as well as unorganized. The political system includes the decision-making bodies of society on both the local and the national levels.

It is through the opinion system that much of public anxiety about dangerous goods transports is expressed. This anxiety is not necessarily motivated by the actual conditions of these transports, however even if the anxiety is not well motivated it can be of such magnitude that measures to reduce it can still be regarded as necessary. It is reasonable to make a closer study of the extent to which the expressions in the opinion system reflect people's anxiety, and to try to establish the degree to which this anxiety has an effect on various parts of society. It would also be appropriate to investigate what actual conditions relating to transports of dangerous goods could call for greater safety measures. Taking such measures can lead to diminishing or removing the anxiety of the population, as well as increasing safety.

What distinguishes dangerous goods from ordinary goods is that in the event of an accident during the transfer of dangerous goods, acute injury and damage to people and surroundings can result. Accordingly, the interaction of the transfer process with the environment has to be studied. Note that various components in the society system and also some external systems are involved, as is shown in Figure 2. The environmental influence caused by transport of dangerous goods, e.g. noise and pollution, is the same as for all other transports, so this aspect is not included in the study.

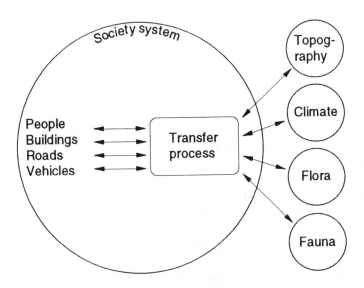

Figure 2: The interactions of the transfer process.

Two aspects of the interaction between the transfer process and the affected systems are of special interest. One is manifested when the interaction results in a malfunction in the process – an accident. The other is the effect of an accident on the systems – the consequence.

This is a problem that can be studied from three different coherent viewpoints. The first is to study the anxiety of the population and the perception of the risks involved in the transport of dangerous goods. The anxiety might be of such a magnitude that corrective measures ought to be taken regardless of whether the real risk is of the same degree as the perceived one. The second of the three is to consider the transport system and its customer, the industry system (among others), and how these systems work regarding dangerous goods transports. The third is to study the interaction of the transport process with the surrounding systems. These three are not independent of each other, particularly not when it comes to suitable measures that should be taken.

In this study only the third viewpoint, the transports of dangerous goods as a process interacting with the surroundings, is being dealt with (Figure 3).

3.   QUALITATIVE PERSPECTIVE - OCCURRENCE OF DAMAGE

3.1 Basics

A dangerous substance loaded in a tank must escape to be able to inflict harm or damage to the environment. This can happen in different ways: by the opening of a valve, by the causing of a crack or a hole in the tank wall or by the bursting of the tank in two or more parts. A leak in the tank wall can occur by the degeneration of the wall, e.g. caused by corrosion, to the degree that it cracks. A hole in the tank wall can also be made by penetration or tearing of the wall by external objects. For the tank to rupture, it is necessary for the internal pressure to become so high that the tensile stress of the tank wall reaches the ultimate strength of the material, which can be caused by extreme heating of both tank and contents. At high temperatures, the material of the tank wall is weakened which can precipitate a rapture (Droste and Schoen, 1988a and 1988b). The tank can also rupture from severe external impact.

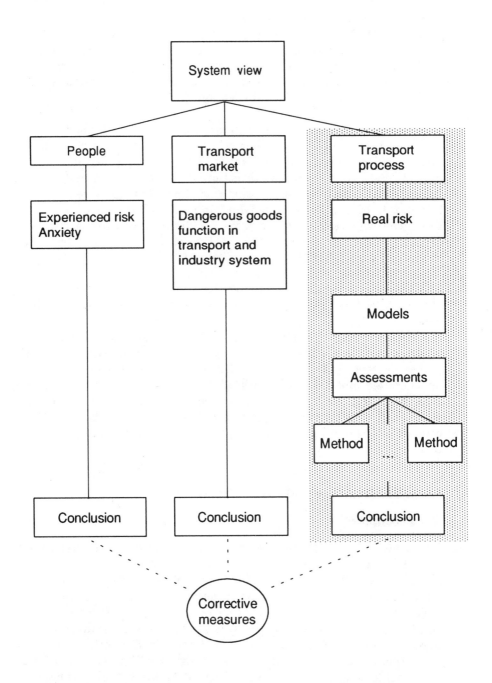

Figure 3: Three areas of study, derived from a system overview.

External forces are required to break a valve or to make a hole in a tank wall, and heat to open a safety valve or to rupture the whole tank, so that the transported substance can affect the environment. Such energy transfers to the tank can happen in accidents (traffic), the occurrence of which are influenced by the subsystems topography, climate, roads, vehicles and people.

An accident occurs when the progress of a vehicle cannot be brought to a stop in a controlled manner on the road before it is stopped by another object, when a vehicle cannot be brought to avoid an obstacle on the roadway or when the direction of motion cannot be brought into line with the direction of the road. Furthermore, a vehicle can be run into by another vehicle which in its turn has "failed" in one of the three ways above. Aside from accidents due to impact, other causes can include, for example, heating of a bearing or a defect in the fuel system, which can start a fire.

The failure of a vehicle to stop, avoid an obstacle or keep to the road may have several causes, e.g. that the friction forces between the wheels and the road surface are less than the lateral forces acting on the wheels so that the vehicle hits an object at the side of the road or another vehicle on the road and is thus damaged. When friction and speed are great enough, a vehicle will overturn and stop in an uncontrolled manner as a result of the friction between the side of the vehicle and the ground or be stopped by an object on or at the side of the road. This could be due to inattentiveness of the driver, to other human shortcomings or to a deficiency in the man-machine interface.

4.   QUANTITATIVE PERSPECTIVE

4.1 Probability

The probability of dangerous goods accidents on a specific route cannot be assessed directly with an acceptable (meaningful) degree of accuracy. The documented history of accidents is far from complete, but even if the records on accidents with tankers involved were complete during the last 30 (or even 50) years, which they are not, the conditions have not remained the same during these years. Vehicle design, road construction and traffic intensity are continuously changing, as is the quantity of dangerous goods transported, both in total and in regard to individual shipments. The problem of limited statistical material on dangerous goods transports is mentioned in some studies (e.g. Meslin, 1981,

Lautkaski, 1986, and Abkowitz et al., 1988).

Recently, methods have emerged for assessing the true accident rate, i.e. the anticipated number of accidents after a certain number of accidents during a given period have been observed. Brüde and Larsson (1987a) use a method called the Empirical Bayes Method (the EB method) together with a variation of it for which prediction models of the number of accidents are used. The problem seems to be to estimate correctly the precision of the prediction model selected.

Junghard (1989) starts out from the assumption that the accidents in an intersection or road section have a Poisson distribution which means that the expected values for accidents in a population of intersections or roads sections of the same type are gamma distributed. This leads to the assumption that accidents randomly selected from a population of intersections or road sections have a negative binomial distribution. The conditioned expected value for accidents in such an intersection or road section, given the observed number of accidents, is calculated with the help of a parameter that is estimated by the use of cross-validation. The prognostic capability is enhanced primarily when the number of accidents, calculated with prediction models (Brüde and Larsson, 1987b) is extremely high or low.

To use these methods, a number of accidents that is not too small must have occurred for every type of object studied, which is seldom the case for tankers with dangerous goods, as usually no such accidents are registered.

4.2 Analytical Methods

Another way of approaching the problem is to take the influencing factors as a starting-point for trying to assess the probability of an accident involving a vehicle loaded with a dangerous goods.

Although the number of subsystems mentioned is not especially large, they are systems in themselves and some are very complex ones. To determine which components of the subsystems are essential to study in connection with the interaction of each of the systems with the transport process is a very elaborate task, all the more so as the systems influence each other. Some components of the different subsystems that influence the transport process are discussed next, however this is not to say that these are the only components worth studying and nothing is said about their probable synergistic effects. Any previous attempt to solve the problems from the starting point indicated here has not been found, but there are studies in some areas within this complex.

The human being has long been studied from physical, psychological and physiological points of view and the results of these studies should be useful in this context, however ergonomics is beyond the scope of this work.

Nowadays, it should be quite possible to assess the probability of vehicle malfunction with adequate accuracy, although this requires an immense amount of work. This also implies participation of car-manufacturers (which should not necessarily be taken for granted).

The influence of weather on traffic safety is dealt with by Björketun (1982), Schandersson (1986) and Möller (1988). All of these writers make reservations concerning the use and generalization of the results. The results can not be used to quantify the change, e.g. due to snow, in the probability of accidents to an acceptable degree of accuracy.

Different aspects of the influence of topography and road design parameters on accident frequency are investigated in studies by Brüde et al. (1980), Polus and Dagan (1987), Lamm and Choueiri (1987), and Arnberg et al. (1978).

4.3 Restrictions

While it is now possible to calculate the influence of curve, width, inclination and intersection of roads on the frequency of traffic accidents it is not yet possible to calculate the influence of the remainder of the parameters. However, it is doubtful whether the accuracy of the available assessments is sufficient; this will be considered later. The inaccuracy is probably due to the fact that the statistical material that forms the base for the determination of the model coefficients is insufficient, but it could also be due to other factors, which is shown by the following statement about the factors influencing the accident frequency in an intersection (Brüde and Larsson, 1986):

> "For many factors that are mentioned in the following there are more or less contradictory research results about how the traffic safety is influenced. This indicates partly great methodological problems, partly the complexity of the interaction of different factors."

Furthermore, calculation methods for the proportion of dangerous goods vehicles in a given number of accidents, as well as for how many accidents result in the dangerous goods affecting the environment, do not exist.

4.4 A Sample Case

The following calculation example has as a starting point two fictitious roads that are not very different from each other but could be alternative roads for transports of dangerous goods. The layout of the roads with bends and intersections is shown in Figure 4. As there are no hills, inclination will have no influence on the outcome of an accident. A summary of the characteristics of the roads is shown in Table 1 (alternative 1 corresponds to road 1 in Figure 4).

Table 1: Characteristics of alternative roads.

	Alt. 1	**Alt. 2**
Road width, m	9	9
Surface	asphalt	asphalt
No. of bends	7	8
No. of intersections		
3 way	3	6
4 way	2	1
Total length, km	16.53	21.80
Velocity, km/h	70	70
No. of axle-pairs	6000	4000

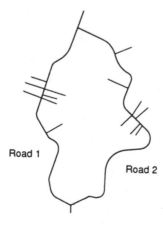

Road 1

Road 2

Figure 4: Two alternative roads.

The number of expected accidents on each road is calculated using the prediction models from Brüde et al. (1980) and Brüde and Larsson (1987). These models take into account the influence of bends, widths, inclinations and intersections. The calculations are carried out for two scenarios: first, all dangerous goods transports use alternative 1, and second, these transports are moved to alternative 2.

With the material that is the basis of the models used, it is almost impossible to calculate a correct confidence interval for the assessments, but $P(O) \pm 2 \cdot \sqrt{P(O)}$, where $P(O)$ is the estimated number of accidents, is given as an approximate 95% prediction interval in Brüde and Larsson (1981). In Figure 5 the total number of accidents calculated for each of the roads and their approximate prediction intervals are shown. These models show that no significant difference in numbers of accidents between the two roads can be observed. It might even be the case that alternative 2 is more "accident prone" than alternative 1.

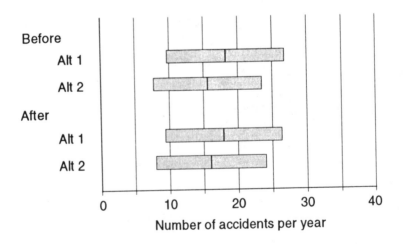

Figure 5: Approximate 95% prediction intervals.

So far, only the expected number of traffic accidents has been calculated. In order to assess the frequency of dangerous goods transport accidents another two steps have to be taken. First, the frequency of dangerous goods vehicle involvements in accidents has to be determined (regardless of whether the dangerous goods reach the environment) and, second, a determination of the extent to which accidents involving dangerous goods lead to the escape of dangerous substances that can affect the environment has to be carried out. However, as there are not yet any models for these two steps, there is no way of completing this example, i.e. of assessing the probability of dangerous goods accidents, other than to make assumptions (or guesstimates) and these assumptions will increase the already great uncertainty.

5.   CRITERIA FOR ROUTING

Starting with some proposed criteria for selecting a route for dangerous goods transports, the feasibility of using these and the results that could be achieved will be discussed. The most common criteria in the literature as well as in the public discussion are:

- that the road environment shall be exposed to minimum risk, in total, regarding health, life and property (risk criterion);

- that the most severe consequences (worst-case) of a single accident shall be as limited as possible (consequence criterion); and

- that certain selected objects must not be damaged or harmed (selective object criterion).

5.1 Risk Criterion

In selecting roads for transportation of dangerous goods, when the purpose of routing is to reduce the risk to the environment, the risk must be quantified to enable a comparison of the roads. One way to do this would be to calculate the frequency from existing statistics and from that, assess the probability. However, it has already been observed that the event of a dangerous goods vehicle meeting with an accident such that the dangerous goods cause damage to the environment occurs with such low frequency that the statistics for these events cannot be used as a basis for probability assessment. This means that other methods of assessing the probability have to be developed.

Another way to quantify risk is to use the factors (vehicles, climate, roads etc.) that influence the transport process, primarily in the physical sense, as a starting point for the probability calculus. As noted above, the relations between the influencing factors and the transport process are known to a rather limited extent, which could be explained by the complex relations and interactions. It is also difficult (if not impossible) to conduct empirical experiments that can lead to generalized theories. Nevertheless, some investigations have been performed in these areas, although several of these studies provide only a basis for qualitative conclusions. While some studies give quantitative results, these cover only the influence of some road parameters such as intersections, road width and so on (e.g. Brüde et al., 1981 and Brüde and Larsson, 1987).

As shown in the example above, these prediction models give relatively large uncertainty intervals but, most important, the link between traffic accidents and dangerous goods accidents (i.e. accidents where the dangerous substance escapes from its container or reacts in a way that causes damage to the environment) is lacking. This means that, even if the influence from the limited number of factors discussed earlier could be regarded as enough, a refinement of the prediction models is not sufficient for the probability of dangerous goods accidents to be assessed in an acceptable way, since the important step from traffic accident to dangerous goods accident is lacking.

The consequences can be calculated with relatively good accuracy, provided the prerequisites are precise enough. Along the road, however, the conditions for an accident to cause great or small damages are varying. The substance that is involved in the accident is also of importance. The conditions at the given moment of an accident are often unforeseeable. However, it is possible to identify, if not the most severe case, at least one of the most severe cases, by using a relatively limited analysis.

Thus the risk cannot be quantified, as the probability of a dangerous goods accident cannot be assessed satisfactorily. This means that the effect of the use of routing cannot be established in terms of risk level changes. From this, the conclusion can be drawn that the routing of dangerous goods is of dubious value and, thus, there is not necessarily any need for this measure. This should be seen in relation to the first criterion, that the road surroundings shall in total be exposed to the minimum risk regarding health, life and property.

5.2 Consequence Criterion

The criterion that the most severe consequences (worst-case) of a single accident shall be as small as possible is quite applicable in practice, as the consequences can be calculated relatively well. What will be achieved by this is not primarily a reduction of the number of accidents but a redistribution of the consequences and (probably) an avoidance of the most severe consequences. A reduction of the number of accidents might be achieved but the possibility of the result being an increase of the number is not small, since the probability of an accident is not part of the assessment according to this criterion.

By a straightforward use of this criterion, it is possible to bring about more accidents and this increase in the number of accidents can involve a greater total extent of damage than there would have been if the fewer, more severe accidents had been allowed to occur. This is of course not a fact at every use of this criterion. The chance that this negative result may occur can be diminished by the use of a careful qualitative probability assessment, but it cannot be eliminated.

This quite qualitative reasoning might appear axiomatic but it is important and not only for the sake of completeness. It has been noted that these facts are easily forgotten, which is why a comment on the possibility of a negative result in total, and over a longer time perspective, is judged to be important.

An absolutely objective opinion of the need and value of routing according to the consequence criterion is not possible within the scope of this study. Possibly such an opinion could be formed based on a study from the first point of view mentioned in subsection 2.2

5.3 Selective Object Criterion

The criterion that certain selected objects are not to be damaged contains implicitly that other objects can be damaged instead. The criterion is applicable in practice and no calculation methods are needed. On the other hand it requires some degree of consensus as to which objects are to be regarded as worthy of, or in need of, special consideration.

Perhaps it is not possible to protect a given object at any price, which is why an analysis of the situation, as it would be after rerouting a transport, ought to be performed. In connection with this it could be worthwhile to find out if a relocation of the object worthy of protection, as an alternative to the transfer of the dangerous goods transports, is possible.

According to this criterion, routing does have value. The existence of a specific need for routing could probably also be regarded as a fact, but the extent of the need cannot be established without knowing what objects (types) are to be considered worthy of protection and the number of these.

6.   EXPECTED DEVELOPMENT

Irrespective of what criterion is selected, methods of implementation are always needed so that the routing will have the desired effect. This problem is dealt with here and, in addition, some options that will emerge in the near future are presented.

6.1 Implementation of Routing

In order to implement routing of dangerous goods transports, there must be, in addition to a method for determining what roads are suitable, a way of informing the drivers in question of the roads that are to be used. The most effective method today is probably to use road signs and/or signposts, but this has its disadvantages. The number of signs along the roads is already quite substantial and their capacity to transfer messages/information to the road-users diminishes with the increased number of signs. Furthermore, the information is, of necessity, short and simplistic and it is not likely that transports of different substances and volumes can be dealt with in individual ways.

The means available today are relatively unrefined when it comes to the practical implementation of routing. This has not been noticed much in connection with routing of dangerous goods vehicles, however the blunt means that are used today to direct traffic flows in general have been noticed. There are some projects in progress with the objectives of using the relatively great technological advances of the past few years to direct traffic so that the infrastructure can be better utilized. To the extent that this succeeds (the prospects are generally considered as good) the opportunity to overcome the directing problems in connection with the routing of dangerous goods transports will be increased.

6.2 Dynamic Traffic Control Systems

The traffic situation is influenced by both the demand for and the availability of roads for transport. The trend of the demand for road transports is growing, while the short-term variation (e.g. during the day) can be very great, especially in the cities. The congestion of traffic is so severe today that great resources in time and capital are lost and, in view of the expected increase in demand, corrective measures will have to be taken. Adapting to the peaks of short-term variation by extending the road network may no longer be possible. However there remain three alternatives: to reduce the demand, to lower the short-term variation and to utilize the existing infrastructure more efficiently. The first measure, to reduce demand, is economically and politically very difficult to carry out and, in addition, a very long-term measure as almost the whole structure of society would have to be changed. The other two measures can be regarded as feasible in a much shorter perspective and then they could be carried out with the help of a central traffic control system.

One task for a traffic control system will be to by various means distribute the demand for road resources in time and the other will be to optimize the traffic flow in the infrastructure for transport of people and goods, both collectively and individually. It is possible to supplement such a system with components especially for the guidance of dangerous goods vehicles.

With the type of system described above, the selection of roads could be done dynamically, in contrast to the way it is done today, and the selection could be dependent on several factors such as the traffic situation at large, time of the day, special public arrangements in the vicinity of the road and, not least, the substance that is being transported. The vehicle, equipped with a computer, would receive information about the way to take depending on the type of the vehicle and the goods.

When it is no longer necessary to calculate some form of average risk level covering a whole year (or more), it is quite possible that the requirements of the probability assessments can be lowered and more qualitative assessments of the probability can be used. When great differences between various possible consequences exist, minor differences in the probabilities mean less to the differences in risk. This means that two roads that cannot be differentiated as regards risk on a yearly basis can very well show such great differences in possible consequences at different times of the day that, in spite of the lack of quantitative assessments of the probabilities, an acceptable qualitative risk assessment can be made and the roads can be selected for dangerous goods transports alternatively depending on the time of day.

6. REFERENCES

ABKOWITZ, M., & CHENG, P.D-M. (1988). Developing a Risk/Cost Framework for Routing Truck Movements of Hazardous Materials. *Accident Analysis and Prevention*, 20 (1), 39 - 51.

ARNBERG, P.W., CARLSSON, G., & MAGNUSSON, G. (1978). *Inverkan av vägojämnheter*. VTI Meddelande nr 95.

BJÖRKETUN, U. (1982). *Samband mellan vägbeläggningar, väderlek och trafikolyckor 1977*, VTI Meddelande nr 317.

BRÜDE, U., & LARSSON, J. (1986). *Faktorer som för korsningar påverkar antal olyckor, olyckskvot och skadeföljd-olyckskostnad*. VTI PM 1984-06-21 rev. 1986-10-27.

BRÜDE, U., & LARSSON, J. (1987a). *Användande av prediktionsmodeller för att eliminera regressionseffekten*. VTI Meddelande 511.

BRÜDE, U., & LARSSON, J. (1987b). *Förskjutna 3-vägskorsningar på landsbygd*. VTI Meddelande nr 544.

BRÜDE, U., LARSSON, J., & THULIN, H. (1980). *Trafikolyckors samband med linjeföring*. VTI Meddelande nr 235.

DROSTE, B., & SCHOEN, W. (1988a). Full Scale Fire Tests with Unprotected and Thermal Insulated LPG Storage Tanks. *Journal of Hazardous Materials*, 20, 41-53.

DROSTE, B., & SCHOEN, W. (1988b). Investigations of Water Spraying Systems for LPG Storage Tanks by Full Scale Fire Tests. *Journal of Hazardous Materials*, 20, 73-82.

JUNGHARD, O. (1989). *Prognosmodeller för olyckor och skadeföljd behandlade med korsvalidering och bootstrap*. VTI Meddelande 579.

LAMM, R., & CHOEIRI, E.M. (1987). Recommendations for Evaluating Horizontal Design Consistency Based on Investigations in the State of New York. *Transportation Research Record 1122*, TRB, Washington D.C.

LAUTKASKI, R. (1986). *Use of Risk Analysis in Enhancing the Safety of Transporting Hazardous Liquefied Gases*, Transportation Research Board, pp 59 - 63, Washington D.C.

MESLIN, T.B. (1981). Assessment and Management of Risk in the Transport of Dangerous Materials: The Case of Chlorine Transport in France. *Risk Analysis*, 1 (2), 137 - 141.

MÖLLER, S. (1988). *Beräkningar av olyckskvot vid olika väglag vintertid med hjälp av schablon.* VTI Meddelande nr 584.

POLUS, A., & DAGAN, D. (1987). Models for Evaluating the Consistency of Highway Alignment. *Transportation Research Record 1122*, TRB, Washington D.C.

SCHANDERSSON, R. (1986). *Samband mellan trafikolyckor, väglag och vinterväghållningsåtgärder.* VTI Meddelande nr 496.

Establishing Credible Risk Criteria for Transporting Extremely Dangerous Hazardous Materials

R. A. Sivakumar
Corporate Research and Development
United Airlines
Chicago, IL 60666 U.S.A.

R. Batta
M.H. Karwan
Department of Industrial Engineering
State University of New York at Buffalo
Buffalo, NY 14260 U.S.A.

ABSTRACT

A measure of risk that is more appropriate while routing extremely dangerous hazardous materials is to minimize the expected risk, given that an accident occurs. This paper introduces this alternate measure of risk, and discusses efficient solution procedures for determining an "optimal" path for a single shipment for such a hazardous material.

1. INTRODUCTION

The past two decades has seen our society recognize the fact that the transportation of Hazardous Materials (HM) is a necessary evil that has to be endured. This has led researchers from the Operations Research community, among others, to develop models to ship these commodities in the least risky manner. Here, risk could be considered to be the number of fatalities, or dollar damage that is caused due to the occurrence of an accident. Since, in most of the cases, risk is encountered only when an accident occurs--a probabilistic event --it is a random variable, and it makes sense to talk about expected risk. The primary concern facing the HM transportation planner is to minimize the risk, in the event of an accident *en route*. Almost all of the literature in this field has tended to reflect this view, and has focused on determining route(s) that minimize the *a priori* expected risk (see, for example, Batta and Chiu, 1988).

The models that minimize the *a priori* expected risk implicitly assume that the shipments of the HM in question will continue on the same route, even after the occurrence of an accident. With the accident probabilities being very low (these are estimated to be of the order of 1 *per* million truck miles), the route that minimizes this objective is *expected* to be the best route over these million miles, or so--typically comprising a very large number of shipments. However, this model is highly inadequate for the transportation of those HM which, when involved in a single accident, can prove to be highly catastrophic. This inadequacy stems from the fact that two *a priori* events of equal probability can have very different total costs if one of them involves the possibility of a catastrophic event. Also, there is often a complete re-evaluation of the entire routing scenario after the occurrence of the accident, possibly resulting in a revision of the routing parameters.

Any model that is considered for extremely dangerous hazardous materials--those that typically result in a large number of casualties when involved in an accident--must make use of the fact that the current routing scenario ends at the occurrence of the first accident. Further, all shipments of these materials must be made through route(s) that minimize the risk if an accident occurs *en route*. In other words, the problem is to determine a path that minimizes the expected risk, given that an accident occurs. In this paper, we introduce and analyze this model. Glickman and Sherali (1991) have introduced a similar model, where the objective is to determine a route that minimizes the expected number of fatalities given that the number of fatalities exceed a certain number. Before proceeding further we note that our approach could be enhanced by the consideration of uncertainty; for instance by the use of an extra variable such as the probability of nothing happening (safe journey) or the unconditional root-mean-square risk. This is a suggested future research topic.

After formally developing the model to route extremely dangerous hazardous materials, we proceed to discuss some solution procedures.

2. PROBLEM DEFINITION

2.1 Framework

Let $G=(N,A)$ be an undirected, planar transportation network, where N is the set of nodes (intersections) and A is the set of links (roads). Let O and D be the origin and destination nodes respectively. If P is a path from O to D, P can be represented by a sequence of unique links $(l_1, l_2, ..., l_m)$. Let p_i be the probability that an accident occurs in link l_i, and r_i be the risk from such an accident.

The *a priori* expected risk along the path P, is given as follows:

$$ER(P) = p_1 r_1 + (1 - p_1) p_2 r_2 + ... + (1 - p_1)(1 - p_2) ... (1 - p_{m-1}) p_m r_m \quad (1)$$

In the above expression, each term gives the expected risk along that particular link. Clearly, an accident can occur at a link, only if the previous links were traversed in a safe manner. It is this probability of safe traversal that is given within the parantheses. Similarly, the probability of an accident occurring somewhere along P can be given as follows:

$$PR(P) = p_1 + (1 - p_1)p_2 + ... + (1 - p_1) ... (1 - p_{m-1}) p_m \quad (2)$$

Now, the expected risk given that an accident occurs somewhere along the path P is the ratio of the expressions given in equations (1) and (2). Since this provides the expected risk, conditioned on the event of an accident, we also refer to it as the conditional risk. It is statistically known that the accident probabilities typically are in the order of 10^{-7} truck miles. Thus the $(1-p_i)$ terms and their higher order products can be ignored without sacrificing much accuracy. After doing this, the conditional risk along P, which we call $CR(P)$, reduces to the following:

$$CR(P) = \frac{p_1 r_1 + p_2 r_2 + ... + p_m r_m}{p_1 + p_2 + ... + p_m} \quad (3)$$

The two measures discussed above--the *a priori* expected risk, and the conditional risk--are illustrated *via* a simple numerical example, shown in Figure 1. Each link in the network has *two* measures--the number of fatalities, and the probability of accident. The path 0--1--3--D is the least *a priori* expected risk path, but it has the highest conditional risk. On the other hand, the path 0--1--2--3--

D has the least conditional risk, but the highest *a priori* expected risk.

The results in Figure 1 have a simple intuitive explanation. Taking path 0--1--3--D causes one to traverse links that have a relatively high consequence given an accident occurs on them but have a low accident probability. On the other hand, taking path 0--1--2--3--D causes the reverse to be true. This brings out an interesting fact.

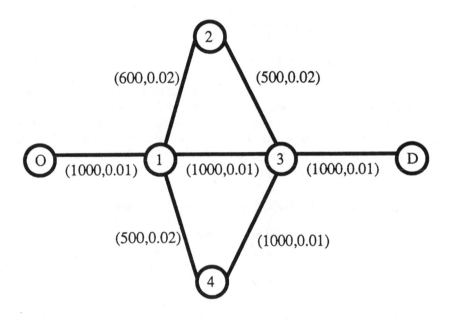

Path	P[Accident]	Expected Risk	Expected Risk Given an Accident
0-1-3-D	0.03	30	1000
0-1-4-3-D	0.05	40	800
0-1-2-3-D	0.06	42	700

Figure 1: Example network.

Taking path 0--1--2--3--D is certainly good from a perspective of our modeling scenario where the routing ends given the first accident, when one takes only consequence into consideration. However, taking the path 0--1--2--3--D is likely to cause this accident to occur after a fewer number of shipments, since the accident probability while traversing the path is relatively high. Therefore, a compromise situation is probably warranted, where a constraint is put on the path's total accident probability. The mathematical formulation presented in the next section does not do this, but it is our intention to incorporate this constraint in our future work.

2.2 Mathematical Formulation

Given the above definition of conditional risk along a specific path, the objective is to determine a path P^* that minimizes $CR(P)$ over all loopless paths. The network formulation of the Conditional Risk Problem (CRP), thus can be written as follows:

$$(CRP): \text{Minimize} \quad \frac{\sum_i \sum_j e_{ij} x_{ij}}{\sum_i \sum_j p_{ij} x_{ij}}$$

$$\text{Subject to} \quad (I) \quad \sum_j x_{ij} - \sum_j x_{ji} = \begin{cases} 1 \text{ if } i = 0 \\ -1 \text{ if } i = D \\ 0 \text{ otherwise} \\ \text{for } i = 1...|N| \end{cases}$$

$$(II) \quad \{ x_{ij} \mid x_{ij} = 0 \text{ or } 1, \sum_j x_{ij} \le 1\}$$

In the above formulation $e_{ij} = p_{ij} r_{ij}$ is the expected risk along link (i,j), and the binary variable x_{ij} is 1 if link (i,j) lies in the solution path, and 0 otherwise. Constraint sets (I) and (II) are the well known loopless path constraints; strictly speaking, sub--tour elimination constraints (see Nemhauser and Wolsey, 1988) have to be added to the above formulation to prevent cycles from occurring outside the path. It is necessary to explicitly restrict the paths to be loopless, as there can be instances where it is mathematically preferable to loop, just to reduce the conditional risk. The desire to loop can be seen from our example, where looping around 1--2--3--4 causes a decrease in the objective of conditional risk---clearly such a strategy, which increases the accident probability by looping on a road segment, is not feasible to implement. We note, however, that it is possible to construct examples in which looping can cause an increase in the objective.

3.   SOLUTION PROCEDURES

The formulation presented in the previous section for the conditional risk problem falls into the category of "ratio problems" in the
operations research literature (see Gondran and Minoux, 1984).
Determination of a loopless path that minimizes the ratio is an NP--
Hard problem, and hence, for a general case, cannot be said to have
been solved unless a complete enumeration has been performed.

A naive approach that immediately comes to mind is to determine
the shortest loopless paths in the increasing order of the *a priori*
expected risk (see Yen, 1971), and from these choose the path that
minimizes the conditional risk. However, for a general case, there
is no correlation between the *a priori* expected risk and the
conditional risk, and hence a complete enumeration has to be done
before one can claim to have solved the conditional risk problem--
an approach that can be very prohibitive even for moderately sized
networks. Another alternative is to relax the binary restriction on
the x_{ij}'s (by converting the constraint set (II) to $0 \le x_{ij} \le 1$), and then
use the Charnes and Cooper approach (Bazaraa and Shetty, 1979) to
convert the resultant fractional program into a linear program. A
branch and bound procedure can then be employed, by branching on
fractional variables. The advantage of this procedure is that it provides a lower bound on the (CRP). A raw branch and bound procedure, like this one, may be expensive based on previous experience.

A bisection search procedure was used to heuristically solve the
(CRP). A successful application of a similar procedure has been
reported for the problem of determining the least ratio cycle in a
graph (Lawler, 1981). Briefly speaking, this procedure starts off by
determining a valid lower and upper bound for the optimal ratio. An
estimate for the optimal ratio is then 'guessed' by choosing the
midpoint in this interval. This estimated ratio is then compared to
the optimal ratio by solving a shortest path problem with modified
cost coefficients. The new interval is then determined based on
whether the estimated ratio is more than, or less than the optimal
ratio, based on the sign of the length of the shortest path. This procedure is repeated until the interval cannot be pruned any further, or
the optimal ratio is achieved. An extensive explanation of this
procedure can be found in Sivakumar, Batta, and Karwan (1991).

The bisection search procedure was tested on a network
representing Albany County of the State of New York. The risk and

probability of accident occurrence were randomly generated for each link to obtain 100 'realistic' data sets. The histogram of the frequency distribution of the gap between the final upper bound (best feasible solution) and the final lower bound given in Figure 2.

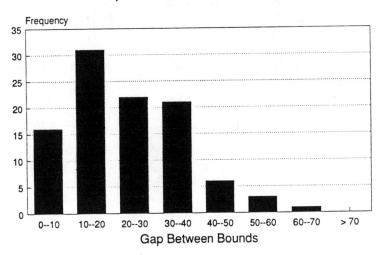

Figure 2: Performance of bisection search procedure.

4. REFERENCES

BATTA, R., & CHIU, S.S. (1988). Optimal obnoxious paths on a network: Transportation of hazardous materials. *Operations Research*, 8, 84-92.

BAZARAA, M.S., & SHETTY, C.M. (1979). Nonlinear programming: Theory and algorithms. New York: John Wiley and Sons.

GLICKMAN, T.S., & SHERALI, H.S. (1991). Catastrophic transportation accidents and hazardous materials routing decisions. *Probabilistic safety assessment and management conference*, Beverly Hills, California.

GONDRAN, M., & MINOUX, M. (1984). Graphs and algorithms. Chichester: John Wiley and Sons.

LAWLER, E.L. (1976). Combinatorial optimization: Networks and matroids. New York: Holt Rinehart and Winston.

SIVAKUMAR, R.A., BATTA, R., & KARWAN, M.H. (1991). A network-based model for transporting extremely hazardous materials. *Working Paper*, Department of Industrial Engineering, State University of New York at Buffalo.

YEN, J.J. (1971). Finding the K shortest loopless paths in a network. *Management Science*, 17, 712-716.

Chapter 3: Application of Simple Risk Assessment Methodology

A Simple Consequence Model Based on a Fatality Index Approach

Lars H. Brockhoff
Department of Chemical Engineering
Technical University of Denmark, Building 229
DK-2800 Lyngby
Denmark

H.J. Styhr Petersen
Institute for Systems Engineering and Informatics
Commission of the European Communities
Joint Research Centre Ispra, T.P. 321
21020 Ispra, (VA)
Italy

ABSTRACT

In this paper, a simple and transparent consequence model for various class 2 substances (e.g., chlorine, ammonia and LPG) and class 3 substances (e.g., gasoline) is proposed. The basic assumption in the model is that the number of fatalities from a given release will increase with an increased amount of substance released --- the fatality index concept. The parameters for the model can be estimated from historical accident data on releases and avoid a large number of the assumptions necessary in traditional consequence models.

The model proposed estimates consequences for three different population density classes: rural, semi-urban (or industrial) and urban. Unfortunately, the accident data do not permit this estimate, and the overall average of the fatality index is therefore used as a first approximation for semi-urban (industrial) areas. The fatality index for the other two classes is estimated based on population density statistics.

1. INTRODUCTION

Traditional consequence models, including release, dispersion and vulnerability calculations, can be useful for estimation of the risks related to fixed installations storing, producing or using hazardous materials, if care is taken when making the necessary assumptions, but large uncertainties are found in the results (Amendola et al., 1991).

For accidents related to transportation of hazardous materials, further (uncertain) assumptions have to be made, because the location of the accidents are not known in advance. This introduces even larger uncertainties in results related to transportation. In this paper a simple model for human consequences (number of fatalities) based on historical data is presented. The model avoids some of these uncertainties, and is believed to be more transparent than traditional consequence models.

The limits of this approach are that phenomena like the wind speed and the wind direction, the release duration and other accident specific characteristics are not taken into account.

For accidents concerning transport, it is clear that the wind speed and direction, for instance, are very important for the emergency team and for the surrounding population. However, when averaging over all accident locations, it seems a reasonable assumption that the population is not clustered in one given direction from the transport corridor.

The parameters of the model are estimated for toxic releases of chlorine and ammonia and for flammable liquefied hydrocarbons (class 2.3) and flammable hydrocarbons (class 3 and 3.3). The consequence model is thus useful for the vast majority of the substances transported and is used in a decision support system for routing transports of dangerous goods. The relevant models and an adherent computer program is developed during a Ph. D. study undertaken at the Joint Research Centre of the European Community in collaboration with Department of Chemical Engineering, Technical University of Denmark.

2. ELABORATING ON THE PROBLEM

The consequences of a total release from a transportation tanker will depend on a number of parameters e.g., the type and quantity of the substance and the population density in the vicinity of the accident. For flammable substances ignition is essential for the possibility of severe consequences to humans. The event tree shown in Figure 1 shows a possible structure of the problem, and facilitates estimation of the parameters. The risk of transporting dangerous goods can thus be expressed by the probabilities of the single events in the event tree shown in Figure 1.

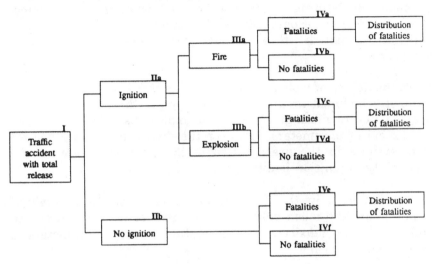

Figure 1:   Event tree for analysis of human consequences of releases from transportation accidents.

In this paper the releases of the toxic gases chlorine and ammonia, the flammable liquefied gases (LPG, propane etc.), the highly flammable liquid hydrocarbons (gasoline, jet fuel etc.) and the moderately flammable liquid hydrocarbons (diesel oil, domestic fuel etc.) will be analyzed.

Obviously the number of fatalities will depend on the number of people in the vicinity of the accident. In the study of toxic releases (Brockhoff et al., 1992) the semi quantitative distinction U (urban), I (industrial, semi-urban) and R (rural) was introduced. This distinction will be adopted also in this paper and is discussed in section.

The probability of ignition (level II in Figure 1) was for toxic releases of e.g., chlorine and ammonia assumed to be zero (Brockhoff et al., 1992). In parallel it will be assumed that releases of the above mentioned flammables will only result in fatalities if the substance is ignited.

The distinction between fires and explosions (level III in Figure 1) was made because they a priori were thought to have different consequences, and because the three types of substances were thought to differ in their willingness to explode. It is however difficult to distinguish between the two processes. Not only are observers of an accident believed to report "an explosion" if at any time during the accident there was a "boom". Also on a theoretical level there are problems in defining what is a fire and what is an explosion. Furthermore a major part of the accidents are believed to be hybrids, which are initiated with a fire and later produce an explosion, or vice versa.

Since the degree of violence with which the three substances burn is implicitly included in the two later steps of the event tree, it is decided not to distinguish between explosions and fires in the estimation of parameters. However the distinction is included for completeness in the presentation of the data. In section 6 the probability of ignition is discussed.

The parameters in the distribution function of the number of fatalities are the mean value and the variance for each substance and population density class. Estimation of these parameters is presented in section 4.

3. THE DATA

A list of 882 transportation accidents (Haastrup and Brockhoff, 1992) and 1190 accidents at fixed installations has been compiled as background of the project. Of these, 93 included chlorine and 77 included ammonia. For the current purpose it is necessary as a minimum to know the number of fatalities. The available data for calculation of fatality indices for chlorine and ammonia divided in the three population density groups are summarized in the Tables 1 and 2.

The analysis of fire and explosion accidents includes road and rail transportation accidents involving diesel oil, fuel oil, gasoline, jet fuel etc. and combustible liquefied gases (LPG, LNG, propane etc.) extracted from Haastrup and Brockhoff (1992). In total 178 accidents from this list are included (from now the JRC-list).

In France very comprehensive lists of accidents are published every year (since 1988) (France, 1988 90), which are thought to be very complete in the degree of reporting. Three volumes (a total of 1538 accidents) were available for the current analysis, including 51 relevant accidents (from now the MT-list).

Table 1: The 36 accidents involving chlorine for which the number of fatalities was recorded. Fourteen of these resulted in one or more fatalities.

	Accidents with recorded number of fatalities	Tons released (if 1 or more fatalities)	Number of fatalities
U	4 of 7	98	14
I	1 of 3	30	19
R	1 of 3	29	1
?	8 of 23	124	121
Total	14 of 36	281	155

Table 2: The 16 accidents involving ammonia for which the number of fatalities was recorded. Six of these resulted in one or more fatalities.

	Accidents with recorded number of fatalities	Tons released (if 1 or more fatalities)	Number of fatalities
U	4 of 7	151	37
I	1 of 2	180	2
R	0 of 6	–	–
?	1 of 1	2	2
Total	6 of 16	333	41

Tables 3 and 4 show how the accidents in the two lists were distributed according to transportation mode, type of consequence and type of substance.

Table 3: The 178 relevant accidents from (Haastrup and Brockhoff, 1992). The numbers in parentheses are the number of accidents resulting in one or more fatalities.

	ROAD			RAIL		
	spill	fire	expl.	spill	fire	expl.
Diesel	16 (2)	2 (2)	1 (0)	5 (1)	5 (3)	2 (0)
Gasoline	19 (0)	20 (11)	8 (6)	6 (1)	6 (4)	5 (1)
LPG	3 (0)	8 (5)	23 (22)	4 (1)	15 (0)	30 (9)

Table 4: The 51 relevant accidents from (France, 1988 90). The numbers in parentheses are accidents resulting in one or more fatalities.

	ROAD			RAIL		
	spill	fire	expl.	spill	fire	expl.
Diesel	22 (0)	0 (0)	0 (0)	0 (0)	0 (0)	0 (0)
Gasoline	18 (1)	2 (0)	2 (1)	1 (0)	1 (0)	0 (0)
LPG	2 (0)	0 (0)	0 (0)	2 (0)	1 (0)	0 (0)

4.   THE DISTRIBUTION OF THE NUMBER OF FATALITIES

As previously mentioned the distribution of the number of fatalities is characterized by a mean value and a variance. The fatality index approach offers a possibility to include the relevant aspects in the estimation of the mean number of fatalities. The fatality index was introduced by Marshall (Marshall, 1977) and the possible use of it as an approach to human consequence models was elaborated on by the authors in a previous paper (Brockhoff et al., 1992). The fatality index is defined by the equation:

$$N = \beta \times W^n = W \times [\beta \times W^{n-1}]\tag{1}$$

where N is the number of fatalities, W is the amount released and $[\beta \times W^{n-1}]$ is the index.

The equation expresses that the number of fatalities depends on a constant β and on the amount of substance released (to an extent determined by the exponent n). The value of β will depend on the substance and on the population density at the location of the accident. Ideally the values of the different factors should be extracted from the data. However, data are sparse as can be recognized from the previous section, and the values will therefore be estimated, based partly on theoretically reasoning and partly on data supporting the reasoning.

4.1 The exponent n

For toxic releases it was shown that n could be either 0.75 or 1, which gave results with insignificant differences. The value 1 was adopted because of the simplicity gained (Brockhoff, et al., 1992). The data given in Tables 6 and 7 show that the number of fatalities increases with increasing quantities released, but do not permit an estimation of the value of n. Marshall (1977) argues that theoretically the value should be 0.67 for explosions, but finds that 0.5 is closer to accident data. Therefore n = 0.5 is adopted, and it should be noted that this value (since it is a synonym of the square root) preserves some transparency.

4.2 The effect of population density

The value of ß is, as already mentioned, a function of the population density at the site of the accident. The ratios between the three population density classes should ideally be estimated from the accident data. The sample is however far too small for such an estimation. As a consequence it was chosen to base the estimate on general population density estimates.

For toxic releases it was argued that the ratio in mean number of fatalities between urban and rural areas should be approximately 1:64 (Brockhoff et al., 1992). A ratio of 1:100 (and thus 1:10:100) is however thought to be within the uncertainty of the population density data and provides a more transparent model (Funtowicz, 1991). It was therefore chosen to use the ratio 1:100 for toxic releases.

The ratio for fires and explosions is a priori expected to be smaller as mainly the population at the road, rather than in the vicinity matters. Various scenarios can be (and have been) constructed to analyze the effect of the population of the road e.g.:

- The road width is either 10 m or 30 m.
- Three possible "damage" distances: 50 m, 150 m and 300 m.

- In urban areas the population of the road is either: 1 car per 10 m, with 2 persons and 1 pedestrian per meter (i.e., 12 persons per 10 meters) or 1 car per 20 m and 1 pedestrian per 5 meter (i.e., 3 persons per 10 meters). The general population density in urban area is 1600 pers/km^2.

- In rural areas the population of the road can be expressed: 1 car per 20 m, with 2 persons and 1 pedestrian per 10 meter (i.e., 2 persons per 10 meters) or 1 car per 20 m and no pedestrians (i.e., 0.67 persons per 10 meters). The general population density being 25 pers/km^2.

The ratio in population density estimated from the scenarios given above, imply a ratio between urban and rural areas of a factor between 2 and 25 as can be seen from Table 5 .

Table 5: The ratio in population density for three "damage radii", two different width of the road and different population density at the road.

Width of road	Damage radius	Persons per 10 m urban road	Persons per 10 m rural road	Persons in urban	Persons in rural	Ratio
10 m		12	0.67	131	7	19
	50	3	2	41	20	2
30 m	meters	12	0.67	128	7	19
		3	2	38	20	2
10 m		12	0.67	467	22	21
	150	3	2	197	62	3
30 m	meters	12	0.67	458	22	21
		3	2	188	62	2
10 m		12	0.67	1190	48	25
	300	3	2	650	127	5
30 m	meters	12	0.67	1170	47	25
		3	2	631	127	5

Based on Table 5 and considering the transparency, a ratio of 1:10 (and thus 1:3:10) is proposed for the fire/explosion model.

4.3 The value of β

The minimum criteria for using an accident for assessment of the value of ß, is that both the amount released and the number of fatalities can be estimated. Section showed that 14 accidents involving chlorine and 4 accidents involving ammonia fulfil these criteria. Tables 6 and 7 show that for fire and explosions both the parameters were estimated by the source for 18 LPG accidents, 4 gasoline accidents and zero diesel accidents.

Table 6: The recorded liquefied petroleum gases accidents (see Table 3) for which both the amount released and the number of fatalities were recorded.

Date	Mode	Event	Amount (ton)	Fatalities
280659	rail	explosion	18	26
250169	rail	explosion	63	3
220172	rail	explosion	65	1
190774	rail	explosion	69	7
240278	rail	explosion	45	25
300579	rail	explosion	30	1
180143	road	fire	8	5
231070	road	fire	4	2
310190	road	fire	<18	6
196065	road	explosion	14	10
250762	road	explosion	14	10
150770	road	explosion	7	1
090372	road	explosion	18	2
210972	road	explosion	14	4
010273	road	explosion	19	9
110778	road	explosion	22	216
150778	road	explosion	36	100
160778	road	explosion	23	15

Table 7: The recorded gasoline accidents (see Table 3) for which both the amount released and the number of fatalities were recorded.

Date	Mode	Event	Amount (ton)	Fatalities
240977	road	fire	25	7
140280	road	fire	30	1
150981	road	fire	30	5
060787	road	fire	25	1

For the remaining 19 LPG, 19 gasoline and 8 diesel oil accidents (where the number of fatalities were recorded, but not the quantity) it was assumed that a release during rail transportation involved 60 tons and during road transportation 25 tons.

The three ß values (for U, I and R) cannot be estimated from the data. For toxic releases it was argued that I would equal the overall average value of , and that R and U should be 8 times smaller and bigger respectively. The approach is adopted in this paper, using the ratios 10 and 3 for toxic releases and fires/explosions respectively, as discussed in the previous section.

The ß average values are calculated, using the formula:

$$\text{ß average} = \frac{\sum\limits_{i=1}^{r} N}{\sum\limits_{i=1}^{r} W^n} \qquad (2)$$

where r is the number of accidents, N is the number of fatalities in each accident, W^n is the quantity released in the power n.

For chlorine and ammonia the $ß_{average}$ values are estimated to be 0.55 and 0.12 respectively (based on Tables 1 and 2). For LPG the estimated $ß_{average}$ is 2.9, for gasoline 1.3 and for diesel $ß_{average} = 1.3$.

The ß values for the five substances can now be estimated for each of the three population density classes. These values are shown in Table 8.

Table 8: The ß values of the five fatality indices estimated for each population density class.

	Urban	Industrial	Rural
Chlorine	5.5	0.55	0.055
Ammonia	1.2	0.12	0.012
LPG	8.7	2.9	1.0
Gasoline	3.9	1.3	0.43
Diesel	3.9	1.3	0.43

4.4 The Variance

A detailed analysis of the distribution of the values was made for the toxic releases where the population density is known for a large fraction of accidents (Brockhoff et al., 1992) and the parameters of a S-shaped curve were estimated. It was however found that a more simple model could be used instead. The simple model expresses that 1 out of 10 accidents have consequences 10 times more severe (than average) and 1 out of 10 consequences are 10 times less severe.

The data for fires and explosions do not reject the application of the same model to these types of consequences. It is therefore proposed to use the discrete distribution of 83.3% for the average (estimated) ß, 8.3% for 10 x ß and 8.3% for 0.1xß.

5.   THE PROBABILITY OF (ANY NUMBER OF) FATALITIES

The previous section analyzed the distribution of the number of fatalities, under the restriction that the release resulted in (any number of) fatalities. The probability of getting an accident which results in fatalities will be analyzed in this section.

The probability of fatalities is assumed to be a function of both the population density and the toxicity/flammability of the substance. Population density was however rarely reported for fire and explosion accidents, but significant differences between road and rail accidents were observed. These differences are assumed to reflect differences in the mean population density of these two transportation modes.

Since the two above mentioned factors are also determining the value of ß it is proposed that the probability of fatalities is a function of the value of the fatality index.

For chlorine and ammonia, the probabilities of fatalities are read directly from Tables 1 and 2 since the samples are regarded representative of accidents involving these two substances (Brockhoff et al., 1992). For flammables, the problem is slightly more complex. The relevant data available (from the JRC-list,Table 3) for flammables are shown in Table 9.

Table 9: The probability of fatalities, given an accident with release and ignition (i.e., fire or explosion).[1]

		JRC-list (all accidents)		JRC-list (large release accidents)	
		Road	Rail	Road	Rail
Diesel,	Observations	2 of	3 of	- -	- -
fuel	Percentages	367%	743%		
Gasoline	Observations	17 of 28	5 of 11	4 of 9	1 of 1
	Percentages				
LPG etc.	Observations	27 of 31	9 of 45	12 of 13	7 of 10
	Percentages				

[1] The second row includes only those accidents for which the amount was recorded (and was large).

The probability of fatalities is assumed to be a function of both the population density and the toxicity/flammability of the substance. Population density was however rarely reported for fire and explosion accidents, but significant differences between road and rail accidents were observed. These differences are assumed to reflect differences in the mean population density of these two transportation modes.

Recognizing the number of observations in each class, it is concluded that the probability of (any number of) fatalities for both diesel oil and gasoline is 60% for road and 45% for rail accidents respectively. For LPG accidents the probability for road is estimated to be 90%. For railway accidents, there seems to be a tendency to report even small releases (without explicitly giving the quantity released). A coherent estimate would therefore be a 70% probability of accidents with fatalities. The differences between road and rail accidents are assumed to originate from the different composition of population density areas passed through.

The JRC-list on which these estimates are based is however not regarded complete. More complete data could be extracted from the MT-list (Table 4), but the level of detail would be lost. The overall average probability of fatalities (given an ignition) can however be estimated to be 17% (1 of 6 accidents). In comparison the average values estimated from Table 9 are 50% (63 of 125 accidents) and 73% (24 of 33 accidents) respectively. The estimated probabilities of fatalities for each of the three substance classes are thus between 3 and 4 times too large. It is conservatively chosen to reduce the estimates to approximately 50% of their values.

Tables 10 and 11 show the estimated probabilities of fatalities together with the estimated fatality indices. Based on the tables it is possible to express the probability of fatalities as a function of the fatality index.

Table 10: The probabilities (percent) of fatalities for a toxic release of 20 tons, divided in population density classes.

	Urban		Industrial		Rural	
	Prob. fat	ß	Prob. fat	ß	Prob. fat	ß
Chlorine	57%	5.5	33%	0.55	33%	0.055
Ammonia	57%	1.2	- -	0.12	0%	0.012

Table 11: The probabilities (percent) of fatalities for the three flammable substances (20 tons release), for road and rail accidents.

	Road		Rail	
	Prob. fat	$20^{-0.5}$ x ß	Prob. fat	$20^{-0.5}$ X ß
LPG	45%	0.65	35%	0.65
Gasoline	30%	0.29	20%	0.29
Diesel	30%	0.29	20%	0.29

The estimate of the probability of (any number of) fatalities, can now be related to the fatality index. Figure 2 shows the probability of fatalities as a function of the fatality index (remembering the definition of fatality indices), assuming 20 ton release as a median of releases.

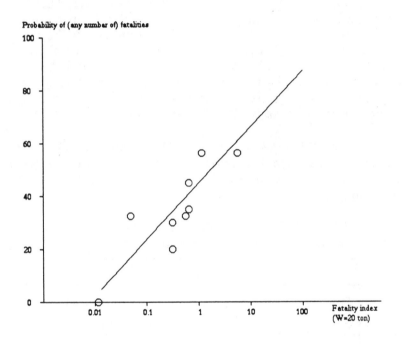

Figure 2:   The probability of fatalities as a function of the fatality index, assuming 20 ton release. A possible straight line is indicated.

Though recognizing that the function described by the points should be a S-shaped curve, the more simple solution of a straight line is proposed. The line shown in Figure 2 is characterized by the equation:

$$P\ (fat) = 44 + 22 \times log_{10} (\beta \times W^{n-1})\ for\ 0.01 \leq \beta \times W^{n-1} \leq 350 \qquad (3)$$

6.   THE PROBABILITY OF IGNITION

As already mentioned it is assumed that the probability of igniting the gases chlorine and ammonia is zero.

For the flammables an evaluation of the fraction of accidents resulting in ignition, should ideally be performed using a complete list. The MT-list (Table 4) which is regarded relatively complete does however only include 49 accidents and is thus not suitable for observing differences between the three substance-types. The JRC-list however serves this purpose. Table 12 show the relevant data.

Table 12:   The relevant data for assessment of the probability of ignition, given an accident with a large release.

	MT-list			JRC-list		
	Road	Rail	Total	Road	Rail	Total
Diesel, fuel	0 of 22	0 of 0	0 of 22	3 of 19	7 of 12	10 of 31
Gasoline	4 of 22	1 of 2	5 of 24	28 of 47	11 of 17	39 of 64
LPG etc.	0 of 2	1 of 3	1 of 5	31 of 34	45 of 49	76 of 83

The available data do (as already discussed) not facilitate a distinction between different amounts released or different types of areas to which the release occurred (though both factors influence the probability of ignition). Furthermore the data in Table 12 do not expose significant differences between road and rail accidents.

The (relatively complete) MT-list implies that approximately the same overall number of accidents involving diesel and gasoline are observed whereas LPG accidents are observed 4 to 5 times more seldom. Furthermore this list implies that the probability of ignition in an gasoline accident is approximately 20% (5 ignitions of 24 accidents), which is assumed to determine the absolute level.

In the JRC-list the number of diesel accidents is half as large as gasoline accidents (31 versus 64). As the MT-list is far more complete, this is interpreted as if they are reported (and not occurring) more rarely and that the 30 "missing" accidents were without consequences. The proportion between the three types of substances is thus 16% : 61% : 92% or approximately 1 : 4 : 6.

Assuming an absolute level determined by the MT-list of 20% ignitions for gasoline accidents, and a ratio of 1 : 4 : 6 between the three substances, Table 13 can be compiled.

Table 13: The probability of ignition, given an accident with a large release of one of the three substance types.

	Probability of Ignition
Diesel	5%
Gasoline	20%
LPG etc.	30%

7. DISCUSSION

The paper presents a model for estimation of human consequences of a transport accident involving hazardous goods. The parameters are estimated for the toxic gases chlorine and ammonia and for flammable hydrocarbons, class 2.3 (LPG etc.), 3.3 (gasoline etc.) and 3 (diesel, fuel etc.). These parameters are summarized in Table 14.

An overview of the differences in the level of consequences for the different types of substances, is better provided by a frequency-consequence curve (fN-curve). Figure 3 shows the fN-curves for chlorine, ammonia, LPG and gasoline assuming that an accident with a 20 ton release has occurred (in an area which in probabilistic terms consists of 20% urban, 30% industrial and 50% rural area).

The fN-curves are thought to reflect reality reasonably well. However, the model is still regarded as a pilot model and further data (i.e. accident reports, where the type and quantity of the substance, the number of casualties and the population density are all reported) are required. These may be provided if agreement on the proposal from the European Communities (EC, 1991) which requires detailed accident reports is reached.

Table 14: A summary of the parameters estimated for the human consequence model.

	URBAN		INDUST		RURAL	
	P (N>=1)	Fat. Index	P (N>=1)	Fat. Index	P (N>=1)	Fat. Index
Chlorine	0.60	5.5 x W	0.38	0.55 x W	0.16	0.055xW
Ammonia	0.46	1.2 x W	0.24	0.12 x W	0.012	0.003xW
Class 2.3	0.15	8.7 x $W^{1/2}$	0.12	2.9 x $W^{1/2}$	0.09	1.0 x $W^{1/2}$
Class 3.3	0.086	3.9 x $W^{1/2}$	0.064	1.3 x $W^{1/2}$	0.044	0.43 x $W^{1/2}$
Class 3	0.022	3.9 x $W^{1/2}$	0.016	1.3 x $W^{1/2}$	0.011	0.43x $W^{1/2}$

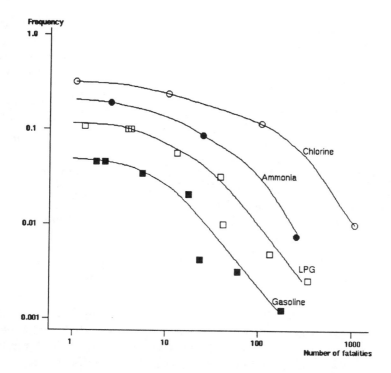

Figure 3: The estimated fN-curves for chlorine, ammonia, LPG and gasoline, assuming a 20 ton release.

The consequence model presented is a part of an integrated model consisting of an estimation of the probability of a traffic accident, an estimation of the probability of a tank release and finally the consequence model. The integrated model will be presented in the corridor application context of this conference.

The model described in this paper has been developed to support policy decisions in the risk management of dangerous goods transports, and should not be considered suitable to predict single accidents in specific locations.

The strength (and to some extent the weakness) of the model is the rather close link to real accident data. This link means that the prediction of the average number of fatalities, and the upper and lower limits, are closely related to past accident experience, in terms of human fatalities. Thus if no accidents with a given substance have happened in the past, the model is not suitable, unless the model users are prepared to make a top level assumption concerning the similarities with another (known) substance.

It is also clear that the model is not suitable for predicting the outcome of specific technical changes in the transport means, like installations of safety valves or similar. For predictions of this kind, more detailed traditional risk analysis models are needed.

Both the simple fatality index model and the complex dispersion models (in combination with vulnerability models) are based on numerous assumptions. Furthermore consequence models seem to be inherently uncertain (Amendola et al., 1991). The fatality index approach, which is proposed in this paper as a model for human consequences recognizes this! However, it is the easiest to use, it gives results closer to actual observations (Brockhoff et al., 1992). and it is more transparent. The authors therefore recommend using the proposed model for public policy related risk assessments since the decision makers will be better able to understand the limitations of the results--limitations which are present in both approaches.

Acknowledgments The authors would like to thank P. Haastrup, Joint Research Centre, Ispra for his contribution to the development of the approach presented. S. Funtowitch is acknowledged for his ability to point out uncertainties and possible simplifications.

8. REFERENCES

AMENDOLA, A., CONTINI, S., & ZIOMAS, I. (1992). Results of a European benchmark exercise. *Journal of Hazardous Materials*, 28.

BROCKHOFF, L., PETERSEN, H.J., & HAASTRUP, P. (1992). A consequence model for chlorine and ammonia based on a fatality index approach. *Journal of Hazardous Materials*, 29, 405-425.

E.C. Proposal for a council directive on the appointment of an officer for the prevention of the risks inherent in the carriage of dangerous goods in undertakings which transport such goods, and on the vocational qualification of such officers. Official Journal of the European Communities, No. C 185/5, 1991.

FRANCE (1988-90). Inventaire des pollutions accidentelles et accidents industriels en 1987, 88 and 89. Technical report, Ministere de l'Environnement et de la prevention des risques technologiques et naturels majeurs, Service de l'Environnement industriel, France.

FUNTOWICZ, S.O. (1991). Private communications.

HAASTRUP, P., & BROCKHOFF, L.H. (1992). A list of transportation accidents involving hazardous chemicals. Technical Report EUR 14 549 EN, European Communities.

MARSHALL, V.C. (1977). How lethal are explosions and toxic escapes? *Chem. Eng.* London, 323, 573.

A Simple, Usable, Empirical Risk Model for Small Communities

Eugene R. Russell, Sr.
Civil Engineering Department
Kansas State University
Seaton Hall
Manhattan, Kansas
66506-2905, USA

ABSTRACT

This paper describes a model that was written for communities with populations less than 50,000. Its primary objective is to alert officials of these communities of the threat to life, property and environment from the transportation of hazardous materials. It could be called an "awareness model". Its primary purpose is to alert small-town, local officials to their qualitative level of risk via a simple yet thorough process. The process itself is the key to awareness. A secondary objective was to keep it simple. It is intended primarily for communities whose officials have little or no technical expertise in hazardous materials transportation. The model takes local officials through the process of acquiring and analyzing data to determine their risk and their vulnerability. To those knowledgeable in traditional risk analysis, risk is usually defined as the probability of an incident times the consequence of the accident/incident. The model leads one through this process with the use of surrogate variables and empirical charts and tables, without even mentioning probability or consequences that would be analytically developed by a technical analyst. The user is told in a step-by-step manner what data to get, where to get it and how to use it. He/She uses charts and tables to get risk and consequence subfactors that are summed to the communities' risk.

1. INTRODUCTION

In the USA, there are close to 21,000 communities, both urban and rural. About ninety-eight percent of these communities are under 50,000 population. Together, they comprise almost half of the population that resides in communities. More striking is the fact that close to seventy percent of all communities are rural, defined as having less than 2,500 population.

Hazardous materials transportation incidents can occur anywhere. Although the consequences may not affect as many people, these small communities can be devastated by a major incident. These small communities are the least prepared to cope with a hazardous materials incident. Many may have no awareness of the danger. Their officials, police and fire departments (many times volunteers) may have little or no training or even the knowledge to cope with the problem or may not even understand that they are at risk.

A risk model was developed to address the needs of the thousands of small communities in the USA. It was primarily aimed at those who had no experience in this area. One of its main objectives was that it be kept simple, easy to understand, yet practical and usable by local officials. The process of using the risk model is intended to be an awareness exercise that, in itself, could be more valuable than the risk model result.

The model has been criticized by some experts in risk modeling as lacking rigorous mathematical analyses. Keep in mind that this model is not for those communities with risk modeling expertise available. It is for small communities with little or no risk modeling expertise. However, for those without expertise, it gives reasonable results with limited data and resources. In addition, the process is educational and definitely an awareness exercise. It could be called an "awareness model". Its primary purpose is to alert small-town, local officials to their qualitative level of risk via a simple yet thorough process. The process itself is the key to awareness, and may be more important than the numerical result. It is felt that the basic structure of the model is sound and meets the objective of having a model that can be easily used by local officials with no special education or training, no special expertise in mathematical analysis, and no source of good data bases to work with. It should not be used for other purposes without considerable modification.

2.   BACKGROUND OF MODEL DEVELOPMENT

2.1 General

The development of a simplified risk analysis model for local communities to use in assessing their risk from the transportation of hazardous materials followed very definite steps. These steps included the following:

1. collection of literature on risk analysis,

2. synthesis of collected literature,

3. discussion of this literature,

4. decision making on the variables to be included in the model,

5. writing drafts of the model,

6. evaluation of the model by research staff, local officials and professionals throughout the country, and

7. revision of the model based on the comments and suggestions received.

The initial model developed was essentially an additive model. It consisted of empirical values on eight variables which were added to obtain a total risk index. It had a very simple structure but some values appeared to be arbitrary. Comments received on early drafts indicated that the initial structure of the model was not well received because it did not fit into the traditional methodological form of traditional risk analyses, i.e., risk = probability of occurrence x consequences of an occurrence. To give the model a more rational basis, it was reorganized into a multiplicative form by a series of empirical equations which attempted to fit better into the mainstream of risk analysis, as most people perceive it.

2.2 Model Philosophy

The model is strictly empirical and, although attempts were made to insure that it was consistent with theory, it was not developed from a theoretical basis nor was any new theory developed. Although empirical equations were developed to support all steps in the process, the manual user sees no equations. The user is led step-by-step through a series of tables. The equations could be used by persons with modeling expertise for calibration and adjustments to the tables or for sensitivity analysis.

2.3 Previous Model

Griffith and Gabor at the Disaster Research Center at the Ohio State University developed a simple model (Gabor & Griffith, 1979). This model attempted to assess community vulnerability and was highly simplified. The model, which required great subjectivity on the part of local officials, employed a ten point weighting system and included four major factors:

1. Density,

2. Proximity,

3. Transportation, and

4. Forms of Threat

The simplicity of this model was very appealing and the author used this model as a starting point. It served as a basis for the evolution of subsequent models.

2.4 Early Decisions

It was decided that a "how to" manual was in order, capable of guiding any local person with concern for hazardous materials transportation through a simple step-by-step procedure and process using only local resources and data readily available to them. Any equations or theory within the process would not need to be apparent to the user. The resulting manual was to be strictly empirical, with tabular values adjusted to give results that the staff and local advisory group considered to be reasonable values on test communities. It should be noted that the tables could be recalibrated for communities with different characteristics.

Although a set of equations were developed, it was felt that the user should not be aware of any equations and should be led instead through a series of steps using tabular values. A series of worksheets are provided to users for all steps in the process. Input from a broad-based, knowledgeable local advisory committee convinced the author that this was the best approach. The use of these tables would eliminate equations and minimize computation. All mathematical symbols and decimals were avoided wherever possible. The latter point was stressed by our local advisory team.

2.5 Data

The next important decision made by the research team concerned data availability of hazardous materials flowing through communities.

It was decided that the best approach was locally obtained data. Two sources were decided upon:

1. field or survey identification of manufacturing and storage facilities having hazardous materials and

2. field identification of vehicles carrying hazardous materials through the community by a road survey or placard counts.

Road surveys performed by Zajic and Himmelman consisted of taking fifteen minute interval counts every two hours for twenty-four hour periods on Monday, Thursday and Saturday (Zajic & Himmelman, 1978). This approach was tested on Manhattan, Kansas. It was concluded that placard counts should be at least directly proportional to the actual amounts flowing within a community.

It was concluded from these road placard counts in Manhattan, that small local communities, generally with limited personnel and resources, should make continuous road counts on a weekday--probably Tuesday, Wednesday, or Thursday--between 6:00 a.m. and 6:00 p.m. If the community was on a major through highway or it suspected for some reason that there was a great deal of night traffic, twenty-four hour counts might be necessary.

In regard to quantity, a weighting system had to be developed which assigned various weights to different vehicle sizes, based on vehicle sizes and their capacities. It was assumed all vehicles were full.

An additional problem raised at the time the road surveys were taken in Manhattan was how to identify the hazardous materials carried by other modes--rail, air, pipelines, and barges. It was decided by the research team that local officials would have to consult with appropriate officials, agencies, or companies to determine the hazardous material carried, the quantity, and the frequency of shipment.

2.6 Maps

Base maps were needed for local officials to identify transportation routes and the various consequence variables to be selected. It was decided that USGS topographic maps which are available for almost every part of the United States were suitable. For determining population distributions and other consequence variables, standard government aerial photographs, which are readily available for almost any part of the United States, are suitable.

2.7 Variables

The final decision made before the risk analysis model was first drafted was on the selection of pertinent variables. It was decided that the variables used in the Disaster Research Center's (DRC) community vulnerability model--density, proximity, forms of transportation, and forms of threat--were a good starting point (Gabor & Griffith, 1979).

3. ADDITIVE MODEL - FIRST DRAFT

The first approach to the model was based on the DRC's model. This model was modified and expanded to meet the objectives of this project. This model employed a ten point weighting system as shown in Table 1 (Gabor & Griffith, 1979).

Table 1: Community vulnerability model table.

Factor		Maximum Weight (Points)
Density		1
Proximity		1
Transportation	a) Road	1
	b) Rail	1
	c) Barge	1
Forms of Threat	a) Major Fire	1
	b) Explosion	1
	c) Toxic Release (air)	1
	d) Toxic Release (water)	1
	e) Acute Corrosion	1
		TOTAL 10
		Risk Index = r*/10

* where r is the total of the weights actually assigned.

The model was sent out to several reviewers. Comments were very unfavorable. Reviewers said it was confusing and too much emphasis was on fixed facilities. A second version was developed with a scale of 0 to 100. Emphasis was directed toward transportation routes.

The model below indicates that the total community risk value was put on a 0 to 100 scale. In addition, the emphasis of the model was directed exclusively towards the transportation routes by various modes. It was felt by this research team that the 0 to 100 scale was superior since people seem to have an intuitive feel for ratings on this scale. It also tends to avoid fractions or decimals.

4.   MULTIPLICATIVE MODEL

Based on the critiques received from the twenty-six reviewers and input from the advisory panel, a major decision was made by the research team to make a significant change in the basic structure of the risk assessment model. It was concluded that simplicity could be maintained and still have the model follow empirical equations that were closer in form to traditional risk analysis theory, i.e., Risk = Probability of an incident x Consequences.

Revision centered on establishing a model on a set of empirical equations. Formulas were developed and a new factor (Risk Factor) was developed. Risk factor is a surrogate for the probability of an incident. The consequence factor is a surrogate for consequences and is the sum of four evenly weighted variables: population density, environment, property, and manufacturing and storage facilities. Table 2 illustrates the multiplicative risk analysis model.

Charts and tables were developed for the users. The final risk index is the risk factor times the consequence factor.

5.   SUMMARY OF MODEL DEVELOPMENT

The evolution of this Risk Analysis centered around developing a simplified risk analysis model that, although not theoretical in nature, did not violate accepted theory or tests of reasonableness. It was concluded early in the development that simplicity and reasonableness should be the most important criteria.

Table 2: The final risk model.

	Maximum Value
Risk Factor (RF)	1
Sub-Factors	
Twelve Hour Average Density (THAD)	
Average Forms of Threat (AFT)	
Consequence Factor (CF)	100
Sub-Factors	
Population Density	25
Environment	25
Property	25
Manufacturing and Storage Establishments	25

$$RI = (RF)(CF)(CAL)$$

where   RI = Risk Index (0 to 100)
        RF = Risk Factor (0 to 1)
        CF = Consequence Factor (0 to 100)
        CAL = Calibration factor if needed (usually = 1)

Empirical equations, relationships, and tables were developed so that results could be calibrated and adjusted and be better tested for sensitivity and reasonableness. These are presented in Appendix A so that those persons with analytical interest could better understand all aspects of the model so that calibrations and future refinements could be made. These equations and analytical relationships are neither shown nor mentioned in the manual itself and local officials are not asked to compute mathematical equations. They are only asked to perform a series of steps which guide them through a series of tabular values, and simple mathematics, i.e., addition and multiplication. Worksheets are provided to aid them in the process.

6.   EXPLANATION OF VARIABLES AND MODEL USE

The manual is presented as a step-by-step process. To explain the model, the variables and the use of the model, it will be explained in order of these steps.

The following steps are listed in the manual:

Step 1.   Obtain maps and aerial photographs.

Step 2. Conduct manufacturing and storage establishment surveys.

Step 3. Obtain data on pipelines, barges, air and rail.

Step 4. Plot one-mile route segment corridors.

Step 5. Plot manufacturing and storage data.

Step 6. Conduct highway traffic survey.

Step 7. Determine risk sub-factors.

Step 8. Determine risk factor.

Step 9. Determine consequence sub-factors.

Step 10. Determine consequence factor.

Step 11. Determine risk index.

Step 12. Determine level of preparedness.

Step 13. Determine community vulnerability.

Step 14. Select a response plan.

This paper will concentrate on steps 4 through 11.

6.1 Step 4. Plot Corridors

It is suggested that U.S. Geological Survey (USGS) topographical maps be used. They are easily obtained, large-scale, and show all detail necessary for the model.

Lines are drawn one-half mile on each side of each transportation route where hazardous materials is a concern. This results in a one-mile wide corridor. Each route is then marked into one-mile segments. The result is a series of one-mile square segments along the transportation routes. Figure 1 shows an example of a plotted highway corridor.

6.2 Step 5. Plot Manufacturing and Storage Facilities

Businesses and storage facilities falling within the one-mile segments are marked on the maps. They have to be known from existing data or a survey.

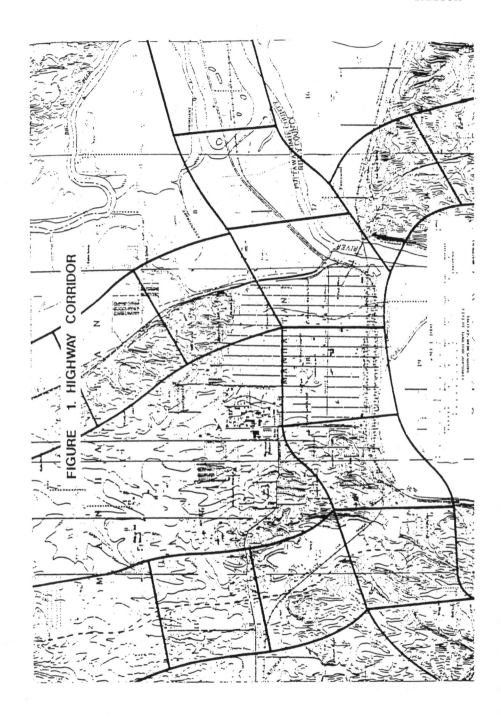

Figure 1: A typical plot of highway corridors.

6.3 Step 6. Conduct Highway Traffic Survey

Conduct a highway traffic survey of hazardous materials along each designated highway route. It has been assumed that any applicable data on rail, barge or air transportation has been obtained by contacting the company involved. (A field count can be made if necessary).

These field counts are basically a hazardous materials placard count. A 12-hour count, 6 a.m. to 6 p.m., on an average weekday is used. Forms are provided to check off each placard observed.

6.4 Step 7. Determine Risk Sub-Factors THAD and AFT

6.4.1 THAD

In this step, total vehicle counts for all modes are combined and the community's twelve-hour average density (THAD) value is determined from a table that uses total vehicle counts (all modes) and total route mileage (all modes) as input. Road, rail, waterway and air are number of vehicles or carriers; for pipelines, one count per one-tenth mile is used. The THAD value is the total vehicle count divided by the total route mileage. The user gets the value from Table 3 for various ranges of carrier count and route mileage.

6.4.2 AFT

Since quantity and type of hazardous material are important factors, adjustments were made to the counts. The average forms of threat sub-factor considers carrier size and material threat. Two tables are used, one to adjust for quantity and one to adjust for the type of threat. The figures in Table 4 are used to increase the vehicle count to adjust for quantity.

The values in Table 5 are used to adjust common placards to account for some materials being more hazardous than others. For example, if a material is a threat from fire, explosion and toxic release, the count is multiplied by three. Poison gas and explosives are multiplied by six.

The tables are used first to adjust the placard count for quantity. Then the forms of threat table is used to adjust the quantity-adjusted values to account for the form(s) of threat, i.e., magnitude of the potential consequence. An AFT value is obtained from Table 6.

Table 3: Twelve hour average density (THAD).

	0-5	6-10	11-15	16-20	21-25	26-30	31-35	36-40	41-45	46 or more
				Total Route Mileage (TRM)						
0-5	10	3	2	1	1	1	1	1	1	1
6-10	30	10	6	4	3	3	2	2	2	2
11-15	50	17	10	7	6	5	4	3	3	3
16-20	70	23	14	10	8	6	5	5	4	4
21-25	90	30	18	13	10	8	7	5	5	5
26-30	110+	37	22	16	12	10	8	7	6	6
31-35	130	43	26	19	14	12	10	9	8	7
36-40	150	50	30	21	17	14	12	10	9	8
41-45	170	57	34	24	19	15	13	11	10	9
46-50	190	63	38	27	21	17	15	13	11	10
51-55	210	70	42	30	23	19	16	14	12	11
56-60	230	77	46	33	26	21	18	15	14	12
61-65	250	83	50	36	28	23	19	17	15	13
66-70	270	90	54	39	30	25	21	18	16	14
71-75	290	97	58	41	32	26	22	19	17	15
76-80	310	103	62	44	34	28	24	21	18	16
81-85	330	110	66	47	37	30	25	22	19	17
86-90	350	117	70	50	39	23	27	23	21	18
91-95	370	123	74	53	41	34	28	25	22	19
96 or more	390	130	78	56	43	35	30	26	23	21

*These values were derived from the formula $\text{THAD} = 10\left(\dfrac{\text{TCC}}{\text{TRM}}\right)$ where

TCC = range midpoint of total carrier count

TRM = range midpoint of route mileage

+NOTE: If THAD is 100 or more, set your Risk Factor at 1.00.

6.5 Step 8. Determine Risk Factor

The risk factor table values are obtained by the relationship

$$RF = \frac{(\text{THAD})(\text{AFT})}{100} \tag{2}$$

The user, however, gets the value from Table 7.

For very small vehicle counts (THAD below 10), Table 8 is used instead of Table 7.

Table 4: Vehicle table.

Mode	Vehicle Type	Adjusted Placard Count
Truck	Non-Tanker	1
	Tanker	2
Rail	Box Car	1
	Non-Box Car	3
Air	Large Aircraft	2
Water	Barge	10
Pipeline	1/10 of a mile	2

Table 5: Forms of threat table.

Placard	Threat			Count Per Placard
Chlorine			Toxic Release	1
Combustible	Fire			1
Corrosive			Toxic Release	1
Dangerous	Fire	Explosion	Toxic Release	3
Explosives		Explosion		6[a]
Flammable	Fire	Explosion	Toxic Release	3
Flammable Gas	Fire	Explosion		2
Flammable Solid	Fire			1
Nonflammable Gas		Explosion		1
Organic Peroxide	Fire			1
Oxidizer		Explosion	Toxic Release	2
Poison			Toxic Release	1
Poison Gas	Fire		Toxic Release	6[a]
Radioactive			Toxic Release	1

[a] Because explosives and poison gas are exceptionally hazardous.

It can be noted that at high values of THAD and/or AFT, the risk factor becomes one which means that the consequence governs the risk. The authors believe this is logical for the small towns but may need some adjusting or calibrating for large towns.

Table 6: Average forms of threat table.

Total Adjusted Placard Count[a]	AFT Value
1-20	1
21-40	2
41-60	3
61-80	4
81-100	5
101-120	6
121-140	7
141-160	8
161-180	9
181 or more	10

[a] adjusted for vehicle type (quantity) and form(s) of threat (potential consequence level)

Table 7a: Risk factor (RF).

	Average Form of Threat (AFT)									
	1	2	3	4	5	6	7	8	9	10
1-5*	.03	.05	.08	.10	.13	.15	.18	.20	.23	.25
6-10*	.08	.15	.23	.30	.38	.45	.53	.60	.68	.75
11-15	.13	.25	.38	.50	.63	.75	.88	1.00	1.00	1.00
16-20	.18	.35	.53	.70	.88	1.00	1.00			
21-25	.23	.45	.68	.90	1.00					
26-30	.28	.55	.83	1.00						
31-35	.33	.65	.98	1.00						
36-40	.38	.75	1.00							
41-45	.43	.85	1.00							
46-50	.48	.95	1.00							
51-55	.53	1.00								
56-60	.58	1.00								
61-65	.63	1.00								
66-70	.68	1.00								
71-75	.73	1.00								
76-80	.78	1.00								
81-85	.83	1.00								
86-90	.88	1.00								
91-95	.93	1.00								
96-99	.98	1.00								

* For the range   THAD 1-10 use Table 8.

NOTE: RF = 1.00 for all values below or to the right of the solid line = 1.00.

Table 7b: Small THAD risk factor.

THAD	Adjusted Placard Count	
1-5	201-220	.28
	221-240	.30
	241 or more	.33
6-10	201-220	.83
	221-240	.90
	241 or more	1.00

Table 8: Population density table.

Average Population Density	Population Density Value
0	0
1-80	1
81-160	3
161-240	5
241-320	7
321-400	9
401-480	11
481-560	13
561-640	15
641-720	17
721-800	19
801-880	21
881-960	23
961 or more	25

6.6 Step 9. Determine Consequence Sub-Factors

The consequence factor is made up of four components or sub-factors:

1. population density
2. environment
3. property value, and
4. manufacturing and storage

Each component is equally weighted rated from 0 to 25. The consequence factor is the sum of these four numbers and can have a value between 0 and 100.

6.6.1 Population density

Population density is basically the average population density within all the one-mile square route segments. A population is estimated for each segment and the average of these per-square-mile estimates is considered to be the average population density affected. Table 8 is used to obtain the population density sub-factor.

6.6.2 Environment Sub-factor

To keep the model simple, it was decided that the threat to water supply or waterways was a reasonable surrogate for a small town environmental factor. The model uses the percent of open water or channels within the one mile route segments. The percentages for each segment along the route are added. Then Table 9 is used to determine the environment sub-factor.

Table 9: Environment table.

Average Percent of Waterway	Environment Value
0-10	5
11-20	10
21-30	15
31-40	20
41 or more	25

6.6.3 Property Sub-factor

Land use maps or aerial photos can be used to determine land use. Total land use values are determined for each one square mile segment and then averaged over all segments. Table 10 is used to determine property sub-factor.

Table 10 could be revised or weighted to reflect a community's priorities. Special uses such as schools, high-rise, etc., could be accounted for.

Table 10: Property sub-factor.

Land Use Type	Land Use Values	Property Factor
Open space	1	5
Agriculture	2	10
Commercial	3	15
Industrial	4	20
Residential	5	25

6.6.4 Manufacturing and Storage

The manufacturing and storage facilities having hazardous materials within the one square mile route segments have been accounted for by data base or survey and marked on the maps. The average number per route segment is calculated. Table 11 is used to determine the manufacturing and storage sub-factor.

Table 11: Manufacturing and storage sub-factor.

Average Number of Establishments with Hazardous Materials	Sub-factor Value
1-5	5
6-10	10
11-15	15
16-20	20
21 or more	25

If there are a number of establishments close to each other, or if there is high employment at an establishment, adjustments are suggested according to Table 12.

Table 12: High employment--high concentration adjustment.

Category	Adjustment
High Employment	add 1 factor for each 25 employees
High Concentration	add 1 factor for each pair of establishments within 500 feet of each other

6.7 Step 10. Determine the Consequence Factor

The consequence factor is the sum of the values arrived at for the four sub-factors.

6.8 Step 11. Determine Risk Index

To determine the risk index, the risk factor is multiplied times the consequence factor. The result is just a number and may not mean much to small community officials. Table 13 suggests a more meaningful risk level in more understandable, subjective terms.

Table 13: Level of risk table.

Calculated Risk Index	Level of Risk
1-35	Low
36-70	Medium
71 or more	High

The above breakpoints were arrived at empirically but not arbitrarily. Eleven towns were studied. They were selected by the project advisory panel. The panel's expert judgment was used to subjectively rate the towns as low, medium or high risk. Then the model was applied and the risk factor numbers correlated to the panel's judgment. Towns judged to be low risk all had RI values less than 35; those judged to be high risk all had RI values greater than 70.

An attempt was made to judge a community's preparedness through a questionnaire relating to emergency response training and equipment. Twenty questions were scored 5 points each. "No" answers were negative, e.g., no equipment, no training, etc. Table 14 was developed to estimate preparedness.

Table 14: Preparedness table.

"NO" Answer Score	Level of Preparedness
0-35	High
36-70	Medium
71 or more	Low

Based on previous work, a community vulnerability rating was suggested (Zajic & Himmelman, 1978). These are presented in Table 15.

Table 15: Community vulnerability table.

Preparedness			
Risk	High	Medium	Low
Low	Low	Low to Medium	Medium
Medium	Low to Medium	Medium	Medium to High
High	Medium	Medium to High	High

It should be emphasized that the preparedness table and the community vulnerability table are quite subjective. The manual had a section suggesting a type of response plan for various levels of risk and vulnerability.

7. DISCUSSION

The author has argued over the years that the process of following the manual steps leading to a risk index is more important than the risk index itself. Many professionals in risk assessment find the model over simplified or too subjective. The answer is that if a town has risk assessment expertise, they don't need this model. On the other hand, if they have none, or have never assessed their risk or given it any attention, they can benefit greatly from the process of applying the model. The author believes that the model is simple, usable and can be valuable to small towns.

This risk model was developed to be easy to understand and simple to use by local officials with no particular expertise or knowledge of hazardous materials. It was decided that the model would not be mathematically complex. Also, data requirements were kept to a minimum. The prevailing philosophy was that an empirical model with step-by-step guidelines and minimum computations would be better than a theoretical, analytical model.

The resulting model was strictly empirical, adjusted to give results that the staff and a knowledgeable advisory group felt were reasonable values for several small communities. The model was written in a "how to" manual format capable of guiding any local person through simple step-by-step procedures. Worksheets were developed to clarify the procedure. Although equations were developed, the user is not aware of them and is led instead through a number of worksheets using values from a series of tables and charts.

The author's closing comments are to reiterate the primary purpose of the model and to emphasize the point with an example. It is intended as an awareness model for small communities with little or no risk modeling expertise. It should not be used for other purposes without considerable modification. Eleven small towns in Kansas used the model. The results were very good. In all cases, the expert panel with first-hand knowledge of these towns agreed with the model results which varied from low to high risk.

Acknowledgment The research reported in this paper was conducted under the sponsorship of the USDOT/Federal Highway Administration, Office of the Secretary, University Research Program.

The complete report and manual that was developed is available through the National Technical Information Service (NTIS), Springfield, VA/USA/22161 (Russell, et al., 1986). The Association of Bay Area Governments (ABAG) adapted the model to a large metropolitan area (Jackson, 1983; City of Union City Hazardous Materials Transport Risk Assessment, 1983).

8. REFERENCES

City of Union City Hazardous Materials Transport Risk Assessment (1983). Association of Bay Area Governments, Berkeley, CA.

GABOR, T., & GRIFFITH, T.K. (1979). *The assessment of community vulnerability to acute hazardous materials incidents*. Disaster Research Center, The Ohio State University, Columbus, OH.

JACKSON, K. (1983). *A report on the use of a modified Kansas State University risk assessment model in a San Francisco Bay Area*. Association of Bay Area Governments, Berkeley, CA.

RUSSELL, E.G., et al. (1986). *Risk assessment/vulnerability users manual for small communities and rural areas*. DOT/OST/P-34/86/043, Springfield, VA, NIST.

ZAJIC, J.E., & HIMMELMAN, W.A. (1978). *Higher hazardous materials spills and emergency planning*. New York: Marcel Dekker, Inc.

The Risks of Handling vs. Transporting Dangerous Goods

J. R. Kirchner
W. R. Rhyne
H&R Technical Associates, Inc.
P. O. Box 4159
Oak Ridge, Tennessee
37831-4159 USA

ABSTRACT

Attention is focused on the transportation of dangerous goods because transportation routes frequently pass through or near areas in which many people live and work. However, it is not uncommon for industrial facilities at which the dangerous goods are loaded or unloaded to be located near public access areas with significant population densities. In such cases, the risks to the public posed by handling activities may be significant, and they may even be the dominant risk contributors. Three risk assessments addressing both the transportation and handling of dangerous goods are reviewed. The results of these risk assessments show that risks posed by the handling of dangerous goods can be much higher than the associated transportation risks, even when very long transportation routes are involved.

1. INTRODUCTION

In recent years, much attention has been focused on the risks associated with transporting dangerous goods by rail, highway, and waterway. Transportation accidents involving dangerous cargo are of concern primarily because they may occur in areas of high population density where they have the potential to affect large numbers of people. Yet, many facilities where these same dangerous goods are loaded or unloaded are located close to populated areas. Thus, accidents involving loading, unloading, or other handling activities may have off-site consequences just as high as postulated transportation accidents. In fact, when the risks of transportation and handling activities are compared, handling is often the higher-risk activity. This situation is generally attributed to the low accident rates (usually on the order of 1×10^{-6} accidents / mile traveled) used in predicting transportation accident risks and the relatively high probabilities of handling errors (10^{-4} - 10^{-2}/operation).

Three risk assessments performed recently by the authors consider both the handling and transportation of dangerous goods. In the remainder of this paper, we examine the results of these risk assessments.

2. RISK ASSESSMENT OF HYDROGEN FLUORIDE CYLINDER HANDLING AND TRANSPORT

This study considers the movement of 850-lb cylinders of anhydrous hydrogen fluoride (HF) by flatbed truck across a large plant site. Loading and unloading activities at a cylinder storage area and at two process areas are also considered.

Common carriers deliver the cylinders to the site in covered tractor-trailers. Each shipment may contain up to eight cylinders. Each cylinder is clamped inside a rigid steel "cage" for shipment. This cage prevents the cylinder from contacting other cylinders or large objects during shipment. The cage has slots in its base to allow the insertion of forklift tines.

An electric forklift is used to unload the cylinders from the tractor-trailer. The forklift moves each cylinder across the loading dock to a fenced storage area.

When a cylinder is needed at one of the two process areas, a forklift picks up a cylinder (in its cage) in the storage area and loads it onto a flatbed truck (with no side walls or tailgate). The truck driver straps the cylinder to the truck bed and transports it to the requesting process area. One of the process areas is about 0.6 miles away (road distance), and the other one is about 1.1 miles away.

At Process Area 1, the cylinders are housed in a supply shed (open on the front) with an attached loading dock. The cylinders sit on scales inside the shed, one behind the other, and they are connected to a supply manifold by flexible tubing. An electric monorail hoist is provided to lift the cylinders and move them between the scales and the front of the loading dock. When a cylinder is delivered to this area, the truck driver parks a few feet from the loading dock, removes the tie-down straps, picks up the cylinder using an electric forklift, and places the cylinder (in its cage), on the end of the loading dock. Operators use the monorail hoist to lift the cylinder out of its cage and move it to one of the scales. Operators lower the depleted cylinder into a cage using the hoist before it is loaded onto a flatbed truck by forklift for return to the cylinder storage area.

At Process Area 2, only one cylinder is used at any given time. The cylinder scale in this case is located at the back of a deep (~ 30 ft), covered loading dock. A monorail hoist is provided for lifting cylinders into and out of their protective cages and moving them between the scale and the front of the loading dock. The actions which take place when a cylinder is delivered to this area are basically the same as those described for Process Area 1.

We identified a total of 19 accident initiating events associated with handling activities. The initiating events considered include forklift-induced damage to a cylinder, forklift collisions with other vehicles, hoist failures, and operator errors in connecting a cylinder to a hoist or in operating the hoist. We constructed an event tree for each of the initiating events. Using the event trees, we developed 90 handling accident scenarios involving HF releases. We estimated the expected frequency of each scenario using subjective estimates of handling error probabilities, published human error probabilities, known frequencies of the handling operations, and analytically determined failure thresholds for the cylinder. We predicted concentrations of HF in air downwind of the postulated releases using state-of-the-art source term models (Fauske and Epstein, 1988) and atmospheric dispersion modeling techniques. Isopleths for severe health effects were defined using toxicity data for HF. We conservatively assumed that the wind would be blowing toward the area of highest population density at the time of a release.

We broke the on-site transportation route into three segments for analysis (Figure 1). The segments range from 0.6 miles to 1.2 miles. Accident forces considered as cylinder failure mechanisms included impact, crush, puncture, and fire. We identified transportation accident scenarios involving these forces by reviewing a detailed fault tree analysis of chlorine cylinders (Andrews et al. , 1980). In total, we analyzed 192 scenarios. We estimated expected frequencies for these scenarios using accident

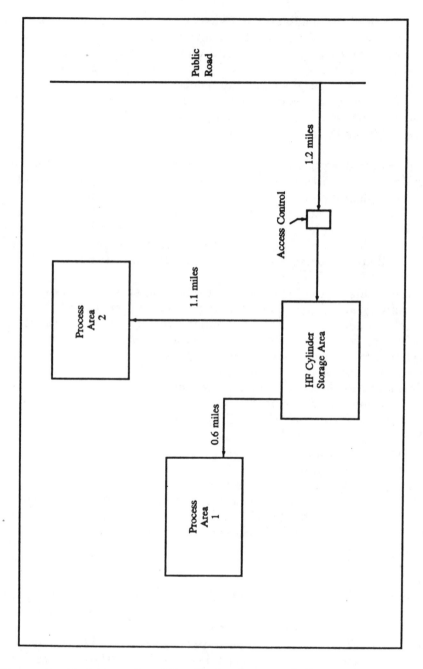

Figure 1:   Hydrogen fluoride cylinder transportation route
            segments.

force distributions reported by Clarke et al. (1976) and highway accident rates based on data published by Jovanis et al. (1989) and Harwood et al. (1990). Table 1 presents the accident rates suggested by these two references. The route segment between the public road and the cylinder storage area is similar to a rural two-lane road, so we used an accident rate of about 2 x 10^{-6} accidents/mile. The other two segments are within the restricted area of the plant and are best characterized by a local street (except that plant traffic is slower) or an urban one-way street (except that plant traffic is slower and moves in two directions). These characterizations suggest an accident rate of about 1 x 10^{-5} accidents/mile for the segments connecting the cylinder storage area to the process areas. However, the accident rate for these segments should be considerably lower because of the low speed limit and the exclusion of the general public. Rhyne et al. (1988) easily justified a factor of 15 reduction in published accident rate for the case of an escorted vehicle convoy on a military base with already restricted traffic. Jovanis et al. (1989) suggest an accident rate reduction factor of 7.5 for cases of controlled route access. For this analysis, we used a factor of 10 reduction for the two storage area-to-process area segments, yielding an accident rate of 1 x 10^{-6} accidents/mile.

Table 1: Accident rate data.

Road Type	Accident Rate (accidents/10^6 mile)
Controlled access	3.80[a]
Non-controlled access	28.40[a]
Local street	15.60[a]
Rural two-lane	2.19[b]
Rural multilane undivided	4.49[b]
Rural multilane divided	2.15[b]
Rural controlled access	0.64[b]
Urban two-lane	8.66[b]
Urban multilane divided	12.40[b]
Urban multilane undivided	13.90[b]
Urban one-way street	9.70[b]
Urban controlled access	2.18[b]

[a] Taken from Jovanis et al. (1989).
[b] Taken from Harwood et al. (1990).

We predicted the consequences of these scenarios using the same methods applied to the handling accident scenarios. The worst-case accident location (location closest to the most people) for each route segment was assumed for purposes of estimating the accident consequences.

Figure 2 is a frequency-consequence curve showing the results of this study. Note that the handling accidents only cause severe health effects for a few people, whereas some of the transportation accidents have the potential to injure large numbers of people. Yet, the total expected frequency of all postulated transportation accidents is less than 1×10^{-5} per year. In contrast, the sum of the predicted handling accident frequencies is almost 1×10^{-1} per year. Handling accidents represent >99% of the total estimated risk.

3.   RISK ASSESSMENT OF LOW-LEVEL RADIOACTIVE WASTE HANDLING AND TRANSPORT

This study considers the proposed movement of low-level radioactive waste from three temporary storage areas to two proposed disposal sites. One of the proposed disposal sites (Waste Site 1) is a large, in-ground trench for waste burial. The other proposed disposal site (Waste Site 2) is a group of above-ground concrete casks sitting atop concrete pads. At Waste Site 2, personnel place steel waste boxes containing waste of higher radioactivity inside the casks, which are then sealed. When the concrete casks are all filled and sealed, they are covered with earthen mounds.

Two basic waste forms would be moved in this effort: solid waste in sealed drums (taken to Waste Site 1) or boxes (taken to either Waste Site 1 or 2) and uncontainerized solid waste hauled in dump trucks to Waste Site 1. In addition, liquid leaching out of Waste Site 1 is automatically collected and transported to one of the temporary storage sites by tank trucks for treatment.

At a storage site, waste boxes or drums are removed from temporary storage areas by a forklift. The forklift driver then places the container onto a flatbed truck for transport to the appropriate disposal site. At the disposal site, personnel unload the containers from the flatbed truck using a forklift. In the case of Waste Site 1, the flatbed truck is usually driven down into the trench for unloading. Sometimes, personnel may unload a truck near the trench and use a mobile crane to lower the containers into the trench. Dump trucks hauling waste to Waste Site 1 always drive down into the trench to dump their loads.

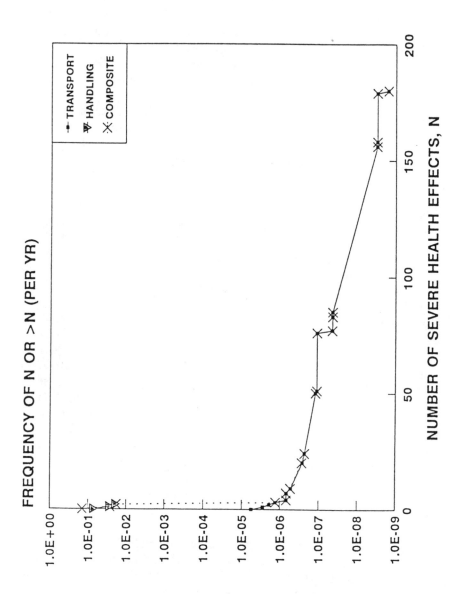

Figure 2: Frequency - consequence curve for severe health effects.

We identified seven accident initiating events associated with waste handling. These initiating events include forklift-induced damage to waste containers, forklift collisions with other vehicles, and fires in the trench at Waste Site 1. We constructed an event tree for each of the initiators to develop accident scenarios. From the event tree analysis we identified a total of 20 handling accident scenarios. Next, we estimated the expected frequencies of these scenarios using the expected frequencies of waste movements, subjective estimates of handling error probabilities, and crude estimates of container failure thresholds (based on analyses performed as part of other studies). We estimated the amount of radionuclides made airborne in a given scenario using projected radionuclide concentrations in the waste along with empirical release fractions for powder spills (Sutter, et al., 1981) and for the burning of combustible materials with powder contamination (Halverson et al., 1987). We employed a simple Gaussian dispersion model to estimate concentrations of the various radionuclides in air downwind of the postulated releases. Doses to exposed populations were then estimated using a standard breathing rate of 1.2 m^3/h (Shleien and Terpilak, 1984) and data provided by Eckerman et al. (1988). Finally, we estimated the expected excess cancers over a 70-year period using the calculated doses and cancer risk estimates reported by the International Committee on Radiation Protection (1977).

We considered five different routes in the analysis of low-level waste transportation (Figure 3). Note that uncontainerized solid waste is only transported between Storage Site 2 and Waste Site 1. We constructed an event tree (Figure 4) to evaluate the various modes of release in a truck accident. Because of the preliminary nature of this study, we made two major simplifying assumptions: (1) all mechanical forces can be grouped together and (2) containers that do not fail as a result of mechanical forces applied in an accident are damaged such that their resistance to fire is reduced by 50%.

The roads connecting the storage sites to the waste disposal sites are rural, two-lane roads; therefore, we assigned an accident rate of 2 x 10^{-6} accidents/mile (Harwood et al., 1990) for all five routes. The probability of mechanical forces occurring in an accident was taken from Fischer et al. (1987). We took the assigned failure probability for drums or boxes subjected to mechanical forces in truck accidents from Geffen et al. (1981). The tank truck failure probability is based on data reported by A. D. Little, Inc. (1989). We assumed a value of 0.8 for the probability of waste material dispersal given that mechanical forces act on a dump truck.

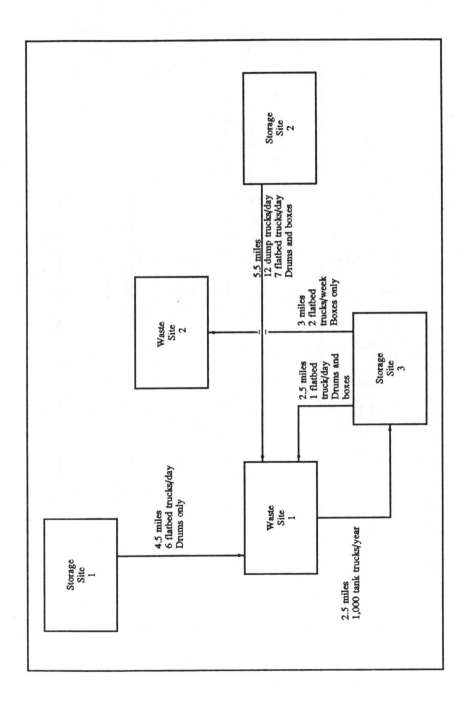

Figure 3: Low-level waste transportation routes.

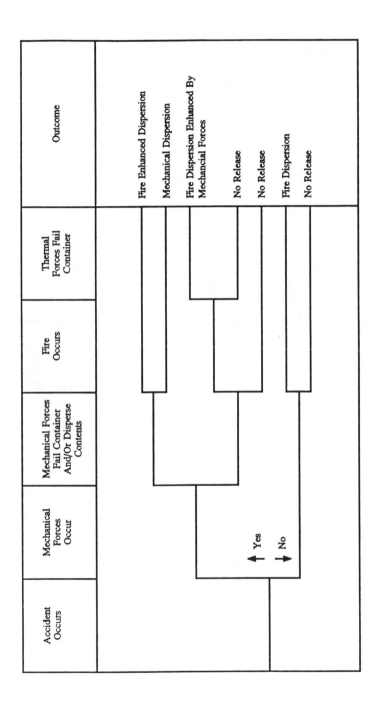

Figure 4: Simplified event tree for low-level waste transportation.

The assigned probability of fire in cases where mechanical forces occur and the assigned container failure probability, given a fire occurs, are based on data given by Clarke et al. (1976). We assumed that a fire lasting 30 minutes or longer is required to fail a drum, box, or tank truck. This assumption is based on a detailed analysis of a gasoline tank truck by Rhoads et al. (1978). We used the probabilistic distribution of truck fire durations reported by Clarke et al. (1976).

We predicted the consequences of the postulated transportation accident scenarios involving airborne releases using the same methods and assumptions applied to the analysis of handling accidents. Spills of leachate from tank trucks, especially spills not involving fire, have the potential for contaminating groundwater and any nearby surface water bodies. We used simplified transport models to predict possible human intakes by these pathways.

The total risk for handling accidents is estimated to be $< 1 \times 10^{-13}$ health effects/year. About 98% of this risk is attributed to container damage by forklift tines or drum handler jaws. These scenarios are expected to occur frequently; however, the amount of material released is small. In contrast, the total risk associated with transportation accidents is $< 1 \times 10^{-14}$ health effects/year. Almost 60% of this risk is contributed by spills from leachate tank trucks during a vehicle fire. Several accident scenarios involving solid waste are also significant contributors to transportation risk. Overall, handling accidents contribute 94% of the total estimated risk. It is worth noting that some of the transportation accident scenarios having relatively low expected frequencies are significant risk contributors because of the larger amount of waste available for release compared to handling accidents.

4. RISK ASSESSMENT OF HANDLING AND TRANSPORTATION FOR A COMPRESSED GAS CYLINDER SAMPLING PROGRAM

This risk analysis focuses on a large-scale program to collect and sample unlabeled or mislabeled compressed gas cylinders from several locations on a large plant site.

A specially trained team picks up the cylinders for transport to a central sampling and rebottling facility on-site. This facility is located inside a vapor-containment tent so that any gas releases would not threaten nearby workers or the public. Cylinders in good condition are sampled, labeled, and taken by truck to a storage area for eventual reuse. Cylinders in poor condition are mounted inside a large, sealed tank. A drill-type device then punctures the cylinder. The gas is collected in the tank and transferred to a new cylinder. Personnel label the new cylinder before it is taken to the storage site by truck.

At a cylinder pick-up location, a specially trained team in protective clothing collects cylinders one at a time and moves them to a flatbed truck using a hand cart. The flatbed truck is equipped with side panels, a cylinder storage rack, and a lift gate. The technicians secure larger cylinders in the truck's cylinder storage rack; however they place the smaller cylinders in plastic drums partially filled with vermiculite. They strap these drums to the truck stakes at the forward end of the truck bed. Upon arrival at the sampling facility, the truck is driven into the vapor containment tent. Personnel seal the tent before unloading commences. Then the team (still in protective clothing) removes the cylinders from the truck one at a time and places them in a storage rack to await sampling. When a cylinder is needed for sampling, a technician in protective clothing moves it to the sampling cabinet a few yards away. If the cylinder is large, the technician uses a hand cart to move it. Otherwise, the cylinder is hand carried to the sampling cabinet. After sampling, the technician removes the cylinder from the cabinet, labels it, and moves it to yet another storage rack. Periodically, personnel load a truck with processed cylinders and transport them to a central cylinder storage area for eventual reuse. The handling procedures for reloading the truck at the sampling facility and unloading it at the storage area are the reverse of the procedures described for the handling of unprocessed cylinders; however, the team members do not wear protective clothing when handling processed cylinders.

We analyzed each handling operation to identify actions which, if performed incorrectly, could result in damaging a cylinder. Based on this exercise, we identified 22 accident scenarios which could result in cylinder failure and gas release. The accident scenarios involve cylinder drops from truck beds and tip-over of cylinders during handling on the ground. We estimated expected frequencies for these scenarios using (1) subjective, yet conservative estimates of handling error probabilities, (2) cylinder failure thresholds based on simple and very conservative energy balance calculations for typical large and small cylinders, and (3) information regarding the number of cylinders to be sampled. Because of the limited time and resources available to complete this study, no formal consequence analysis was performed. However, we made a rough estimate of the likelihood of experiencing serious health effects, given a gas release, based on the expected fraction of cylinders containing toxic gas and the number of cylinder pickup locations in close proximity to populated areas.

Old, unidentified cylinders are located at many different locations on the plant site. Analysis of all possible transportation routes was not feasible in this case, so we assumed that a 5-mile trip on a rural, two-lane road is representative of a typical trip from a pickup location to the cylinder sampling facility. We used an accident rate of 2×10^{-7} accidents/mile for this route. This rate is based on a rate suggested by Harwood et al. (1990), but we reduced the reported rate by a factor of 10 in consideration of the training and the cautious nature of the cylinder transport team. This lower accident rate is often realized by trucking firms with good safety programs (Modern Bulk Transporter, 1990). The storage area for processed cylinders is 2 miles from the sampling facility. The road connecting these facilities is not open to the public, so we used an accident rate of 6×10^{-8} accidents/mile. This accident rate is derived from a rate of 6×10^{-7} accidents/mile published by Harwood et al. (1990). As before, we reduced the published rate by a factor of 10 to account for the high degree of safety consciousness demonstrated by the cylinder transport team.

We estimated cylinder failure thresholds from simple energy balance calculations for typical large and small cylinders. We used data reported by Fischer et al. (1987) and Clarke et al. (1976) to estimate the probabilities of experiencing various accident forces.

As mentioned earlier, we performed no formal consequence analysis; rather we made a crude estimate of the probability of experiencing serious health effects, given a gas release, based on available information about the cylinder processing campaign.

The expected number of severe health effects for the entire campaign is 1.5. Greater than 99% of this value is attributed to handling accidents. We identified two simple modifications that would reduce the expected number of severe health effects to about 0.2. The project team implemented these modifications shortly after completion of the risk assessment.

5. DISCUSSION

Table 2 summarizes the results of the three studies described above. These studies vary considerably in the degree of analytical detail employed, but in all three cases, handling risks are greater than transportation risks.

Table 2: Summary of risk assessment results.

Study	Handling Risk	Transportation Risk
HF cylinder handling and transport	4.4×10^{-2} health effects/year	2×10^{-5} health effects/year
Low-level radioactive waste handling and transport	$< 1 \times 10^{-13}$ health effects/lifetime	$< 1 \times 10^{-14}$ health effects/lifetime
Gas cylinder sampling program	1.5 health effects/campaign	4.7×10^{-5} health effects/campaign

In all three studies we adopted the expected number of health effects per unit time as a measure of risk. Many decision makers are primarily concerned with preventing high-consequence accidents; they place little emphasis on potentially more frequent accidents that would affect only a few people. For these decision makers, the expected number of health effects per unit time does not convey the needed information. Yet, other decision makers seek to minimize all adverse effects resulting from their operations involving dangerous goods. For these decision makers, the expected number of health effects per unit time is an entirely appropriate measure of risk.

Another important fact that should be considered is that all three of these studies consider short transportation routes of a few miles. The risk assessment of chemical munitions disposal alternatives performed by the U. S. Army and its contractors (Baronian, 1988) is an example of a study which addresses long transportation corridors and numerous on-site handling operations. For one alternative addressed in that study (ship all munitions from their current storage sites across the continental United States to a single disposal plant), transportation accidents account for about 95% of the estimated risk to the public. The transportation risk contribution is lower (59%) for the case of disposal at several regional plants; this option involves shorter transportation distances. Referring back to the low-level radioactive waste study, if the transportation routes were about 100 times longer, transportation accidents could be the dominant risk contributors. In contrast, the other two studies summarized herein predict handling risks that are several orders of magnitude higher than the associated transportation risks. In these cases handling risks would likely be greater than transportation risks even if the transportation routes were 1,000 miles long and the accident rate were increased by a factor of 10. Obviously, the population density along the transportation corridor is an important factor. However, the populations at risk in these cases were not small. (Hundreds of people could be affected in some accident scenarios considered.)

Based on these insights, we conclude that decision makers should not neglect the risks associated with on-site handling of dangerous goods because these risks are often higher than the associated transportation risks.

6. REFERENCES

A.D. LITTLE, INC. (1989). *Risk Assessment for Gas Liquids Transportation from Santa Barbara County* (draft report). Cambridge, MA: Arthur D. Little, Inc.

ANDREWS, W.B. ET AL. (1980). *An Assessment of the Risk of Transporting Liquid Chlorine by Rail*, PNL-3376. Richland, WA: Pacific Northwest Laboratory.

BARONIAN, C. (1988). *Chemical Stockpile Disposal Program Final Programmatic Environmental Impact Statement*. Aberdeen Proving Ground, MD: U. S. Army Program Executive Officer - Program Manager for Chemical Demilitarization.

CLARKE, R.K. ET AL. (1976). *Severities of Transportation Accidents*, SLA-74-001. Albuquerque, NM: Sandia National Laboratories.

ECKERMAN, K.F. ET AL. (1988). *Limiting Values of Radionuclide Intake and Air Concentration and Dose Conversion Factors for Inhalation, Submersion, and Ingestion*, EPA- 520/1-88-020. Washington, DC: U. S. Environmental Protection Agency.

FAUSKE, H.K., & EPSTEIN, M. (1988). Source term and two-phase flow phenomena in connection with hazardous vapor clouds. *Proceedings of the 5th Miami International Symposium on Multiphase Transport and Particulate Phenomena.*

FISCHER, L.E. ET AL. (1987). *Shipping Container Response to Severe Highway and Railway Accident Conditions*, NUREG/CR-4829. Washington, DC: U. S. Nuclear Regulatory Commission.

GEFFEN, C.A. ET AL. (1981). *An Analysis of the Risk of Transporting Uranium Ore Concentrates by Truck*, PNL-3463. Richland, WA: Pacific Northwest Laboratory.

HALVERSON, M.A. ET AL. (1987). *Combustion Aerosols Formed During Burning of Radioactively Contaminated Materials - Experimental Results*, NUREG/CR-4736. Washington, DC: U. S. Nuclear Regulatory Commission.

HARWOOD, D.W. ET AL. (1990). A truck accident rate model for hazardous materials routing. Paper No. 890458 in *Proceedings of the Transportation Research Board 69th Annual Meeting*. Washington, DC: Transportation Research Board.

INTERNATIONAL COMMISSION ON RADIOLOGICAL PROTECTION (1977). Recommendations of the International Commission on Radiological Protection. *Annals of the ICRP*, Vol. 1, ICRP Publication 26. New York, NY: Pergamon Press.

JOVANIS, P.P. ET AL. (1989). A comparison of accident rates for two truck configurations. Paper No. 880406 in *Proceedings of the Transportation Research Board 68th Annual Meeting*. Washington, DC: Transportation Research Board.

MODERN BULK TRANSPORTER (1990, June). Usher Transportation includes customers, others in overall safety program. *Modern Bulk Transporter*, 36-43.

RHOADS, R.E. ET AL. (1978). *An Assessment of the Risk of Transporting Gasoline by Truck*, PNL-2133. Richland, WA: Pacific Northwest Laboratory.

RHYNE, W.R. ET AL. (1988). Probabilistic source term for accidents associated with the transport of chemical munitions. Presented at the American Institute of Chemical Engineers Annual Meeting, Denver, CO, August 21-24, 1988.

SHLEIEN, B., & TERPILAK, M.S. (1984). *The Health Physics and Radiological Health Handbook*. Olney, MD: Nucleon Lectern Associates, Inc.

SUTTER, S.L. ET AL. (1981). *Aerosols Generated by Free Fall Spills of Powders and Solutions in Static Air*, NUREG/CR-2139. Washington, DC: U. S. Nuclear Regulatory Commission.

Transportation Also Begins and Ends with Risks

Brian J. Griffin
F.G. Bercha and Associates (Alberta) Limited
Robert G. Auld
R.G. Auld and Associates Limited
#250, 1220 Kensington Road N.W.
Calgary, Alberta
Canada T2N 3P5

ABSTRACT

In estimating the risks of transporting dangerous goods by truck, the analysis often focuses on the truck movements and not the loading and unloading operations. Appropriate methodologies for analyzing the loading and unloading operational risks are discussed using a recent case study involving the transportation of· high vapour pressure liquids. Relevant results from this case study are presented to demonstrate the format and type of information contained in the results and to describe why the five analytical methodologies used in the case study were appropriate. General conclusions regarding the detailed evaluation of dangerous goods truck loading and unloading risks are presented.

1. GENERAL INTRODUCTION

In estimating the risks of transporting dangerous goods, the analysis often focuses on the movement of dangerous goods to the detriment of loading and unloading operations. This may be partly due to the historical record of where incidents have occurred and partly due to the increased public exposure en route, as compared to the end points. The subject of this paper is the potential risk of loading and unloading dangerous goods trucks. Why should these risks be evaluated? How should these risks be evaluated? What benefits can result from the evaluation?

The complete transportation system should be addressed when estimating risks and if warranted, the loading and unloading operations can be specifically analyzed using appropriate methodologies. A recent case study involving the transportation of high vapour pressure (HVP) liquids will be used to describe these risk analysis methodologies. The general objective of the case study was to analyze the risk of a fire or explosion from a natural gas liquid (NGL) truck transportation and handling system. Although the loading, transport, and unloading operations were all analyzed in the study, the topic for the present paper focuses on the loading and unloading operations.

Why evaluate the loading and unloading risks? The case study transportation system is similar to many HVP liquid transportation systems operating in Canada. Both loading and unloading facilities in the case history were unmanned and remote from the feed operations and therefore were operated by the truck driver and not a plant operator. Should HVP liquids be released during loading or unloading, the fire and explosion hazards pose an immediate safety concern for nearby people, in addition to the potential destruction of facilities. A review of historical accidents showed that although most risk occurred during transportation, there were also severe accidents with both the loading and unloading of HVP liquids. The relevance of this historical record regarding the case history facilities was in question, however, since the designs and operating procedures are not standardized throughout the industry and, in fact, may vary significantly within individual companies. Potentially severe consequences from HVP liquid releases and a general uncertainty of the chances for a release contributed to the decision that a specific analysis of the loading and unloading operations was warranted.

Following a general description of risk analysis methods in Section 2, the case study is presented in Section 3, the selection of risk analysis methods for the case study is given in Section 4 and case study risk results are presented in Section 5. General conclusions regarding the evaluation of dangerous goods truck loading and unloading risks are given in Section 6.

2.   GENERAL DESCRIPTION OF RISK ANALYSIS METHODS

When analyzing dangerous goods transportation risks, historical accident rates are often calculated based on available statistics for both accidents and spills resulting from accidents. In general, the historical performance can include incidents associated with the loading, transportation, and unloading operations which can be compared to exposure terms such as the number of loading/unloading operations carried out or the kilometres travelled. For HVP liquid truck shipments, the reliability of historical data was limited due to both the quality of information recorded and the integration of data from different sources based on different specifications. In particular, estimates of accident and spill rates based on generic performance of loading and unloading facilities can be misleading since the designs and operating procedures associated with these facilities are not standardized. Careful selection or adaptation of data is therefore necessary for application to a specific operation.

Historically, the hazards associated with the HVP liquid truck transport industry have been controlled through engineering standards which have evolved from experience. This experience includes the lessons learned from historical accidents or near misses within the industry and within other similar industries. HVP liquid loading and unloading facilities are generally designed to American Petroleum Institute (API) codes and American Society of Mechanical Engineers (ASME) Boiler and Pressure Vessel Codes. In-house standard practices pertaining to spacing, isolation and fire protection are also applied. Although these codes and practices are standardized, they have evolved over many years and therefore older facilities may not meet newer standards such as the requirements of API 2510 issued in 1989. The implementation of these codes and practices through design, construction, operating, maintenance, and management practices within the industry is varied and can have a direct impact on potential risks. Risk assessment methods therefore should not be limited to code compliance methods but should also include methods for predictive analysis based on site specific parameters.

A large variety of risk analytic methodologies have been developed in order to characterize both the probability of hazard occurrence and the magnitude of consequences or impacts. Different methods have been developed for application to one or more steps in the risk analysis process from identification of hazards to evaluation of risk mitigation measures. Depending on the application, risk characterization may include qualitative description, definition of the relationship among causal factors, definition of the sequence of consequence effects, quantification of a risk measure, and specification of uncertainties in risk estimates. The following is a limited selection of the many available methodologies: a) Use of

engineering codes and practices; b) Safety reviews and audits; c)
Evaluation of risk indices; d) Consequence analysis; e) What-If
analysis; f) Hazard and operability studies; g) Network analysis
such as fault and event trees. General descriptions of these
methodologies are given elsewhere [Lees, Centre for Chemical
Process Safety, 1985, 1989. Burton et al., and Fischoff et al.]

A number of methods are specifically used to measure risk.
General risk measurement methods include statistical methods
[Bercha et al., 1989], probabilistic methods [Rasmussen], Delphi
methods [Salem et al.] and Bayesian methods [Burton et al.]. These
techniques are particularly useful in providing a measure of
overall system risk and information for subsequent risk modelling
techniques such as network and simulation methods. Network risk
analysis methods include reliability diagrams [Henley et al.],
event trees [Pat -Cornell], fault trees [Bercha, et al. 1983], and cause-
consequence diagrams [NARS]. In general these methods model the
interaction of events which determine the potential hazard
occurrence and/or its consequence effects. Simulation methods
include Monte Carlo or stochastic modelling [Henley et al.]. This
technique is used to characterize risk based on a complex number of
uncertain inputs or causal factors. Finally, consequence effects
may be measured through modelling of the hazard migration and
potential human response.

Given the large shopping list of risk analysis methodologies, a case
study will be utilized to illustrate an optimum mix for application to
transportation risk analysis. One method generally does not
answer all questions and both the selection and adaptation of
methods may be specific to the risk system being analyzed.

3.   CASE STUDY FACILITIES

The case study facilities which will be addressed in this paper
include three NGL truck loading facilities located from 100 to 240
km from a common unloading facility. Approximately four tanker
trucks per day transport NGLs from any of the three production
facility storage tanks to the unloading facility where it is stored and
subsequently transferred to an existing pipeline. NGL is a highly
flammable hydrocarbon gas compressed at approximately 1700 KPa
within the handling system in order to store it in liquid form. If
pressurized liquid NGL is released to the atmosphere, it will
vaporize and form a hazardous cloud which if ignited will cause a
fire or explosion with potential harmful consequences to both

exposed people and equipment. Given this hazard, it was assumed that a spill or leak was not tolerable and instead of including a detailed consequence analysis, the study was restricted to analyzing the potential for releases.

Similar loading facilities are installed at two locations as shown in the schematic flow diagram presented in Figure 1. A legend for the symbology used in all facility flow diagrams is given in Figure 2. As may be seen, two lines connect the tanker truck to the loading facility; namely, a pressure equalization line connected to the vapour space and a liquid NGL transfer line. When the tanker truck is connected for loading, liquid NGL is pumped through line 1 from the storage vessel to the tanker truck using one of two available pumps. Check valves are located downstream of the pumps while excess flow valves protect the liquid system on the storage vessel discharge and upstream of the breakaway valve. High pressure is controlled by valves on both lines 1 and 2. A breakaway valve is installed only in line 2. In an emergency the shut-down control will close the storage vessel outlet valve and stop both pumps. There are no on-site operators and therefore the truck driver operates the facilities alone. Operations include connecting the truck, establishing the load volume, loading, and disconnecting the truck.

The third loading facility was more recently constructed from a different design as shown in corresponding schematic flow diagram, Figure 3. Compared to the other loading facilities, this is a one pump system with additional safety protection devices installed. Pressure safety valves are located for thermal relief and potentially blocked-in lines. The NGL line pressure control valve was changed to a differential pressure control valve and included in the emergency shutdown system. Both low flow and pressure switches activate pump shut-down. Breakaway valves are also installed in both lines 1 and 2. Although high pressure vapour in line 2 cannot be diverted to a flare, the storage tank pressure safety valve is used as an atmospheric release location. Again, the truck driver is responsible for loading, however, a site operator controls the procedure with a key interlock.

The common unloading facility was located on an existing processing plant site and a schematic flow diagram is given in Figure 4. As may be seen, lines 1 and 2 are the NGL liquid and equalizing vapour connections to the tanker truck. Line 3 is an NGL supply line from the plant and line 4 is the transfer line to the pipeline. The unloading facility is more complex than the loading facilities previously described. Operation of the facility will not be

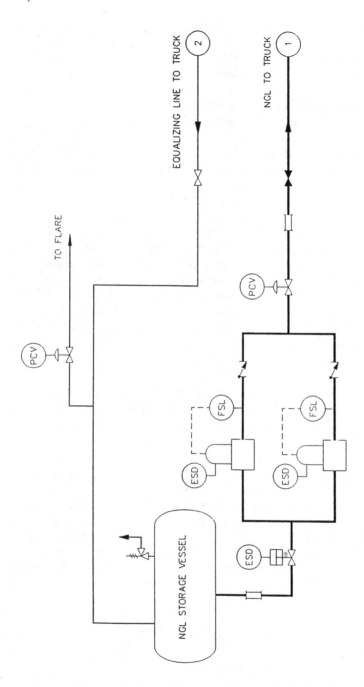

Figure 1: Schematic flow diagram of two truck loading facilities.

further described, but a number of safety devices included in the design will be briefly highlighted. Excess flow and check valves protect the storage vessel from emptying in the event of equipment ruptures. Pressure safety valves are located in lines for thermal relief on both storage vessels. Breakaway valves are installed in both tanker truck connection lines 1 and 2. A pressure switch prevents the compressor from operating when insufficient vapour pressure is present. In an emergency, the shutdown control will close all valves indicated, however, there is no valve in line 3.

SYMBOL	DESCRIPTION	SYMBOL	DESCRIPTION
	PRESSURE SAFETY VALVE	(FSL)	FLOW SWITCH (LOW)
	CHECK VALVE	(PSH)	PRESSURE SWITCH (HIGH)
	EXCESS FLOW VALVE	(PDV)	DIFFERENTIAL PRESSURE VALVE
	BREAKAWAY VALVE	FT, PT, TT	FLOW, PRESSURE, TEMPERATURE TRANSMITTER
	MANUAL VALVE	(DI)	DENSITY INDICATOR
	CONTROL VALVE (HYDRAULIC)	(ESD)	EMERGENCY SHUT DOWN
	CONTROL VALVE (PNEUMATIC)	————	NGL LIQUID FLOW
	PUMP	– – – –	INSTRUMENT CONNECTION
	COMPRESSOR	————	NGL VAPOUR FLOW
CSC	CAR SEALED CLOSED		

Figure 2: Flow diagram legend.

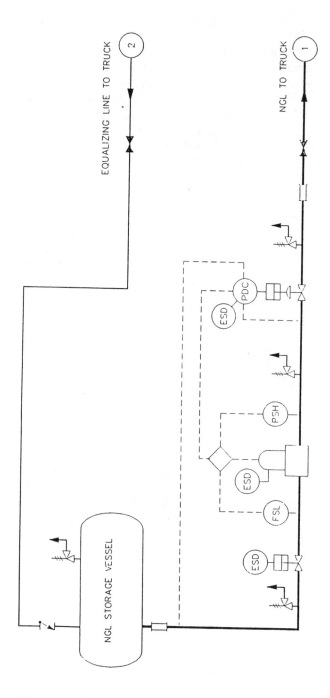

Figure 3: Schematic flow diagram of third truck loading facility.

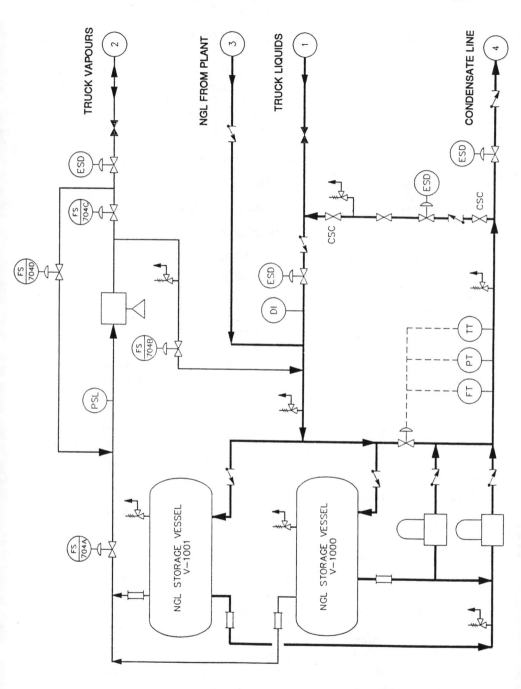

Figure 4: Schematic flow diagram of truck unloading facility.

4.   CASE STUDY RISK ANALYSIS METHODS

The complete NGL transportation system was analyzed in the case study, however, the topic for the present paper involves only the loading and unloading operations. Of the general risk analysis methodologies described in Section 2, the following were selected for analyzing the loading and unloading risks:

a) **Site visits** and assessment of actual operating conditions;

b) Review of **historical release incident records**;

c) Review of compliance to **standard practices**;

d) Analysis of potential for accidental releases by adapting two standard risk analysis methodologies:

   i) **Knowledge Based Hazard and Operability Study**;

   ii) **Fault Tree Analysis**.

This approach incorporated the historical knowledge base of industry in an evaluation of risks specific to the case study facilities. It was developed to satisfy the objectives of understanding the potential hazards and establishing optimum risk mitigation strategies.

A **site visit** of each facility was undertaken in order to assess the siting of equipment, location of safety devices, and actual operating procedures. During these visits, technical documents such as engineering drawings and operating procedures were collated for further analysis. Photos taken during the visits were used to update technical information where required in order to analyze current operations.

Historical data on the truck transportation of HVP liquids was collated including statistical information on traffic volumes, loading and unloading accidents, spills, and cargo classification. It was assumed that the truck transportation of NGL was associated with risks similar to other HVP liquids such as propane. In addition to evaluating historical accident records, the associated consequences such as release volumes, fires, and explosions were also reviewed in order to confirm the potential consequence severities. Design and operating information from the case study facilities was then compared to the historical data in the development of site specific accident scenarios.

The newer loading facility was of a different design than the older two facilities as previously described in Section 3. A **safety review** was carried out on the three loading facilities to identify deviations from good practice among the different sites. The comparison included both equipment and procedures. Although basic code specifications were met at all sites, the design, operating, and maintenance procedures varied among sites. The objective of the safety review was therefore to ensure that the progress and experience in hazard control implemented at one site was incorporated at all sites.

Following the review of compliance with standard practices, a hazard analysis method was required to identify site specific hazards and evaluate potential accident scenarios. The **knowledge based hazard and operability study** was adapted to meet this study objective. The knowledge base included information from the historical review of release incidents in addition to operator experience and was relied upon to significantly decrease the analytical effort that would have been required with more rigorous hazard and operability approaches. This methodology involved a systematic comparison of the engineering and operational documentation to the established specifications, standards, procedures and operating experience in order to evaluate potential deviations from the design intent. Results complemented those from the safety review with predictions of accidents specific to the case study facilities.

The final risk analysis methodology, **fault tree analysis**, was applied to further define accident scenarios previously identified and establish optimum risk mitigation strategies. The fault tree was used as a model to graphically present a logic network of all important causal factors leading to accidental releases of NGL at the loading or unloading facilities. The level of detail to which the sequence of causal factors is defined is dependent on the objectives of the analysis and for the case study all hazards were included with less than 50 factors. Probabilities of fault tree events are often estimated through historical statistics, however, the available data for HVP loading or unloading facilities were of poor quality. Should more detailed quantification be required, data on component failures within related process industries should be used as an alternative which may have less inherent uncertainty than historical transportation data. For the case study, such detailed quantification was not carried out, however, a qualitative evaluation and a risk sensitivity analysis based on available data provided valuable insights for recommending risk mitigation measures. Fault tree analysis is a flexible methodology which can be adapted for varying levels of analytical effort to answer the questions: "How can an accident occur?" "What are the chances of an accident occurring?" "Would a particular change reduce the chance of an accident occurring?"

5. CASE STUDY RISK RESULTS

Relevant results from the case study are presented to demonstrate the format and type of information contained in the results and to describe why the selected methodology was appropriate. As previously described, the case study risk evaluation used five analytical methodologies to analyze the risk of a fire or explosion from an accidental NGL release. Different consequences from a fire or explosion were not analyzed in this study but rather, any such scenario was assumed intolerable. It was concluded that risks were most sensitive to the transportation operation and in particular to road accidents. Loading and unloading operation risks were not negligible, however. Furthermore, the associated potential release quantities were much greater. Results from each of the methodologies used to analyze the loading and unloading systems are further described in this section.

The **site visits** proved critical to the risk assessment because a number of factors were identified which otherwise may not have been considered. The involvement of operations personnel during site visits ensured that their experience knowledge base was incorporated within the analysis. Engineering and procedural documentation is seldom up-to-date, and often does not reflect changes specifically affecting hazard control. In addition, safety features identified in the documentation may be of limited effectiveness due to the implementation or maintenance carried out on site. For example, although loading instructions for the truck driver were posted at all three locations, the location, readability, completeness, and overall effectiveness varied significantly.

Historical incidents resulting in releases of flammable compressed gas during loading and unloading operations were collated. Although records of such incidents were not comprehensive, relevant information on causes and consequences was presented as shown in the example given in Table 1. Such information provided insight to historical accident scenarios, however, equipment and procedures have generally been improved over the years and therefore some scenarios were not applicable to the case study.

Table 1: Example historical releases of inflammable compressed gas during loading and unloading.

Year	Location	Material Released	Fire	Explosion	Property Loss $	Amount Released (Litres)	Fatalities (F) Injuries (I)	Available Data
RELEASES DURING LOADING								
1973	Nebraska LPG Gas Bulk Plant	LPG	x		8,000		1(I)	Tanker truck ignited while loading.
1977	Kentucky Refinery	NGL	x	x			2(F)	Natural gas tube-type tanker truck exploded during loading.
1986	Alberta	Butane						Filling line ruptured while filling tanker truck. Vapour from line ignited at nearby construction site.
RELEASES DURING UNLOADING								
1978	Florida LP Gas Plant	NGL	x					Tanker truck unloading at the tank farm.
1980	Binger, Ok.	LPG	x				1(F)	Tanker truck moved during unloading and broke transfer hose. Car ignited plume.
1986	West Virginia	LPG	x					Flexible line ruptured.

A safety review of the three loading facilities was carried out to evaluate the **compliance with standard practices**. A number of differences were found among facilities due to such influences as the age of the facility and the modifications to improve original designs. An example of the comparison among designs (not including details) is given in Table 2. Safety protection devices were also specifically compared among facilities and the results were presented as shown in the example given in Table 3. From a safety perspective, design consistency will ensure all facilities benefit from the safest practices. Although all installations met basic design codes, the safety review provided a method for transferring the safety technology among all the facilities.

In order to identify potential accident scenarios, a **knowledge based hazard and operability study (HAZOP)** was carried out at each facility. The study involved input from operators, truck drivers, technical support engineers, and management personnel in addition to risk analysts. Standard HAZOP procedures were adapted to efficiently utilize the knowledge base of both the operators and the historical industry performance. Results from the study were collated in spreadsheets such as the example given in Table 4. As shown in this example, hazard scenarios were identified through systematic questioning of potential deviations in the operations. Any existing protection from an accident scenario in terms of both safety control devices and safety procedures are indicated in the spreadsheet. Finally, remarks on potential risk mitigation measures are noted. Recommendations for reducing risks resulted directly from applying this methodology and are further described at the end of this section.

Table 2: Example of loading facility design comparisons.

Design Feature	Site 1	Site 2	Site 3
Pressure relief on pump discharge	No	No	Yes
Pressure switch on pump discharge	No	No	Yes
Storage Vessel Pressure Relief	Yes	Yes	Yes
Pressure alarm low on storage vessel	No	No	Yes
Sight glass on storage vessel	Yes	No	No
Breakaway valves			
- on NGl liquid line	Yes	Yes	Yes
- on equalizing line	No	No	Yes
Loading Instructions	Inadequate	Inadequate	Complete

Table 3: Example of loading facility safety device comparison.

Function	Site 1	Site 2	Site 3
Pressure relief			
- NGL storage vessel	Yes	Yes	Yes
- Feed line to vessel	Yes	Yes	Yes
- Discharge from pump	No	No	Yes
Flow control			
- Excess flow			
pump suction	Yes	Yes	Yes
pump discharge	Yes	Yes	Yes
NGL vessel feed	Yes	Yes	Yes
- Breakaway valves			
Pump discharge	Yes	Yes	Yes
Equalization line	No	No	Yes
- Facility isolation	Yes	Yes	Yes
System Shutdown			
- ESD			
Pump suction flow	Yes	Yes	Yes
Pump shutdown	Yes	Yes	Yes
Pump discharge	No	No	Yes
- Low pump flow	Yes	Yes	Yes
- High pump discharge pressure	No	Yes	Yes

A **fault tree analysis** was then carried out to decompose the accident scenarios previously identified into a sequence of causal factors. An example of part of a subtree is presented in Figure 5. In this example, a storage vessel release is defined by a vessel failure or a release on the equalizing or NGL side of the vessel (Logic given by OR gate). Both equalizing and NGL side releases require an equipment failure and a protection device failure to occur (Logic given by AND gate). The protective check valve is only located at one facility and therefore equipment failures at the other facilities will lead directly to a storage vessel release. NGL side equipment failures include piping, or control valves, or breakaway valves and or the PSV (Logic given by OR gate). In general, causes identified in the fault tree include mechanical failures, environmental effects, operator problems, and maintenance effects. All historical scenarios were included in the fault tree representation along with scenarios specific to the case study sites. Although the limited

Table 4: Example of knowledge-based HAZOP study results.

| Scenario | Hazard | Protection Provided | | | | | | Remarks |
| | | Loading | | Unloading | | | |
		Control	Procedures	Control	Procedures		
NGL flow fails to shut down when trucker presses ESD button	Plant and personnel not protected from hazard requiring ESD action	• suction to pumps shut down • pumps shut down	• ESD • radio to control room	• fill line shut down • vent line shut down • shipping line shut down	• ESD • phone to control room		Adequate protection
Truck rolls and breakaway valves on NGL line fail to close	NGL release to atmosphere	• excess flow valve	• ESD	• jerk wire	• ESD		• Review adequacy of excess flow valve tight shut-off. • Ensure procedures to ensure jerk wire connected.
Inadequate loading instructions	Improper action by trucker causing release of NGLs	• ESD button		• ESD button			• Provide permanent sign listing procedures. • Define procedures as completely as possible. • Ensure trucker training.
PSV emissions not adequately dispersed	NGL gas pocketing resulting in fire or explosion						• Check for safe release.

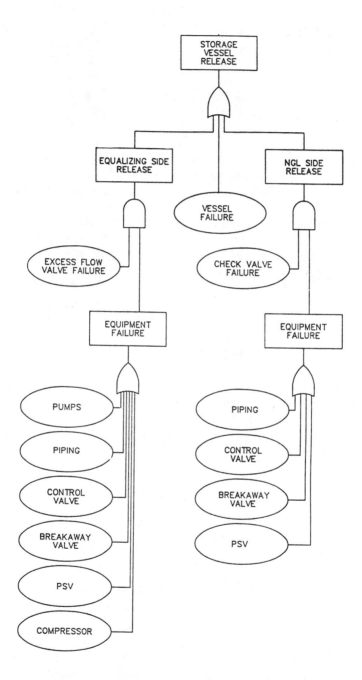

Figure 5: Portion of fault tree describing fire or explosion while unloading NGLs.

information available on historical accidents prevented quantification of base events, quantification of higher level events was possible. Most importantly the development of the fault tree model provided a method to assess risk sensitivity to causal factors. A number of specific risk mitigation measures were recommended based on the fault tree analysis and are described in the following paragraph.

Measures which may be implemented to mitigate the risk associated with the loading and unloading facilities were identified from the safety reviews, the hazard and operability studies, and the fault tree analysis. The combination of methods provided recommendations for transferring safety experience from one site to all sites, for mitigating new hazards which were identified and finally, for optimizing the implementation of measures to effect the greatest risk reductions. Generally, both preventative and protective measures were identified to reduce the potential occurrence of fires or explosions, however, resulting consequences were not analyzed. An example of risk mitigation measures recommended from the case study are given in Table 5. As may be seen, three types of examples are given; namely, two associated with the facility equipment, one associated with safety devices, and one association with procedures.

Table 5: Example risk mitigation measures.

Recommendation	Action Taken
Provide rack to support hose	Provided cantilever swing-away support
Install breakaway valves on equalizing lines	Done and operation jerk lines checked
Vent PSVs at safe height	Vents raised
Upgrade and post permanent loading/unloading procedures	Procedures upgraded to include Pre-loading /unloading safety checks. Large sign displaying trucker's operating procedures installed at each site.

The above results were generated within the required level of effort and met the original objectives of the case study. Risks associated with the transportation system were identified, the effectiveness of potential risk mitigation measures was estimated and results were presented in a format suitable for assisting management decisions. Of the 20 principal recommendations resulting from the study, 14 were effectively implemented shortly after completion of the study, 3 were not implemented because of changes in operating strategy and 3 were not implemented because of a deliberate and conscious decision to accept the risk.

6.   GENERAL CONCLUSIONS

The case study showed that the loading and unloading risks were not negligible and improvements were clearly identified which could be implemented to further mitigate the risks.

It is apparent from this case study that a mix of risk analysis methodologies can provide the optimum procedure to identify, analyze, and mitigate the risks involved in transporting dangerous goods. Specifically, it was found that risk methodologies involving:

a) Site visits and natural operating condition assessments;

b) Historical incident record reviews;

c) Standard practice compliance reviews;

d) Knowledge based HAZOP studies;

e) Fault tree analysis;

were all important for the case study. Each contributed to the successful achievement of the study objectives.

One of the greatest benefits of this case study was the increased awareness obtained by both operations and management of the handling and transportation risks. The risk analysis methodology provided a very understandable explanation of the risk elements involved in handling and transporting NGLs. The fact that 70% of the recommendations were implemented almost immediately attests to the usefulness of the methodology in identifying risk elements and potential improvements and in providing a rational justification for implementing changes.

7.   REFERENCES

BERCHA AND ASSOCIATES LIMITED, LAVALIN OFFSHORE INC., FENCO CONSULTANTS LIMITED (1983). *Oil Spill Risk Assessment*. Final Report for Dome, Esso, and Gulf, supporting Environmental Impact Statement for the Beaufort Sea Development.

BERCHA, F.G. & ASSOCIATES LIMITED (1989, December). *Caroline Project Emergency Risk Analysis*, Final Report to Shell Canada Limited.

BURTON, I. ET AL. (ED.) (1982). *Living with Risk, Environmental Risk Management in Canada.* Institute for Environmental Studies, University of Toronto.

CENTRE FOR CHEMICAL PROCESS SAFETY (AIChE) (1985). *Guidelines for Hazard Evaluation Procedures.*

CENTRE FOR CHEMICAL PROCESS SAFETY (AIChE) (1989). *Guidelines for Chemical Process Quantitative Risk Analysis.*

FISCHOFF, B. ET AL. (1981). *Acceptable Risk.* Cambridge University Press.

HENLEY, E.J., & KUMAMOTO, H. (1981). *Reliability Engineering and Risk Assessment.* University of Houston and Kyoto University.

LEES, F.P. (1980). Loss Prevention in the Process Industries. *Hazard Identification, Assessment and Control*, Vols. I and II.

NARS-Nordic (1975). *Working Group on Reactor Safety Recommendations*, 35.

PATE-CORNELL, E. (1984, September). Fault Trees Vs. Event Trees in Reliability Analysis. *Risk Analysis.*

RASMUSSEN, N.C. (1981). The Application of Probabilistic Risk Assessment Techniques to Energy Technologies. *Annual Review of Energy.*

SALEM, S.L. ET AL. (1980). *Summary of Issues and Problems in Inferring a Level of Acceptable Risk.* The Rand Corporation.

Chapter 4: Uncertainty in Risk Estimation

Analysis of Risk Uncertainty for the Transport of Hazardous Materials

F.F. Saccomanno
Institute for Risk Research
University of Waterloo
Waterloo, Canada, N2L 3G1

O. Bakir
Dept. of Civil Engineering
University of Waterloo
Waterloo, Canada, N2L 3G1

ABSTRACT

An approach is presented for considering uncertainty in the analysis of risks for the transport of hazardous materials, taking into account variations in the risk inputs and in its final estimate. Two types of uncertainty are considered: uncertainty of knowledge concerning reliability in the estimates and uncertainty of process concerning the degree of inherent randomness in the input and risk output variables.

The approach is illustrated through an application to a simple risk estimation problem for an accident-induced release of hazardous materials in road transportation. Most likely point estimates of risk and its inputs are obtained based on a reported sample. Uncertainty of knowledge is considered by establishing confidence intervals about these estimates. Uncertainty of process is considered by establishing a probability density function for the release rate. This function can be used to set tolerance levels for the acceptance of risk for a range of possible values.

1. INTRODUCTION

A number of significant advances have taken place in recent years concerning the analysis of risks for the transport of hazardous materials. With a better understanding of the process and access to more reliable data, researchers are able to obtain more accurate estimates of the potential risks involved. In recent years, models that estimate the risks of transporting hazardous materials have become increasingly more rigorous and complete. These models attempt to incorporate the full complexity of the relationships that govern the transport of hazardous materials by various modes and different shipment conditions.

One would expect that as the level of understanding of the process increases, the resultant estimates of risk would be subject to reduced uncertainty and enhanced confidence. However, research into risk uncertainty has only underscored a general lack of agreement among the research community on the nature and reliability of the reported estimates. Despite controlling for underlying assumptions, risk analysis continues to reflect significant inconsistencies in the estimates as reported by various groups, and contradictory conclusions regarding the most appropriate actions to take.

The problem of obtaining reliable estimates of risk and its inputs is referred as "uncertainty of knowledge". One of the objectives of this paper is to review the extent of uncertainty that is present in current risk estimates for the transport of hazardous materials. Since, as is often the case, these estimates are based on very limited sample sizes, they cannot be expected to fully coincide with the true values they are intended to estimate. Given this uncertainty, it is preferable to report these point estimates as "intervals", that bound the "true value" with a reasonable degree of certainty. Accordingly, the first phase of this exercise is to establish confidence intervals for risk and its inputs, based on the reported sample estimates.

A second type of uncertainty needs to be considered. This is the "uncertainty of process" that results from inherent randomness in risk estimation. For the transport of hazardous materials, risk is generally expressed as a random product of a series of random inputs, e.g. accident likelihood, containment system failure or release fault, rate and volume of material released, hazard areas associated with each potential threat for different releases and materials, and damages to nearby population and environment. The values assigned to anyone of these inputs are viewed as choices from a bounded sample space, with a unique probability density function. The nature of this function is generally unknown and must be inferred based on the behaviourial properties of each of the random inputs. The probability density function for risk is a combination of the input functions for all constituent inputs. One of the major objectives of this paper is to present an approach for establishing a combined probability density function for risk for a given set of inputs and a given risk formulation.

The analysis of "uncertainty of process" is carried out in two steps: 1) obtain a probability density function for each input, based on prior information concerning the nature of the "random process", and 2) combine these expressions into a resultant probability density function for risk (the random output process). Both input and output functions reflect random variations over specific intervals of observation time and/or space. These random variations are often observed in the data, since data are subject to practical time and space limitations, and since risks for the transport of hazardous materials represent very rare events.

2. CONSIDERING UNCERTAINTY OF KNOWLEDGE

2.1 Sources of Uncertainty in the Estimates

One of the problems in obtaining reliable estimates of risk is inherent variability caused by different assumptions and jurisdictional factors. This limits the number of estimates required to ensure a certain degree of reliability. In this section of the paper, several important sources of uncertainty of knowledge are discussed.

Each of the constituent risk inputs requires the specification of separate models, with their unique set of parameters and relationships. Uncertainty concerning the values assigned to these inputs has three major sources (Saccomanno, Stewart and Shortreed, 1991):

1) Structural differences in the risk models.

2) Underlying assumptions of application and

3) Jurisdictional differences concerning the calibration and validation of model parameters.

A number of examples help to illustrate these sources of uncertainty.

1) Controls placed on accidents for vehicles transporting hazardous materials. These controls are factors which either alone or in combination explain significant variations in accident rates, and include: road/track geometry, traffic composition, and environment. Many studies concerning the risks of transporting hazardous materials are based on average input values rather than values that reflect the true exposure to controlled accident risk, for example, road accident rates that fail to account for differences in highway design, traffic and weather conditions. In many cases, rates that have been estimated for one set of control conditions are then applied to other conditions, without appropriate adjustments. Saccomanno, Stewart and Shortreed (1991) have shown that differences in estimates for accident rates, fault and release probabilities and hazard areas can result in variations in risk of several orders of magnitude (Figure 1).

2) Omission of significant factors from the risk models because they are unavailable in the data bases in certain jurisdictions. For example, the omission of intersection accidents for road transport, which can represent a large component of all accidents involving trucks.

3) Frequently, inputs into the risk analysis process are estimated conservatively, creating a bias towards the worst-case position - a position that is not necessarily reflective of reality. Biases in any constituent risk component can produce cumulative biases in the final risk estimates. The range of uncertainty associated with the final risk estimate may be well beyond acceptable limits on the true value. The Toronto Area Rail Task Force (1988) contended that some of their estimates were most likely high by a factor of 3 due to biasing assumptions to the worst case.

Uncertainty may also arise from jurisdictional differences that affect risks and relevant assumptions concerning the application of the process. Frequently, the absence of data in a given jurisdiction (for example, data on releases involving hazardous materials) has necessitated aggregation of data with other jurisdictions for the purpose of calibrating and validating specific risk models. Confining the analysis to one jurisdiction alone would not resolve this problem, since data from the same jurisdiction could still be subject to technological variations over time.

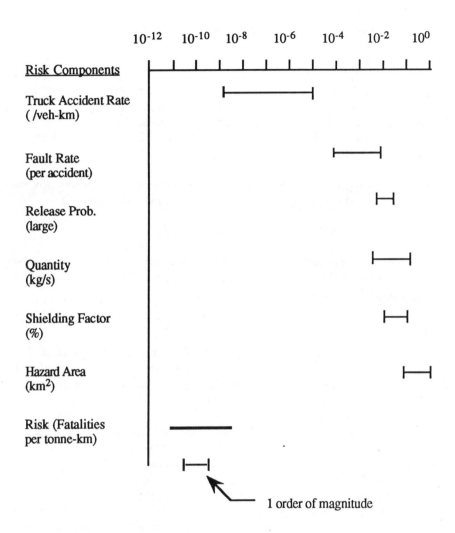

Figure 1: Variations in reported risk estimates.
(Source: Saccomanno, Stewart & Shortreed, 1991).

Many of these assumptions and jurisdictional variations in risk
estimates can be handled through exogenous adjustments. The
purpose of the exercise is to ensure that all estimates apply to a
uniform set of conditions. Nevertheless, even after variations
caused by assumptions and jurisdictional differences are taken
into account, there remains a certain amount of uncertainty in the
estimates that will have to be considered.

2.2 Representing Uncertainty in Sample Estimates

For illustrative purposes in this paper, risk has been defined as a function of two random inputs, i.e., accident involvement and accident-induced release or fault. The most likely value for risk (i.e. its sample mean) is simply a function of the most likely values for its random inputs (ie. their sample means). Since these values are obtained from sample information, they are subject to uncertainty of knowledge.

For a simple product formulation, the mean risk value can be expressed as the product of the sample input means, such that:

$$\mu_R = \mu_X * \mu_Y \tag{1}$$

In a similar manner, a combined risk variances can be estimated from sample estimates of the inputs variances. For two inputs, the expression is of the form:

$$\sigma_R^2 = \mu_Y^2 \sigma_X^2 + \mu_X^2 \sigma_Y^2 \tag{2}$$

Both eqs. 1 and 2 assume that the input variables X and Y are uncorrelated, these expressions can be modified to account for possible correlations among the inputs for a more complex risk formulation (Hahn and Shapiro, 1967).

In addition to reporting the mean and variance of risk and its inputs as single point estimates from a sample, it would be more helpful to express uncertainty in the interval within which their "true value" is expected. The method of confidence intervals offers a way to do this. In this method, we make a statement on the possible range of values for a given variable, based on the results of an experiment (i.e. the reported sample estimates). We can then report with a certain level of confidence (say 95% confidence) that the true value is bounded by the given interval. The method assumes normality in the difference between the sample estimates of a random variable and its true value.

3. CONSIDERING UNCERTAINTY OF PROCESS

Why is it important to consider uncertainty of process for risk and its inputs? Frequently, we are interested not only in the most likely estimate for these variables, but also in the probability that certain values will or will not be exceeded over a given period of time for an assumed set of conditions. For example, the most likely estimate of an event may be found to be acceptable, but given random variations in this variable, there may be an unacceptable chance that a certain critical value will be exceeded over a realistic period of time. This aspect is especially important is setting risk tolerance criteria for the transport of hazardous materials. By considering uncertainty of process, it is possible to assign to each estimate a tolerance measure, essentially, the probability of extracting such a value in a random experiment (e.g. the choice set of sample estimates).

3.1 Probability Density Functions for Risk Inputs

The initial step in this exercise is to obtain, through assumption, realistic probability density functions for the risk inputs. We begin by considering a single risk input (random variable X representing for example accident involvement). Let $f(X;\theta)$ be the probability density function for the random variable X with an unknown parameter, θ. For simplicity, we assume a single parameter for this function; although the approach can be extended to include probability functions with multiple parameters. While the form of this probability density function can be assumed, we will need to estimate its parameter θ, based on observed values of X (accident involvement) from k independent and equally reliable sources. The value assigned to θ is a problem of uncertainty of knowledge in the value of the parameter due to limitations in the sample.

Initially, we define $X_1, X_2 \ldots X_k$ as the observed numbers of accidents in k blocks of n trucks, as obtained by each of k independent and equally reliable sources. The term X_i is an integer random variable defined over the values 0 to n, for a limited time period. The probability density function for X_i can be expressed as:

$$f(X_1, X_2 \ldots X_k \mid \theta) = f(X_i \mid \theta) \qquad (3)$$

for the single input parameter θ.

According to Ochi (1990), a likelihood function can be established for each of the parameter estimates (θ). For the single population case (eq. 3), the expression is of the form:

$$L(\theta) = \prod_{i=1}^{k} f(X_i \mid \theta) \qquad (4)$$

$$= f(X_1 \mid \theta) \ f(X_2 \mid \theta) \ldots f(X_k \mid \theta) \qquad (5)$$

The most likely value for the input parameter (θ) can be obtained by maximizing the likelihood function (partial derivative of eq. 4 with respect to θ and setting the result equal to 0).

To ensure that the accident involvement rate for each vehicle is large, we assume that all trucks in the block travel over a sufficiently large number of kilometres in a given year. Since the average accident rate lies in the 10^{-9} to 10^{-5} range per vehicle - kilometres (Figure 1), we assume an arbitrary distance exposure value of one million vehicle - kilometre per year for each truck in the block.

The above approach assumes a known distribution for the random variable X. In the absence of reliable information on truck accidents, we must establish a probability function $f(X|\theta)$ based on the inherent behavioural properties of the random variable X to account for randomness in the possible values assigned to this variable. For this exercise, we will assume a Binomial distribution, such that for each truck two values are possible: 0 for no accidents and 1 for an accident occurrence. If we observe X accidents in a block of n trucks travelling one million kilometres per year, then the best unbiased estimate for θ is X/n.

Based on these assumptions, the number of trucks X that experience an accident can be expressed as:

$$f(X|\theta) = \binom{n}{x} \theta^x (1-\theta)^{(n-x)} \qquad (6)$$

where $\theta =$ unknown parameter representing the average truck accident rate per million kilometres travelled.

     $f(X|\theta) =$ probability the X trucks experience an accident based on the average accident rate of θ.

From eq. 6, the best estimate for X is given as $n\theta$, where θ is the most likely value for θ from the k source estimates. Again from the Binomial, the most likely estimate for the variance can be expressed as: $n\theta (1 - \theta)$, for a block of n trucks in the sample. The above procedure provides an unbiased estimate of the parameter θ in the probability function $f(X|\theta)$ for the accident input factor. Here we are assuming that the k independent estimates for the number of accidents in a block of n trucks comes from the same or similar blocks.

The adoption of the Binomial to represent accidents in this example is based on at most one accident being allowed per vehicle per observation period. This is for illustrative purposes only. If more than one accident is permitted, the assumption of the Binomial would be incorrect and the Poisson would have to be used.

The above approach can be extended to include any number of constituent risk inputs (e.g. fault occurrence in an accident situation or random variable Y) with their unique probability density functions. In this example, the occurrence of faults in a block of vehicles (random variable Y) has also been represented by the Binomial distribution. The probability density function for Y, defined as $g(Y \mid \phi)$, is an expression similar to eq. 6. Since accident-induced release depends on the prior occurrence of an accident, the random variable Y is defined over a block of trucks experiencing accidents (an integer subset of n).

3.2 Obtaining the Combined Risk Probability Density Function

In this section, a direct approach for estimating the probability density function for risk, $P(R)$, is presented. This approach has been adapted from Hahn and Shapiro (1967) and is applicable to very simple risk formulations. Given numerical difficulties inherent in this direct approach, a parallel approximation is also presented that is based on a system moment generation method.

The direct procedure is initiated by selecting an arbitrary auxiliary random variable Z, that is a function of X (accidents) and Y (release faults):

$$Z = q(X,Y) \tag{7}$$

for the most likely parameter values, θ and $\hat{\phi}$. The nature of this expression is arbitrary and can be modified iteratively depending on the numerical procedure. A wise choice of the auxiliary function can facilitate the calculations considerably. The number of auxiliary variables required by this procedure is one less than the number of inputs in the risk expression (in this case 1 auxiliary function for variables X and Y). As a first step, we select an auxiliary function of the form:

$$Z = X/Y \tag{8}$$

For this formulation the joint probability function for risk is expressed as:

$$p(R \mid \hat{\theta}, \hat{\phi}) = f(X \mid \hat{\theta})\, g(Y \mid \hat{\phi}) \tag{9}$$

Defining the random variables X and Y as functions of R and Z, we obtain:

$$X = h(R,Z) \tag{10}$$

$$Y = 1(R,Z) \tag{11}$$

Substituting $Z = X/Y$ and rearranging terms we obtain:

$$X = \sqrt{RZ} \tag{12}$$

and

$$Y = \sqrt{R/Z} \tag{13}$$

The joint probability function for R and Z become:

$$n(R,Z) = m[X(R,Z), Y(R,Z)]\,[J] \tag{14}$$

$$n(R,Z) = m[\sqrt{Rz}, \sqrt{R/Z}\,]\,[J] \tag{15}$$

where J is the Jacobian of X and Y with respect to R and Z.

We now have a joint probability function incorporating the two combined outputs, R and Z. To rewrite this expression in terms of R alone, it is necessary to integrate the function, $n(R,Z)$ with respect to Z. This yields the probability density function for risk directly as a function of the input distributions for the random variables X and Y. The procedure for isolating the $n(R, Z)$ expression in terms of R is complex for all but the simplest "risk" expression. Hahn and Shapiro (1967) suggests using a system moment generation method to obtain an approximate solution to the above problem.

The moment generating method yields an empirically-determined probability density function for risk, i.e., $p(R|\theta,\hat{\phi})$, given the probability functions for each of its inputs, i.e., $f(X|\theta)$ and $g(Y|\hat{\phi})$. As the name implies, this approximate solution requires the estimation of the first four moments of the random inputs X and Y and the combined variable R. Hahn and Shapiro (1967) has summarized the expressions for estimating the first four moments of risk based on the combined moments of its constituent inputs (i.e., mean, variance, skewness and kurtosis), for any number of inputs and any amount of inter-variable correlation.

To obtain the probability density function for risk, it is necessary to fit selected functions empirically to mathematical expressions described by these moments. We define two terms, β_1 and β_2, as the measures of skewness and kurtosis. Along with the means and variance measures of risk, these parameters can be used to suggest a "best fit" probability function for risk, i.e., p(R). Hahn and Shapiro (1967) suggests a Pearson approximation procedure for this fitting exercise, such that various combinations of β_1 and β_2 define different "feasible" regions for the resultant risk probability functions.

4. DISCUSSION OF THE RESULTS

In this section, a sample set of estimates is obtained for truck accident rates and containment fault rates (given an accident). Several controls have been applied to these estimates to account for different shipment conditions. In this example, we consider a rural freeway situation for a tanker carrying pressure liquefied gas in bulk. Faults reflect failures in the truck tanker containment system that produce a release of hazardous materials, without regard to the size and rate of material released. Various estimates for accident rates and faults were extracted from the literature and these have been summarized in Tables 1 and 2, respectively.

Estimates of the means and variances for accident and fault rates in Tables 1 and 2 are sample estimates, which are subject to uncertainty of knowledge. The 95% confidence intervals for accident and fault rates and for risk (probability of release fault) from this sample are as follows:

Accident Rates		Fault Rates	
Mean	0.61	Mean	5.45
Variance	0.05	Variance	1.10

95% Confidence Intervals (Small samples and the t-distribution)

Mean	0.45 - 0.77	Mean	4.39 - 6.51
Variance	0.02 - 0.17	Variance	0.39 - 9.09

Risk

Mean	3.32
Variance	1.90

95% Confidence Intervals (Sample size n = 50 and normal distribution)

Mean	2.93 - 3.70
Variance	1.30 - 2.88

Table 1: Large truck accident rate estimates (rural roads only).

Freeway	Sources
0.45	Rhyne (1990)
0.40	Harwood (1990)
0.54	Khasnabis, TRR 753 All trucks (1979) 1970-77
0.50	Khasnabis, TRR 753 Michigan trucks (1979) 1972-77
1.12	Khasnabis, TRR 1052 Toll expressways
0.49	Michie (1984)
0.39	Buyco and Saccomanno (1988)
0.76	Harwood and Russell (1989)
0.83	Bercha (1985)
0.59	Saccomanno (1990)

$\hat{\theta} = 0.61$ $\hat{\sigma} = 0.23$

Units: truck accidents per million truck-kilometres

Table 2: Accident-induced faults (rural roads, tanker trucks).

Freeway	Sources
5.16	Harwood California, TRR 1264, 1990
6.90	Harwood Illinois
5.90	Harwood Michigan
5.59	Harwood Weighted Average
3.70	Saccomanno, TRR 1264, 1990

$\hat{\theta} = 5.45$ $\hat{\sigma} = 1.05$

Units: release faults per 100 truck accidents

Notwithstanding the consideration of estimates from similar sources in this simple example, many of the reported risk estimates remain outside the 95% confidence interval. For example, the accident rate from Buyco and Saccomanno (1988) is 0.39 and the fault rate from Saccomanno (et al., 1990) is 3.70. These inputs yield a most likely risk estimate of 1.44, significantly less than the lower bound value of 2.93 for the above 95% confidence interval about the mean.

Given the diversity in the small number of sources reported in our sample, a significant amount of variation in the estimates could still be caused by differences in assumptions and jurisdictional aspects, which have not been accounted for in our analysis. We have assumed that these estimates apply to the same problem, where in fact different problems may have been addressed. It is not surprising, therefore, that many of the estimates were found to lie outside the 95% confidence interval for the mean accident and fault rates. The data set in this example is plagued by an absence of control on the assumptions and few observations. This is reflected in the large confidence interval assigned to variance in the fault rates, where only five observations are involved.

Application of the system moment generating method with a Pearson curve fitting procedure suggests a Normal distribution for the risk output random variable. The results from this simple example are summarised below:

Accident Rates	**Fault Rates**
Density Function:	Density Function:
Binomial Assumed	Binomial Assumed

Risk (Probability of Release Fault)

β_1	0.001	Fitted parameters
β_2	3.12	for Pearson curve.

Density Function:
Normal Fitted

For the control conditions assumed in this example, the cumulative density function for risk can be expressed as standard normal:

$$F(R^*) = \frac{1}{\sqrt{2\pi}} \int_{\infty}^{R^*} \exp{-t^2/2\delta t} \tag{16}$$

where R corresponds to the "standardised" release rate value under consideration, and t is the standardised random variable for risk with a range of possible values from $-\infty$ to R^*.

The cumulative probability distribution for risk, illustrated in Figure 2, should not be confused with the traditional F-N curve representation of risk, wherein the frequency of N or more events (e.g. fatalities) is plotted against N its actual value. The distribution illustrated in this Figure applies to a set of control

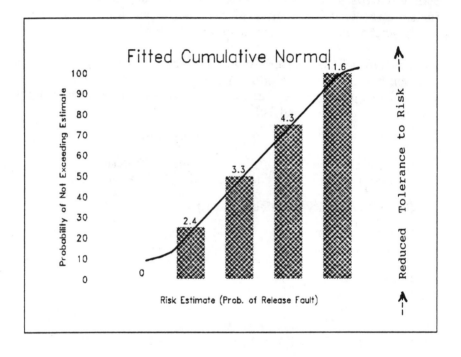

Figure 2: Cumulative probability density function for
risk (release fault).

conditions that fix the extent of consequent damages (i.e., in our example the release fault). This corresponds to the inherent uncertainty of a single point on the F-N curve. The F-N curve can be bounded by any isopleth reflecting the probability that each point on the curve will be exceeded for a given period of time, given the randomness in the process.

For the assumed conditions adopted in this example, the most likely estimate for risk (probability of accident-induced release) is 3.32 releases per million vehicle - kilometres. By itself, this may be acceptable given acceptable risk tolerance levels for the transport of hazardous materials by road. Uncertainty in the process, however, suggests that there is a 10% chance that the risk may be as high as 6.30 releases per year, and this latter value may be unacceptable given the conventional wisdom for this type of activity.

This paper has used a simple example to illustrate how risk can be considered as a random variable with a given probability density function and confidence intervals about each estimated value. The analysis has been hampered by the absence of reliable estimates for a consistent set of assumptions. Nevertheless, the approach has shown that determining the nature of the probability density function for risk (whatever its definition) can facilitate the treatment of uncertainty in risk estimation.

5. BIBLIOGRAPHY

BERCHA, F.G., & ASSOCIATES (1985). Analysis of Transportation Routing Between Edmonton and Swan Hills. Final report prepared for the Alberta Waste Management Corporation.

BUYCO, C., & SACCOMANNO, F.F. (1988, June). Analysis of Truck Accident Rates Using Loglinear Model. Canadian Journal of Civil Engineering, 15, 397-408.

HAHN, G.J., & SHAPIRO, S.S. (1967). Statistical Models in Engineering, John Wiley & Sons, Inc., New York.

HARWOOD, D.W., VINER, J.G., & RUSSELL, E.R. (1990). Truck Accident Rate Model for Hazardous Materials Routing. Transportation Research Board 1264, 12-23.

HARWOOD, D.W., RUSSELL, E.R., & VINER, J.G. (1989). Characteristics of Accidents and Incidents in Highway Transportation of Hazardous Materials. TRR 1245, 23-33.

KHASNABIS, S. (1984). Operational and Safety Problems of Trucks in No-Passing Zones on Two-Lane Rural Highways. TRR 1052, 36-43.

KHASNABIS, S., & ATABAK, A. (1979). Comparison of Accident Data for Trucks and for All Other Motorized Vehicles in Michigan. TRR 753, 9-13.

MICHIE, J. (1984). Large Vehicles and Roadside Safety Considerations. TRR 1052, 1984, 90-95.

OCHI, M.K. (1990). Applied Probability and Stochastic Process in Engineering and Physical Sciences. John Wiley & Sons, Inc.

RHYNE, W.R. (1990). Evaluating Routing Alternatives for Transporting Hazardous Materials Using Simplified Risk Indicators and Complete Probabilistic Risk Analyses. Transportation Research Board 1264, 1-11.

SACCOMANNO, F.F., STEWART, A., & SHORTREED, J.H. (1991). Uncertainty in the Estimation of Risks for the Transport of Hazardous Materials. Paper presented at International Conference: HAZMAT Transport 91. Transportation Center, Northwestern University, 1991.

TORONTO AREA RAIL TRANSPORTATION OF DANGEROUS GOODS TASK FORCE (1988). Final report prepared for Transport Canada, Catalogue No. T44-3/14E.

International Discrepancies in the Estimation of the Expected Accident and Tank Failure Rates in Transport Studies

Philippe Hubert
Institut de Protection et de Sûreté Nucléaire
Commissariat à l'Energie Atomique
CEN / FAR - BP N° 6
92265 Fontenay-aux-Roses Cedex
France

Pierre Pagès
Françoise Rancillac CEPN
CEN-FAR - BP N° 48
92263 Fontenay-aux-Roses Cedex
France

ABSTRACT

Published data on accident and failure rates for both road or rail transport modes on which risk studies are usually based have been compared from one country to another (France, Netherlands, UK, USA) with a view to help reflection in the problem of achieving international standardization. The study addresses the tankers that are used for the transport of gases liquefied under pressure (e.g., LPG and ammonia). Part of the discrepancies observed between figures come from different methodological approaches, such as differences in accident definition or in data collection procedures. But of course actual transport environment has to be put forward as well. A comparison is carried out between observed release rates in road or rail accidents using recent French data and values assumed in the surveyed literature. Although more pessimistic than others, values used in French risk assessment studies are supported by this comparison.

1.   INTRODUCTION

The development of Probabilistic Risk Assessment (PRA) for hazardous material transportation has mainly relied upon the availability of basic data to estimate parameters such as accident and loss of containment frequencies. But a research in the literature (generic studies and PRAs on main toxic or flammable products) shows that in very few cases accident data have led to the setting up of a data base suitable for PRA (Dennis, 1978; Clarke, 1976; Degrange, 1989). In most PRA studies, probabilities are assessed after reviewing heterogeneous data (Appleton, 1988; Lannoy, 1984; HSC, 1991; Degrange, 1984; Meslin, 1979) and are rarely drawn from a single specific data source, or at least with the same consistent definitions of the events likely to occur.

This situation precludes any direct comparison of assessments and consequently any fruitful exchange of results or experience. As a first step and with a view to help in reaching some sort of standardization in this field, the present paper will examine how accident and spill rates have been assessed in generic studies (cf. supra) or in studies dealing with a specific product such as liquefied petroleum gas (Hubert, 1990; Geffen, 1980; TNO, 1983) or ammonia (Castellano, 1989). A second part of this paper is devoted to a closer look at French rail and road statistics, in order to test for the compatibility of field data with international (including earlier French publications) hypotheses used in risk assessment studies.

2.   PROBLEMS IN COMPARING PUBLISHED FIGURES

2.1 Methodological Framework

None of the published figures are straightforwardly comparable. Indeed the definitions for "accident" and "release" are numerous, as well as the methods for assessing the corresponding figures (see Figure 1).

The release frequency (probability of a release per vehicle, per kilometer), the accident frequency and the conditional release probability (probability of a release given an accident) are the key parameters in the risk assessment, but they may not be accessible to direct observation. It may happen that the traffic (vehicle-kilometers) is poorly known, or that hazardous materials accidents are not recorded as such. In some countries, releases, especially large releases, have not been observed.

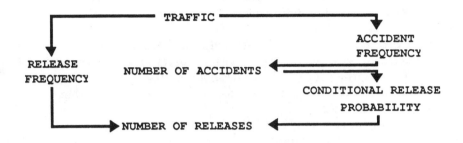

Figure 1: Observable events and estimated parameters.

In the case of road transportation in France, it is possible to use a fairly good record of hazardous material transportation accidents, and the number of "thin walled" tanker releases is high enough for direct estimation of the conditional release probability. However, the traffic is poorly known and it is preferable to deduce the accident frequency from generic lorry statistics. Looking at "thick walled" tankers, releases are scarce, and conditional release probabilities can be determined by applying a deterministic failure analysis to a statistical description of accident environment stresses as suggested in the first Sandia analysis in 1976 (Clarke, 1976). Nevertheless, the small number of observed releases can be used when looking at the plausibility of the theoretical analysis.

An interesting approach has been adopted in Great-Britain, where rail accidents with "thin walled" tankers have been carefully depicted from a deterministic point of view, in order to assess, in every case, a failure probability for "thick walled" tankers. In other words, actual "thin walled" tanker accidents were used as a statistical basis for accident environment description.

Keeping in mind those differences in assessment methodologies, a proper comparison requires the careful definition of the accident and release definition.

2.2 Accident definition

The following definitions have been adopted and the global consistency is not obvious; however, the concepts are usually homogeneous in a given country :

- The Netherlands (TNO, 1983): Accidents with casualties to the lorry's crew and with material damage to the lorry. This is the most restrictive definition.

- USA (Clarke, 1976; Dennis, 1978; Geffen, 1980): For road, accident with casualties, or with property loss in excess to 250 $. For rail, the property loss threshold is raised to 750 $. Note that, on average, 66 wagons are found in a train, 10 of them being involved in an accident. Indeed, the concepts of "train accident" or "wagon accident" are quite intricate, and this situation raises questions when comparing frequencies.

- United Kingdom (Appleton, 1988): Accidents with casualties on the road. Note that the ratio from accidents in general (casualties or material damages) to accidents with health consequences ranges from 3.5 (urban areas and highways) to 1 (rural roads) and averages to 2.2. The frequency is specific of hazardous goods vehicles. Train accidents are "significant" accidents, such as derailments or collisions, or accidents with casualties. The accident rate is computed "per train". The rate "per wagon" has been deduced from US figures.

- France (risk assessment studies): The criterion for inclusion of accidents is casualty or traffic disturbance for road. The quoted figure makes allowance for a higher safety record for hazardous goods vehicles. Train accidents are those involving at least one wagon with material damage.

The figures quoted in the above mentioned studies have been synthesized in Table 1.

Table 1:   Accident frequencies: Occurrence per vehicle (road or rail tanker) and per km.

	Road Accident	Rail Accident
The Netherlands (TNO, 1983)	$4.4 \ 10^{-7}$	$1.4 \ 10^{-7}$
USA (e.g. Geffen, 1980)	$16 \ 10^{-7}$	$9.4 \ 10^{-7}$
United Kingdom (Appleton, 1988)	$6.7 \ 10^{-7}$	$0.7 \ 10^{-7}$
France (Hubert, 1990)	$5 \ 10^{-7}$	$1 \ 10^{-7}$

2.3 Release or failure definition

- The Netherlands (TNO, 1983): A release is considered when the loss of content is above 100 kg for LPG: Instantaneous total release, 7.6 cm diameter holes and small release rates are then distinguished. The same pattern applies to rail and road. Those figures are not given as country specific, rather they rely on some US data.

- USA (Geffen, 1980): As above mentioned, release probabilities are not statistical data, rather they come from a tank failure analysis applied to a statistical description of accident environment (Dennis, 1978). Puncture and breaches are estimated on those bases. Releases that occur without accident are also taken into account, but they are judged as "non significant".

- United Kingdom: The releases that are considered here are associated with punctures; note that the release frequencies (HSC, 1991) have been computed directly and that they are not explicitly linked to the above quoted accident frequencies (Appleton, 1988). Medium spills correspond to 35 kg s^{-1}. They have been associated here with "breach".

- France (Hubert, 1990; Castellano, 1989): The risk analysis is based on major releases (i.e., more than 50% of the content released). The approach is a best estimate after comparing the results of a deterministic method (see above) with a direct statistical approach.

When establishing a comparison, some engineering judgement has to be exercised because the importance of the release may have been defined by the amount released, the release rate or the hole diameter. Eventually, three categories were distinguished (see Table 2).

The distribution of results is quite large. It is even larger when one looks at the release of a significant amount, which is actually the start point in the consequence analysis, small releases being usually neglected.

Table 2a: **Release frequencies for road: Occurrences per vehicle-km (thick walled tankers).**

| | Release | Releases that can lead to consequences | | Minor release (or neglected) |
		Major release (≈ 0.2 m^2 or 100% of content)	Medium release (0.01 to 0.001 m^2)	
The Netherlands (TNO, 1983)	22 10^{-9}	0.66 10^{-9}	1.5 10^{-9}	20 10^{-9}
USA (Geffen, 1980)	180 10^{-9}	43 10^{-9}		137 10^{-9}
United Kingdom (HSC, 1991)	0.5 10^{-9}	0.048 10^{-9}	0.43 10^{-9}	
France (Hubert, 1990)	130 10^{-9}	33 10^{-9}		97 10^{-9}

Table 2b: **Release frequencies for rail: Occurrences per vehicle-km (thick walled tankers).**

| | Release | Releases that can lead to consequences | | Minor release (or neglected) |
		Major release (≈ 0.2 m^2 or 100% of content)	Medium release (0.01 to 0.001 m^2)	
The Netherlands (TNO, 1983)	14 10^{-9}	0.4 10^{-9}	0.98 10^{-9}	13 10^{-9}
USA (Geffen, 1980)	n.a.	42 10^{-9}		n.a.
United Kingdom (HSC, 1991)	2.5 10^{-9}	0.25 10^{-9}	2.25 10^{-9}	$\approx$10 $^{-10}$
France (Hubert, 1990)	n.a.	1.1 10^{-9}		n.a.

2.4 Data sources

- The Netherlands: The road accident frequency is based on generic lorry statistics, using data from traffic and road accident surveys. For hazardous material accident analysis, Dutch records (that are not systematic), US accidents and European rail accidents have been surveyed.

- USA: The studies that are quoted here relied mostly on the same material, i.e., the hazardous material accident file kept by the US department of transportation. In parallel, general traffic and accidents have been surveyed on the road (the data are from 1966 to 1972) and on rail. The accident frequencies are based on events observed for vehicles whose payload is higher than 2 t. The hazardous material accidents served as a basis for a very comprehensive statistical description (conditional on accident occurrence) of accident environment.

- France: In the IPSN and CEPN studies (Hubert, 1988; Hubert, 1990; Castellano, 1990), road accident frequencies have been based on generic data that are relevant for lorry accidents and a correction factor was applied. Data from the national carrier (SNCF) have been used for rail. The national reporting system for hazardous material accidents (about 200 per year) served as a data base for observation on conditional failure probabilities. The raw data on which the quoted studies are based are somewhat old (1976 to 1981). More recent data are mentioned in the discussion and serve as a basis for a critical analysis.

- United Kingdom: Again generic statistics are used for accident frequencies. Note that, for train, hazardous material failure probability are directly observed (i.e., number of failures/traffic).

As mentioned above, accident frequencies and failure probabilities may have been derived from various data sets, even in a given study. On the other hand, some data have been used very often (Clarke, 1976). But the data for accident frequencies are often not consistent with failure analysis nor are they hazardous material specific. This leads most authors to modify raw data.

Other data sources and analyses may be quoted. They are not compared with the previous ones because they were not used as a regular basis for risk assessment studies. In France, EDF, the national electricity producer, set up figures for safety analysis of power plants accidents due to "external events". The LPG producers keep records of their accidents. Also, European figures have been introduced in a computer code, INTERTRAN, which aims at the probabilistic risk analysis of radioactive material transportation (Degrange, 1984).

2.5 Overall comparison

When straightforwardly compared, accident and release frequencies show some discrepancies, but some common features can be delineated (see Figure 2). Road accident frequencies are always in a given country, higher than the rail frequencies. This is generally the case for release frequencies although the United Kingdom is a striking exception.

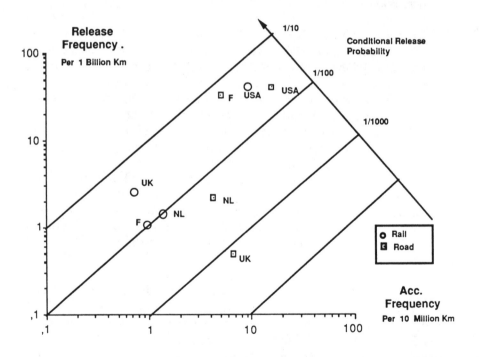

Figure 2: Comparison of accident and release frequencies

It must be noticed that the disparity in international data on accident frequencies is actually quite small. In comparison, in a given country (e.g. France), there is one order of magnitude according to the exact accident definition. The EDF study, (Lannoy, 1984), addressing rural roads, uses a figure which is three times lower than the above quoted one (Castellano, 1989) whereas the LPG carriers, on a damage criterion, assess a figure which is three times higher.

On the other hand, in the case of road, the discrepancies in the release frequencies are very high, and they result mostly from contradictions in the assumed release probability given an accident. In the case of rail, the US figure is isolated. The development, in France, of probabilistic risk assessments for dangerous goods transportation led us to revisit the data, as it was not possible to carry out many studies relying on hypotheses that are contradictory with other international analyses.

3.   CRITICAL ANALYSIS OF RISK ASSESSMENT HYPOTHESES

3.1 Data sources and methodology

In France, as in every country, generic road statistics are available and per kilometer accident frequencies can be derived. More specific is the systematic recording of hazardous material accidents, which allows to gather about 200 accidents per year. Out of those records, a file has been set up after scrutinizing accident reports from 1980 and 1981 (Degrange, 1989). The statistical description of the accident environment had been drawn from this file.

In order to allow a critical analysis, both the raw data from this file and the hazardous material records from 1987, 1988 and 1989 have been used (MTMD).

As regards rail, when the above quoted figures were set up, the recording system was far less developed. To date, it is possible to know precisely the ammonia and LPG traffic (tonnes, wagons, wagons-km, etc.) so that statistical data from 1984 to 1989 may be used in the critical analysis. The main goal of the reanalysis was to test the conditional release probability hypotheses against the present French data.

3.2 Road analysis

Initially, a better safety record for hazardous material transportation had been postulated; the accident frequencies were assumed to be three times lower than for other lorries. In 1983, the traffic has been estimated to about $1.2 \ 10^9$ vehicles x km, and about 240 accidents were observed. It can be deduced that the postulated frequency of $5 \ 10^{-7}$ was an overestimation in spite of this threefold reduction. However, it was not felt that this was a very strong evidence because of the uncertainty in the traffic, and because $5 \ 10^{-7}$ is an average between usual road and highway frequencies. The latter being only $2.5 \ 10^{-7}$, may represent the average traffic better.

The main problem was with the conditional release probability where the French figure (6.6% per accident) is very high when compared to others ($7 \ 10^{-4}$ in the UK, $5 \ 10^{-3}$ in the Netherlands, for example, or even 2.7 % in the USA). Revisiting the French file (1980 to 1981), allowed to confirm the postulated figures : among 45 "thick walled" tankers, 11 (i.e., 26.2%) were found to have leaked, 3 of them being described as presenting a hole. Looking at another criterion, it appeared that 4 of them had lost more than 50% of the content, and actually 3 of them lost 100%. Those observations were found to be fairly consistent with the 6.6% postulated "major release". The 3 and 4 observations do not allow the 6.6% hypothesis to be rejected at the 5 % confidence level, although the hypothesis appeared not to be on the safe side. Table 3 gives observed and predicted releases following a number of accidents (French data) and assuming release rates used in the cited literature. The value given for the confidence level results from the assumption of a binomial distribution of observed releases given a number of accidents and a specified release rate.

Against the two sets of French observations, international hypotheses have been tested, and for none of the periods are they consistent with field data (see Table 3). If allowance is made for the heterogeneities in the definitions, the US figure might be accepted, but it is not believed that the UK and Dutch figures can be compatible with the French reality. The simple fact that 4 total losses of content have been observed in three years prevents from applying international hypotheses to France. On the other hand, the French road tankers are similar to any European road tanker (they comply with the ADR agreement) and the traffic conditions might be slightly worse, but not to that extent.

Table 3:   Observed and predicted "thick walled" tankers
              important releases

	Observed number of accidents	Observed number of releases[1]	Predicted number of releases			
			France (6.6 10^{-2})	US (2.7 10^{-2})	The Netherlands (4.9 10^{-3})	UK (7 10^{-4})
1980-1981 file (Degrange, 1989)	45	4	2.97	1.21*	0.22**	0.032**
Records from 1987 - 1989 (MTMD)	56	5	3.69	1.51*	0.27**	0.039**

1	More than 50% of the content is released
*	Difference between observation and prediction is statistically significant at the 5% level
**	Not compatible with observation at the 1% level

A possible explanation is a better coverage of the French recording systems, which allows the collection of incidents with releases of no consequences (e.g., in a rural area).

3.3 Rail analysis

The French hypotheses for rail relied on observations (1976 and 1977), that were not specific of thick walled tankers and a lot of engineering judgement was exercised (Meslin, 1979). To date, good data are available on rail traffic for LPG and other gases liquefied under pressure (i.e., transported in thick walled tankers) and the accidents and releases are recorded, although, in the latter case, the description of the event is not as systematic as on the road.

From 1984 to 1989, 170.6 10^6 wagon-km were transported, 16 accidents were observed and 27 wagons were stated to be "involved in the accident" (SNCF, 1990). The accident frequency, specific of thick walled tankers, is therefore 1.6 10^{-7} per wagon and per kilometer, a figure which falls in the range of European data. However, on statistical grounds, it is compatible only with the Dutch figure. Based on a Poisson test, the French (at p = 0.015), UK (p = 10^{-4}) and US (p = 10^{-4}) figures can be rejected. All hypotheses, but the US one, are underestimating the reality.

Table 4: Test of international hypotheses against French rail
records (basis 170 10^6 wagon-km of thick walled tankers)
see (SNCF, 1990).

	Observed (1984 to 1989)	Predicted according to hypotheses			
		France	USA	UK	The Netherlands
Accidents	27	17	160	11.9	23.8
Major releases	2 or 1[1]	0.19	7.14	0.42	0.23

[1] One of the releases was a breach on an "empty" wagon, out of which the residual content escaped.

The situation is worse for releases, where European hypotheses underestimate the risk. Should one discard the release that was observed on an empty car, no hypothesis but the US one can be rejected when both releases are considered, the UK hypothesis still cannot be rejected, whereas the French (p = 0.02) and Dutch hypothesis (p = 0.025) can be eliminated.

It must also be mentioned that the conditional release probabilities are underestimated too according to European hypotheses, although to a lesser extent.

In the case of rail, French hypotheses, especially for release frequencies, were rather more optimistic than the one from the literature, in sharp contrast with the situation for road ; however, not only the French, but all European hypotheses proved to be optimistic when compared to actual results. Altogether, those facts are an incentive for a move towards the use of less optimistic figures in the French studies.

4.   CONCLUSION: HOW TO ACHIEVE INTERNATIONAL STANDARDS

An international literature survey has proven that there are important differences in the figures used for hazardous material transportation probabilistic risk assessment. The French hypotheses are among the most pessimistic for road and they are the most optimistic for rail.

The analysis of French hazardous material accident records suggest that the international assumptions (including the French ones) are not conservative, except in the case of the road accident frequency. There are not many data available in the case of rail and the definition and description of accidents and damages must be checked before reaching a definitive conclusion. The case of the conditional release probabilities in road tanker accidents appears to be a real matter of concern : although it is by far the most pessimistic figure in the international literature, the French hypothesis results nevertheless in an underestimation. Statistical tests and engineering judgements both lead to discard the British and Dutch approaches as too optimistic, and a satisfactory explanation has yet to be found. It is hoped that some standardization in the accident and release definition, and intercomparison of the accident recording systems will help in explaining such differences promptly.

5. REFERENCES

APPLETON, P.R. (1988). Transport accident frequency data, their sources and their application in risk assessment. United Kingdom Atomic Energy Authority, SRD R 474. Warrington, UK.

CASTELLANO, S., DEGRANGE, J.P., HUBERT, Ph., PAGES, P., LAMBLIN, J. (1989). Le transport de l'ammoniac anhydre. Analyse et estimation des risques. CEPN N° 159. Paris, France.

CLARKE, R.K., FOLEY, J.T., HARTMAN, W.F., LARSON, D.W. (1976). Severities of transportation accidents, vol. IV-trin. Sandia Laboratories, report SLA74-0001. Albuquerque, USA.

DEGRANGE, J.P., PAGES P. (1984). L'application d'INTERTRAN au calcul des risques accidentels dans le transport des matières radioactives. CEPN N° 75. Paris, France.

DEGRANGE, J.P. (1989). Les données d'accidents routiers disponibles au CEPN, Rapport CEPN N° 146, Paris, France.

DENNIS, A.W., FOLEY, J.T., HARTMAN, W.F., LARSON, D.W. (1978). Severities of transportation accidents involving large packages. Sandia Laboratories, SAND77-0001. Albuquerque, USA.

GEFFEN, G.A. et al. (1980). An assessment of the risk of transporting propane by truck and rail. PNL- 3308, Richland, USA.

HSC (Health and Safety Commission) (1991). Major hazard aspects of the transportation of dangerous susbtances. Advisory group on dangerous substances. Her Majesty Stationery Office, London.

HUBERT, Ph. PAGES, P., CASTELLANO, S., DEGRANGE, J.P. (1990). Le risque associé au transport du propane. CEPN N° 168. Paris, France.

HUBERT, Ph., PAGES, P., DEGRANGE, J.P. (1988). Estimation régionale du risque associé au trafic des matières dangereuses: comparaison d'itinéraires routiers à Lyon. CEPN N° 129. Paris, France.

LANNOY, A. (1984). Analyse des explosions air-hydrocarbure en milieu libre. Etudes déterministe et probabiliste du scénario d'accident. Prévision des effets de surpression. Bulletin de la direction des études et recherches, EDF, série N° 4. Paris, France.

MTMD, Mission du transport des matières dangereuses. Accidents et incidents de transports de matières dangereuses par voies routière et ferroviaire. (Yearly accident statistics). Ministère des Transports, Paris, France.

SNCF (1990). Société nationale des chemins de fer (French national train company). Personal communication and publication in MTMD, Paris, France.

TNO (1983), LPG, a study. TNO report for the Public Ministry of Housing, Physical Planning and the Environment, TNO, Appeldoorn, Netherland.

Perspectives on a Transportation Corridor Risk Analysis

Mark Abkowitz
Kathleen Hancock
Robert Waters
Vanderbilt Engineering Center for Transportation
Operations and Research (VECTOR)
Box 103, Station B
Vanderbilt University
Nashville, TN 37235
U.S.A.

1. BACKGROUND

This paper was prepared in response to a corridor risk assessment specification provided by the International Consensus Conference on the Risks of Transporting Dangerous Goods. The specification was designed to enable a variety of organizations to perform independent risk assessments based on the "same" input conditions, for the purpose of reaching consensus on good transportation risk assessment practice.

To advance the overall objective of understanding the implications of alternative approaches to transportation risk assessment, the authors assembled four teams of researchers, each charged with the same assignment, namely to perform a transportation risk assessment based on the published corridor specification. In the discussion to follow, emphasis was placed on the extent to which various assumptions and evaluation methodologies alter the analysis results.

2. RISK ASSESSMENT RESULTS

For the sake of brevity, the complete risk analysis process undertaken by each team is not reported here. Rather for illustrative purposes, Appendix A at the back of this document presents the detailed approach utilized by Team A to arrive at their risk assessment results. To place the four risk assessments into a common evaluation framework, one of the sets of measures for which each team was asked to present their results considered expected fatality by material type and transport mode (see Table 1). This uniform reporting measure was selected because of its relevance as an aggregate measure of operations safety.

Table 1: Expected annual fatalities.

	Team A	Team B	Team C	Team D
Chlorine				
Truck	0.0	1.12	0.0038	0.447
Rail	0.0036	0.26	0.0058	0.247
LPG				
Truck	0.410	2.26	21.66	2.55
Rail	0.025	0.23	0.14	0.74
Gasoline				
Truck	2.20	1.86	46.79	20.88
Rail	0.07	0.24	0.60	8.47

A quick perusal of Table 1 provides two important observations. First, there are order of magnitude differences in the number of expected fatalities among the teams for the same material and transport mode. From the standpoint of absolute risk assessment, this represents a significant variation in reporting.

Secondly, in the case of chlorine, the teams disagree on the relative safety of truck and rail operations. This is particularly troublesome from the perspective of setting transport policy based on risk assessment results.

The following section begins an exploration as to why these discrepancies occur, with special emphasis placed on how transportation and environmental relationships were represented.

3. ASSUMPTIONS

To investigate the effect of different risk assessment processes on the overall results, the methodology used by two teams, A and B, were selected for more detailed comparative study. In reviewing the work of these two teams, methodological assumptions were divided into two categories, those affecting the representation of *transport operations*, and those related to modeling *environmental conse-*

quences in the event of an accidental release during transport. Table 2 provides a description of the respective assumptions made by Teams A and B relative to transport operations, while Table 3 provides similar information for modeling environmental consequence.

Table 2: Transportation assumptions.

Team A

- National average accident rates on highways were obtained for multilane divided highways for rural and urban corridors.
- Suburban corridor was divided in half with the first half using rural and the second half using urban statistics.
- National average accident rates for rail were obtained for mainline track. This value was used for class 2 track as well.
- No rail sidings were included.
- Corridor specifications (such as grades, intersections, etc.) were not accounted for in the analyses.
- Accident rates are the same for all three substances.
- Release probabilities for trucks were conditional on cargo type, roadway type and location, and packaging.
- Only individual truck and rail car accidents were assumed.
- Truck shipments were assumed one truck at a time. Rail shipments were assumed 10 cars at a time.
- No vehicle accidents were assumed during loading and unloading of trucks.
- Both accident and fixed facility spills were considered for rail loading and unloading.
- Loading and unloading for chlorine and LPG were assumed to be through 4"-diameter hose. No spill hazard was assumed for gasoline.
- Loading/unloading facilities are located adjacent to transportation route.
- No adjustments were made for risk preference or aversion.

Team B

- Accident rates for highways were obtained from the Handbook of Chemical Hazard Analysis Procedures.
- Urban, suburban and rural accident rates were taken as the same.
- National average accident rates for rail were obtained for mainline track. This value was used for class 2 track as well.
- No rail sidings were included.
- Corridor specifications (such as grades, intersections, etc.) were not accounted for in the analyses.
- Accident rates are the same for all three substances.
- Release probabilities for trucks were conditional on cargo type only.
- Only individual truck and rail car accidents were assumed.
- Rail and truck shipments were made one tank or car at a time.
- No vehicle accidents were assumed during loading and unloading of trucks.
- Both accident and fixed facility spills were considered for rail loading and unloading.
- Spill hazard was included for gasoline.
- Loading/unloading facilities were located adjacent to transportation route.
- No adjustments were made for risk preference or aversion.

Table 3: Environmental assumptions.

Team A

- The expected release volume was calculated as the sum of the product of the possibilities of release and the volume of release for 10%, 30% and 100% of the container volume.
- Gaussian dispersion model with instantaneous release, 5 m/s steady-state wind speed and stability class F were assumed for chlorine and LPG.
- Average probabilities for wind direction were determined for 180 degree arc perpendicular to road direction.
- IDLH concentration used for chlorine toxic injury calculation.
- 10% of chlorine injuries assumed fatal due to sensitive populations and inability to relocate quickly.
- ARCHIE fireball model was used for LPG and gasoline induced injuries and fatalities.
- Only blast and fire effects were considered for LPG and gasoline.
- 50% of rural population and 20% of suburban and urban population outdoors are exposed.
- No residents living in right-of-way (ROW).
- Individual resident population risks were based on number of people living within fatal or injurious distance from accident centerline relative to number of people affected by an accident.
- Individual commuter risk based on annual number of commuters relative to number of people affected by accident.
- Injuries or fatalities due to traffic accidents caused by truck accident are not included.
- Population increases due to employment included for 8 hours each day.

Team B

- Material will be released instantaneously and in amounts equaling 10%, 30%, or 100% of payload capacity.
- Dispersion of a vapor cloud was modeled by a Gaussian dispersion model.
- Maximum instantaneous release occurs under wind stability class D and a wind speed of 5 m/s.
- Wind direction will be either N, S, E, or W at the time of release, and will not change direction as plume disperses.
- There are three ways for fatalities to occur: pool fire (gasoline and LPG), toxic cloud (chlorine), and vapor cloud fire or explosion (LPG).
- Residents living in right-of-way (ROW).
- The LC_{50} value for chlorine (fatal to 50% of population) = 500 ppm = .5 kg/m^3.
- Only 20% of the population under the toxic cloud will be exposed (20% are outdoors or have windows open).
- 50% fatalities for the population under the ignited LPG vapor cloud. Estimated concentration for flammability = 5 kg/m^3.
- For LPG, if no pool fire results, all liquid will flash to vapor.
- The hazard presented to life by vapor cloud burn-back is related to the area covered by the cloud at the time of ignition and to the population density, since it can be assumed that anyone caught in the burning plume will be killed. There is little threat to anyone outside the flammable cloud.
- Given a release, a pool fire and vapor cloud fire are equally likely. Immediate ignition occurs in approximately 50% of releases; thus ignition probability does not vary with size of release or population density.
- Population is uniformly distributed, with no time variation.

Three interesting observations emerge when reviewing these exhibits. First, many of the assumptions listed by Team A were implicitly assumed by Team B (e.g., the absence of rail sidings). Secondly, a greater level of operational and environmental detail is associated with the Team A.

Finally, some assumptions, particularly those related to environmental consequence, address the aggregation problem in a different manner, which actually leads to a significant impact on the risk measures reported in Table 1. How this occurred is the subject of the discussion that follows.

4. COMPARATIVE ANALYSIS

The alternative approaches taken by each team affected the direction of each study. The most significant outcome of this deviation in approach was the reversal in finding regarding chlorine transport. Team A concluded that chlorine is safer shipped by truck, actually imposing no "risk" of fatality, whereas Team B asserts that chlorine is safer to move by train.

To investigate why this occurred, the authors examined the differences in the respective team approaches to chlorine transport operations and modeling of environmental consequence. The following was discovered:

1) Team A calculated fatalities based on a single, expected release value, whereas Team B calculated fatalities as a discrete distribution at release volumes of 10%, 30% and 100% of container contents with their corresponding probabilities.

2) Team A considered a right-of-way buffer, which Team B did not.

3) Team B included pool fires as a potential consequence, while Team A did not.

4) Team A assumed ten chlorine cars per train, while Team B considered only one chlorine car per train with ten times as many shipments.

5) Team A considered LC_{50} (chlorine) = 0.0025 kg/m^3 while Team B considered LC_{50} (chlorine) = 0.5 kg/m^3. Team B also used less stable air stability class D in its computations.

6) Team A assumed 50% of the rural population and 20% of the remaining population are exposed to the release, whereas Team B assumed 20% of all people were affected by the plume.

From this list Items 1 and 2 were the major contributors to the difference in outcome concerning the relative safety of truck and rail operations. Item 1 removed the more catastrophic scenarios from the truck risk assessment by averaging the release volume to a less threatening level in contrast to the rail average release volume which is still large enough to cause concern. The difference in consequence assessment is exacerbated by Item 2, which in effect creates a right-of-way buffer for a truck chlorine spill that makes it "impossible" for acute or fatal health effects to reach the general population.

What is, perhaps, most alarming about this conclusion is that the Team A approach has the appearance of being far more comprehensive, and yet one generalized dispersion assumption has the potential to "undo" much of the detail presented elsewhere.

5.   PROBLEMS EXTERNAL TO CORRIDOR SPECIFICATION

To this point in the discussion, attention has been focused on variations in approach within the specification prepared for corridor risk assessment. It is already clear that enough latitude exists for a variety of alternative approaches to emerge which generate widely disparate results.

To further complicate matters, real-world transportation risk assessments face a number of additional uncertainties that were defined as static in the corridor specification. These include the following study limitations:

1) a single container specification to haul each material by each mode.

2) a single routing alternative per mode.

3) consistent and known transport segment geometrics and operating characteristics.

4) consistent and known population distributions.

5) consistent and known land use, hydrology and weather characteristics.

From the authors' experience, these are key considerations, whose variation has a significant impact on risk assessment outcome, similar to that illustrated in the previous section. Furthermore, obtaining this information at the appropriate level of detail can be time-consuming and resource-intensive.

In addition, there are pertinent elements to a transportation risk assessment that were simply omitted from the corridor exercise altogether. For example, the location and mitigation capability of emergency response personnel were not taken into consideration. This is not a condemnation of the intent of the specification, but rather an indication of the breadth of issues that perhaps should be addressed by a transportation risk assessment.

6. CONCLUSION

Transportation risk assessments will be performed more frequently in the coming decade in response to an increased level of regulatory concern and industry self-examination. Consequently, standardization of risk assessment approaches and adoption of "best practice" techniques are desirable goals to seek.

Unfortunately, financial resources and time available to perform the most comprehensive transportation risk assessment possible will be limited in almost all cases. Therefore, the manner in which transport operational and environmental consequence assumptions are made will be a source of uncertainty with the potential to significantly alter the risk assessment findings and their management implications.

The authors believe that, at present, the greatest benefits will come from focusing on the following objectives:

- utilizing information technology to create accurate, detailed transportation networks and associated spatial data defining the geographic area in proximity to the transportation system;

- identifying a subset of environmental consequence models that have a "seal of approval" as being representative of the process they are designed to address; and

- reaching agreement on the most significant factors and of their contribution to risk assessment outcomes.

Beyond the attainment of these goals, the use of sensitivity analyses must become a part of the risk assessment process. Given the influence of the outcome of a transportation risk assessment on government, industry and the public at large, one needs to know how robust the findings are relative to the study recommendations which are being made.

Acknowledgements. The authors would like to express their sincere appreciation for the supporting efforts contributed by Michael Dimaiuta, Emily Goodenough, John Eric Meyer, Michael Pierce, Rajeev Saraf and Douglas Yerkes. Without their assistance, the opportunity to explore important transportation risk assessment considerations described herein would not have been possible. The opportunities presented by the U.S. D.O.T. University Transportation Centers Program and the D.O.E. Environmental Restoration and Waste Management Fellowship Program are also acknowledged.

7. REFERENCES

ABKOWITZ, M.D., LEPOFSKY, M., and CHENG, P.D.M. (to be published in 1992). Selecting Criteria for Designating Hazardous Materials Highway Routes. Forthcoming in *Transportation Research Record*, National Academy of Sciences.

ABKOWITZ, M.D., CHENG, P.D.M., and LEPOFSKY, M. (1990). The Use of Geographic Information Systems (GIS) in Managing Hazardous Materials Shipments. *Transportation Research Record 1261*, National Academy of Sciences.

ABKOWITZ, M.D., and CHENG, P.D.M. (1988). A Risk/Cost Framework for Routing Truck Movements of Hazardous Materials. *Accident Analysis and Prevention*, 20(1).

ABKOWITZ, M.D., and LIST, G.F. (1987). Hazardous Materials Transportation Incident/Accident Information Systems. *Transportation Research Record 1148*, National Academy of Sciences.

CRANE CO. (1988). Flow of Fluids Through Valves, Fittings and Pipe - Technical Paper No. 410. King of Prussia, Pennsylvania.

LAMARSH, J.R. (1983). Introduction to Nuclear Engineering, 2nd Ed. Addison-Wesley, Reading, Massachusetts.

SAX, N.I. and LEWIS, R.J. Sr. (1989). Dangerous Properties of Industrial Materials, 7th edition. Van Norstrand Reinhold, New York.

U.S. DEPARTMENT OF TRANSPORTATION (1989). *Guidelines for Applying Criteria to Designate Routes for Transporting Hazardous Materials*. Research and Special Programs Administration, Office of Hazardous Materials Transportation, Washington, DC.

U.S. DEPARTMENT OF TRANSPORTATION, U.S. ENVIRON-MENTAL PROTECTION AGENCY. *Handbook of Chemical Hazard Analysis Procedures*. Federal Emergency Management Agency, Washington, DC.

U.S. DEPARTMENT OF TRANSPORTATION (1990). *1990 Emergency Response Guidebook*. Report No. DOT-P-5800.5.

APPENDIX A

TRANSPORTATION CORRIDOR RISK ASSESSMENT FOR RAIL AND TRUCK FOR CHLORINE, LPG & GASOLINE

EXECUTIVE SUMMARY

A corridor risk assessment for three hazardous materials (chlorine, LPG and gasoline) transported via two modes (truck and rail) has been performed as an exercise sponsored by the International Consensus Conference on the Risks of Transporting Dangerous Goods. The results of this study indicate that to minimize the annual risk of fatality and injury to the public along the corridor, chlorine should be shipped via truck while LPG and gasoline should be shipped via rail.

The annual risk of fatality and injury for each hazardous material and each mode of transportation are listed in the following table:

| Hazardous Material | Mode | |
	Rail Fatality / Injury	Truck Fatality / Injury
Chlorine	0.0036 / 0.036	0.0 / 0.0
LPG	0.025 / 1.9	0.41 / 19.0
Gasoline	0.067 / 3.0	2.2 / 11.0

Truck transport provides the lesser risk for chlorine because of smaller transported volumes and wider right of ways. Rail provides the lesser risk for LPG and gasoline due primarily to lower accident rates and large shipments requiring less trips. The assumptions, methodology, and results of this assessment are detailed in the following discussion.

ASSUMPTIONS

Risk assessments are value-laden regardless of the objectivity of the analysts and several assumptions are required to perform a risk assessment of this magnitude. The major assumptions made in this analysis are listed below categorized by transportation, and environmental.

Transportation

- Suburban accident statistics are difficult to obtain. Most accident data is for rural and urban. Suburban corridor was divided in half with the first half using rural and the second half using urban statistics.

- Highway contained at-grade intersections which classifies the transport route as multilane divided highway.

- National average accident rates for highways were obtained for multilane divided highways for rural and urban corridors.

- National average accident rates for rail were obtained for mainline track. This value was used for class 2 track as well.

- No rail sidings were included.

- No vehicle accidents were assumed during loading and unloading for trucks. Only fixed facility spills were considered.

- Both accident and fixed facility spills were considered for rail loading and unloading.

- Loading and unloading for chlorine and LPG were assumed to be through 4"-diameter hose. No spill hazard was assumed for gasoline.

- Accident rates for loading and unloading were per use not hours of operation.

- Release probabilities for trucks were conditional on cargo type, roadway type and location, and packaging.

- Release probabilities for rail were conditional on type and location of rail and type of rail car.

- Only individual truck and rail car accidents were assumed. No conditional probabilities for multiple-car accidents were used.

- Loading and unloading facilities are located adjacent to transportation route.

- No adjustments were made for risk-preference or aversion.

Environmental

- Only toxicity effects were considered for chlorine.

- Only blast and fire effects were considered for LPG and gasoline.

- Gaussian dispersion model with instantaneous release, 5 m/winds and stability class F used for chlorine and LPG.

- Average probabilities for wind direction determined for 180 degree arc perpendicular to road direction.

- IDLH concentration used for chlorine toxic injury calculation.

- 10% of chlorine injuries assumed fatal due to sensitive populations and inability to relocate quickly.

- ARCHIE fireball model used for LPG and gasoline injuries and fatalities.

- 50% of rural population and 20% of suburban and urban population outdoors and exposed.

- No residents living in ROW.

- Individual resident population risks based on number of people living within fatal or injurious distance from accident centerline relative to number of people affected by an accident.

- Individual commuter risk based on annual number of commuters relative to number affected by an accident.

- Injuries or fatalities due to traffic accidents attendant to a truck accident are not included.

- Population increased due to employment are included for 8 hours of each day.

METHODOLOGICAL APPROACH

Transportation

The number of vehicles transporting hazardous materials has been calculated from information provided by the corridor risk assessment sponsor. The total amount of each substance was divided by the volume of each respective container for truck and rail. The probability of an accident occurring was obtained from sources listed in the references based on the type and location of road and from general mainline track statistics. The number of vehicles was then multiplied by the accident probability to obtain number of accidents involving hazardous materials.

The number of releases that occur given that an accident has occurred was determined from conditional probabilities. For trucks, these probabilities were based on cargo type, road type and location, and packaging. The normalized values of the release per accident statistics were multiplied together and then by the number of accidents to obtain the number of releases. This number was then multiplied by the release probabilities by volume presented in the Handbook of Chemical Hazard Analysis Procedures. For trucks, the release probability was:

[10%(.6) + 30%(.2) + 100%(.2)]* no. of releases

For rail, the conditional release probabilities were based on the rail type and car type. These values were multiplied by the number of accidents to obtain the number of releases. This number was then multiplied by the release probabilities as above. For rail, the release probability was:

[10%(.5) + 30%(.2) + 100%(.3)]* no. of releases

The results of these calculations were summarized in the analyses.

Fixed facility accident rates, probable release volumes and rates have also been tabulated. These values were obtained directly from the Handbook of Chemical Hazard Analysis Procedures and consist of the number of loadings and unloadings multiplied by the release rate.

Transport and Exposure

Gaussian dispersion assuming instantaneous releases of the expected material volume with 5 m/s winds and stability class F was used to determine the extent of toxic concentrations of chlorine and an explosive cloud of LPG. No cloud was assumed for gasoline. The widths of the toxic or explosive puffs were calculated for several distances from the release site.

The total area below the chlorine puff pathway (with concentration above the IDLH) was used for toxicity calculations. The probability of wind direction for the 180-degree arc on either side of the road was used to determine the likelihood of exposure for the uniform populations on each side of the road. An explosive LPG vapor cloud of explosive concentration does not develop since the volumes of LPG were small for both truck and rail scenarios relative to meteorological conditions and were quickly dispersed. LPG and gasoline fireball scenarios were modeled to determine the expected fatalities and injuries resulting from thermal radiation using the ARCHIE model. Safe distances from fireball thermal radiation were calculated for fatalities and injuries. Residents and motorists within thermal radiation were calculated for fatalities and injuries. Residents and motorists within the respective radius were assumed exposed to this radiation. Thermal radiation levels from a LPG or gasoline fireball are significant and all persons within the distance for fatalities were considered fatalities. All persons between the injury distance and fatality distance were considered to be injured.

Societal and Individual Risk

Societal risks were calculated as the number of deaths or injuries based on the probability of an accident causing a release multiplied by the average number of motorists and residents exposed to the resulting toxic cloud or thermal radiation.

Individual resident risks were calculated by dividing the expected number of resident injuries or fatalities by the average number of residents along the entire segment within the distance from the road/rail that would cause injury or death.

Individual motorist risks in person-trips were calculated by dividing the expected number of motorist injuries or fatalities by the average annual traffic multiplied by the average occupancy rate.

RESULTS INTERPRETATION

Based on risk fatalities and injuries alone, the best mode of transportation for chlorine is truck and for LPG and gasoline is rail. While the least total risk for LPG and gasoline is associated with rail, this is generally due to the high fatalities and injuries of motorists associated with truck transport. These risks may be weighted differently than resident risks in determining the best mode of transportation and have been tabulated in the analyses. The narrower right-of-ways associated with rail transport result in more resident risk with this mode of transport.

As is usually the case, the most significant factor in rail being the lesser risk for LPG and gasoline is reduced number of trips required due to larger loads per shipment. Less trips allow more risk management planning in determining when to make shipments - another advantage of rail transportation.

RISK ESTIMATION CRITIQUE

Several enhancements could be made to this risk analysis to

improve the quality of the results. The following list contains the major enhancements, given sufficient time and resources:

- Actual data should be used whenever available instead of national averages.

- More specific accident rates based on existing traffic and roadway characteristics should be obtained.

- Probabilistic assessments of wind speed, direction and stability class would provide a more realistic assessment of the range of risks.

- Modelling the releases as instantaneous followed by a continuous release at a reduced rate would provide a more realistic assessment of accidents.

- Considering the time of day of the accident would provide a more realistic assessment of the number of people exposed, for example, variations in traffic volume during peak and non-peak times.

- Considering the cost of property damage to nearby buildings and structures due to fire and explosion would provide additional sensitivity to the analysis.

- Considering the chronic effects of materials whose release results in low doses for long periods of time, e.g. contaminated soil or groundwater.

- Developing F/N curves for each mode and material will provide a graphic representation of the relative societal risks associated with transportation of these materials.

The simplifications and assumptions required to perform this study should be taken into consideration when evaluating the reported results.

Hazardous Materials: A Comparison of the Severity of Accidents from Transport and Fixed Installations

Palle Haastrup
Commission of the European Communities
Joint Research Centre, Ispra, Italy

H.J. Styhr Petersen
Department of Chemical Engineering
Technical University of Denmark

ABSTRACT

This paper compares the severity of accidents from transport of dangerous goods with similar accidents related to fixed installations. The comparison is made both in relation to consequences for humans (fatalities) and in relation to environmental consequences, though it is recognised that the definition of environmental consequences is not straightforward.

It is concluded that the frequency of accident with fatalities is approximately half as large for transport accidents compared to fixed installations, and that the consequences are similar in both cases. For accidents with environmental consequences, the corresponding frequency is one third, and the magnitude of the consequences are found to be similar.

An analysis of the reliability of the reporting system showed that this is higher for the fixed installations, and that the uncertainty on the reported number of fatalities is higher for the transport cases.

1. INTRODUCTION

To compare accidents related to transport of dangerous goods with similar accidents related to fixed installations requires a definition of the scales on which to compare. Here there are numerous possibilities, like comparing the total annual number of accidents in the two fields, to compare the total number of fatalities or to compare the number of accidents with fatalities above a certain threshold, etc. The list can be continued nearly infinitively.

It is also necessary to arrive at a common (or at least comparable) definition for accidents in the two areas. For some consequences such a definition may be easy, like comparing the number of accidents resulting in human fatalities or injuries. However, for other types of consequences, like environmental damage, neither the definition of the threshold level for counting an accident, nor the characterization of the consequence is straightforward.

In this paper an attempt to compare accident consequences from transport and fixed installation is made on the following points: Number of accidents leading to human fatalities, number of accidents leading to 10 or more fatalities, number of accidents leading to environmental damage, number of accidents leading to large environmental damage, and finally the reliability of the accident reports in the two fields are compared.

The current study is part of a research program on the transport of dangerous goods carried out by the Commission of the European Communities, by the Joint Research Centre Ispra, which has been underway for the last 5 years in collaboration with the Department of Chemical Engineering at the Technical University of Denmark, among others. Previous results can be found for instance in (Haastrup and Brockhoff, 1990) and (Haastrup and Brockhoff, 1991).

2. ACCIDENT SOURCES

Generally accident case histories are a rich and important source of background data for any risk evaluation or legal initiative. Obviously case histories contain crucial information about what actually went wrong, rather than the "what may go wrong" which is the normal result of a risk assessment, or the "how may we prevent the event" which is the regulatory question.

The Community directive on major hazards (EC, 1982) requests the competent authorities in the member countries to notify the Commission about accidents in chemical industries, and these accidents are collected in a database (Contini et al., 1988). This database now contains 97 accidents and a recent report sums up the lessons learned so far (Drogaris, 1991). However, these accident reports are not available to the public and have therefore not been included in the present analysis, also because the registered accidents relates only to fixed installations. However, the new proposal for a community directive in this area (EC, 1991) requests accident reports to be compiled also from transport accidents.

The accident case histories used for the present study were therefore found in the open literature. More than 10 sources were used of which most consisted of lists of accidents, similar to extract databases. The most important sources were Lees (Lees, 1980), Loss Prevention Bulletin (LPB, 1981 to 1986), Hazardous Cargo Bulletin (HCB, 1986 to 1990), Handbuch Storfalle (HS, 1983) and the French Ministry for the Environment (which has been used as the source for the analysis related to the environment) (France, 1988 to 1990), but several other sources were also used, for instance: Davenport, 1983; Marshall, 1977; Cremer and Warner, 1982; HMSO, 1979; Blything, 1984.

The collection of accident cases is believed to be relative complete in relation to fixed installations, road transport, rail transport and pipeline transport, whereas information about sea and inland water transport are not covered well.

3. ACCIDENTS WITH FATALITIES

In total 2195 accidents of interest were collected, and in Table 1 the distribution of these accidents on fixed installation, loading or unloading and transport is shown.

The content of a given source depends on how the accidents have been chosen by the author of the source. This choice, which is rarely stated explicitly, may influence any subsequent analysis of the accidents. However, as numerous sources have been used it is assumed that any bias concerning transport or fixed installation has been minimized.

Table 1: Classification of 783 accident cases with fatalities.

Class	No. with fatalities	%	No. with 5 or more fatalities	%	No. with 10 or more fatalities	%
Fixed installations	520	66%	157	60%	81	65%
Loading/ unloading	47	6%	14	5%	6	5%
Transport	216	28%	90	34%	38	30%
Total	783	100%	261	99%	125	100%

As seen in Table 1, 28% of the accidents relate to transport with fatalities, a percentage which is only slightly higher for accidents with 5 or more fatalities and for 10 and more fatalities. These results are in agreement with previous results (Haastrup and Brockhoff, 1990), where the similarities related to the consequences of transport and fixed installations were also statistically tested, and with (Appleton, 1988) who reported a slightly higher proportion of accidents related to transport (40%).

Based on this it seems possible to conclude that for every second accident with human fatalities (few or many) in a fixed installation one similar transport accident occur.

An analysis of different time periods showed insignificant variations in this proportion.

4.   ENVIRONMENTAL CONSEQUENCES

To compare environmental consequences of accidents is difficult. However, as a first step it is possible simply to register how many accidents of each type that are available. This has been done based on a very extensive source for France for the years 1987, 1988 and 1989 (France, 1988 to 1990).

An accident was registered as having environmental consequences if one (or more) of the keywords "pollution", "dead fish" or "dead animals" was recorded.

In the following these accidents with consequences for the environment will be denoted "environmental accidents".

Table 2 shows the results of this registration of the environmental accidents. As seen in the table, 21 % of the environmental accidents relate to transport. This should however be considered as a first estimate only, because an evaluation of the consequences is missing.

Table 2: Classification of environmental accidents from 1987-1989 (France, 1988 to 90).

Class	1987	1988	1989	Total	%
Fixed installations	110	84	166	360	79%
Transport	17	31	48	96	21%
Total	127	115	214	456	100%

One option is to use the release size as an indicator for the potential environmental damage. The result of this analysis is shown in Table 3. Here the accident case stories, where the mass is known is shown as a function of transport or fixed installation accidents as a function of the release size.

As seen in Table 3 the proportion of transport accidents is 29 %, somewhat higher than for the total sample. If the release size is used as the criteria for severity of an environmental accident, the fraction of transport accidents thus increases.

Table 3: Transport and fixed installation accidents from 1987-1989 (France, 1988 to 90) as a function of release size.

Class	Less than 10 tons	10 to 100 tons	More than 100 tons	Total
Fixed installations	108	35	11	154
Transport	30	30	3	63
Total	138	65	14	217
Transport in %	22%	46%	21%	29%

A priori it was expected that the fraction of transport accidents would decrease with the magnitude of the release, since transport (by nature) are of a more limited size. However, as seen in table the proportion of transport accidents is highest for the medium size accidents (10 to 100 tons).

Another possible way to evaluate the severity of an environmental accidents is to use expert judgement, i.e. to ask experts in the field to evaluate the severity for the environment of a given accidents. In 92 accident cases a minimum of information related to "where" and "what" was available, and these accidents were evaluated twice by 3 experts. An arbitrary scale from 0 to 100 was used, and each accident was rated accordingly. Table 4 shows the result of this expert rating. Because of the requirements related to the completeness of the information about "where" and "what" the sample used is a sub-sample of the sample used for the classification according to the release size (Table 3).

Table 4:   Transport and fixed installation accidents from 1987-1989 (France, 1988 to 90 as a function of severity based on expert judgement (denoted: Grade)).

Class	Grade less than 10	Grade 10 to 30	Grade more than 30	Total
Fixed installations	30	33	3	66
Transport	16	9	1	26
Total	46	42	4	92
Transport in %	35%	21%	25%	28%

The table shows the accidents according to the grade (below 10, from 10 to 30, and above 30), and the proportion of these accidents related to transport accidents.

It should be noted, that the evaluation of the environmental accidents are not without problems, and that the "severity grade" obtained is uncertain. The results should therefore be used with caution. However, as seen in table 4 the proportion of transport accident varies from 21 to 35 % only, which seem to indicate that the environmental damage is just as severe for transport accidents as for accidents related to fixed installations.

5. RELIABILITY OF CASE HISTORIES

During the compilation of the accident lists used to compare transport and fixed installations it was noted that the same accident often could be found in more than one source, and that important discrepancies between source and types of information on the same level of detail and completeness could be found.

In order to analyse this in more detail, accidents with fatalities described more than once in the sources were extracted for fixed installation and transport respectively.

In Table 5 the number of accident cases described by more than one source is shown as a function of the number of sources. Column 3 indicates how many of the cases related to transport, and column 4 shows the percentage. It is interesting to note that the proportion of transport accidents in the sample of repeated accident reports, is similar to the overall sample (see Table 1). The transport share of the total number of "repeated reports" is however increasing with the number of sources.

Table 5: Number of repeated accident reports from *fixed installation* and transport as function of number of sources.

Number of sources (N)	Number of cases described by N sources	Number of transport cases	Transport cases in percent %
2	85	17	20%
3	27	8	30%
4	19	8	42%
5	9	5	56%
6	1	1	100%
Total	141	39	28%

In the first analysis a discrepancy is defined as (minimum) one source not being in agreement with the others. A discrepancy is thus defined in absolute terms, and two reports indicating 1 and 2 fatalities respectively will be considered in the same way as two reports with 1500 and 1501 fatalities.

It should be noted that discrepancies may also arise because of differences in the definition of "fatalities", for instance in defining after how many days they should be counted. The number of fatalities thus has a built-in uncertainty.

In Table 6 the number of reports with discrepancies are shown as a function of the number of sources describing them together with the proportion of these cases related to transport.

Table 6:   Number of reports with discrepancies as function of number of sources.

Number of sources (N)	Number of cases with discrepancies	Number of transport cases	Transport cases in percent %
2	13	3	23%
3	9	3	33%
4	10	5	50%
5	6	4	67%
6	1	1	100%
Total	39	16	41%

16 of the 39 cases relates to transport, corresponding to 41 %. The relative number of cases with discrepancies is thus nearly twice as large for transport as for fixed installations. Compared to table 5 the proportion of transport accidents in the cases with discrepancies is consistently larger. This seems to indicate that the observation is not due to bias in the sources. One possible explanation is, that the accident reporting system may be better defined for fixed installations compared to transport.

Table 7 shows the mean number of fatalities from the same 141 cases as a function of the number of sources.

Two interesting trends can be identified in Table 7. First of all, the mean number of fatalities related to the cases with discrepancies are consistently higher than for those cases where the sources agree. The total average is approximately 50 times higher for the sources with disagreement. Larger discrepancies are thus found for larger accidents. This also the case for the sub-sample of transport cases, but here the total average is "only" 6 times as large for the cases with disagreement.

Table 7: Mean number of fatalities (road, rail & pipeline accidents in parentheses) as a function of number of sources.

Number of sources	Cases with agreement between all sources		Cases with disagreement between sources	
	Number	Mean number of fatalities	Number	Mean number of fatalities
2	72 (14)	6 (3)	13 (3)	19 (6)
3	18 (5)	6 (6)	9 (3)	938 (14)
4	9 (3)	6 (2)	10 (5)	97 (5)
5	3 (1)	27 (7)	6 (4)	20.6 (10)
6	0 (0)	–	1 (1)	183 (183)
all	102 (23)	6 (3)	39 (16)	256 (19)

Secondly the mean number of fatalities for the transport cases are consistently lower (or equal to) the similar number for the total sample. The average number of fatalities for the transport cases is 3 for the cases in agreement, and 19 in the cases with disagreement, compared to the similar number of 6 and 256 fatalities in the total sample. These figures are however dominated by a few large accidents like Bhopal, in which the uncertainty is also considerable.

6. CONFIDENCE LIMITS

Assuming that the average is the best estimate for the "correct" number of fatalities, given two or more reports from the same accident, then the (absolute) number of fatalities reported in the various accident case histories, can be transformed into numbers indicating the ratio to the average. This makes it possible to do a more quantitative comparison of the uncertainty of reported number of fatalities.

It should be noted that this assumption implies the same percentile deviation on large as well as small accidents, an assumption which it is not possible to test on the current data. However examples like the Bhopal accident, where a large deviation in the reported number of fatalities is observed, supports the assumption.

The transformation was done for the two samples related to transport and to fixed installations, and the resulting accumulated distributions are found in Figure 1.

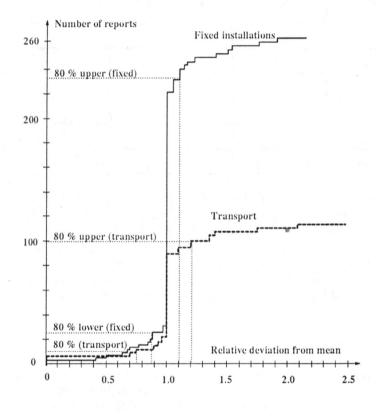

Figure 1: Relative variation of fatalities (accumulated) for transport and fixed installation accidents.

The data used to construct Figure 1 are the sample of transport and fixed installations, actually consisting of 109 and 259 reports respectively. On the figure, 80 % confidence limits are marked for both transport and fixed installation. As seen the confidence limit on the fatalities from a transport accident are wider that for fixed installation. This reinforces the conclusion from the previous section, that the reporting system for fixed installations introduces less uncertainty than for transport accidents.

7. DISCUSSION

This paper compares accidents with hazardous materials related to transport and to fixed installations from a number of different points of view. The findings are summarized in Table 8.

Table 8:   Comparing accidents from transport and fixed installations.

Type of information	Transport accidents compared to fixed installation accidents
Frequency of human accidents	Half as frequent
Human consequences	Similar, also for large accidents
Frequency of environmental accidents	One third as frequent
Consequences of environmental accidents	Similar, also for large accidents
Reliability of reporting system	Lower than for fixed installations
Uncertainty on number of fatalities	Higher than for fixed installations

As seen in the table, the results related to human accidents and environmental accidents are very similar. In both cases transport accidents leads to consequences which are just as severe as for fixed installations. However the frequency of transport accidents are "only" 30 to 50 % of the fixed installation accidents.

However since the quantity of material transported seems to be increasing, this fraction may also be increasing.

Until recently the main focus in Europe was on the risks related to fixed installation, and it was in this area the research and the political initiatives was going on. However the results presented here confirms the need for a close examination also of transport of dangerous goods.

Acknowledgments. The authors wish to thank friends and colleagues at the Joint Research Centre and the Technical University of Denmark for helpful discussions of earlier drafts of this paper. Special thanks go to Lars and Helle Brockhoff, G. Volta and K. Rasmussen at J.R.C. Ispra.

8. REFERENCES

APPLETON, P.R. (1988). Transport accident frequency data, their sources and their application in risk assessment. *Technical Report* SRD R 474, Safety and Reliability Directorate.

BLYTHING, K.W. (1984). In-service reliability data for under ground cross-country oil pipelines. *Number SRD R 326*, page 32. United Kingdom Atomic Energy Authority, Safety and Reliability Directorate.

S. CONTINI, S., AMENDOLA, A., & NICHELE, G.P. (1988). MARS: The Major Accident Reporting System. *I.Chem.E. Symposium Series*, 110, 455.

CREMER & WARNER (1982). *Risk analysis of six potentially hazardous industrial objects in the Rijnmond Area, a pilot study.* A Report to the Rijnmond Public Authority, Technica, Ltd. Reidel Publishing Company. ISBN 90-227-1393-6.

DAVENPORT, J.A. (1983). A study of vapor cloud incidents - an update. *I. Chem. E. Symposium Series*, 80, C1.

DROGARIS, G. (1991). Major accident reporting system. Lessons learned from accidents notified. *Technical Report* EUR 13385 EN, Community Documentation Centre on Industrial Risk, Joint Research Centre, Ispra, Commission of the European Communities.

EC (1982). Council Directive of 24 June 1982 on the major accident hazards of certain industrial activities. *Official Journal of the European Communities*, No. L 230/1.

EC (1991). Proposal for a Council Directive on the appointment of an officer for the prevention of the risks inherent in the carriage of dangerous goods in undertakings which transport such goods, and on the vocational qualification of such officers. *Official Journal of the European Communities*, No. C 185/5.

FRANCE (1988-90). Inventaire des pollutions accidentelles et accidents industriels en 1987, 88 and 89. *Technical Report*, Ministere de l'environnement et de la prevention des risques technologiques et naturels majeurs, Service de l'Environnement industriel, France.

HAASTRUP, P., & BROCKHOFF, L. (1990, October). Severity of accidents with hazardous materials. A comparison between transportation and fixed installations. *J. Loss Prev. Process Ind.*, 3, 395.

HAASTRUP, P., & BROCKHOFF, L. (1991, August). Reliability of accident case histories concerning hazardous chemicals. An analysis of uncertainty and quality aspects. *Journal of Hazardous Materials*, 27, 339-350.

HCB (1986-90). Incident Log. *Hazardous Cargo Bulletin*. Various numbers from 1986 - 1990.

HMSO (1979). *HMSO, Advisory Committee on Major Hazards, Second report.* Health and Safety Commission, HMSO, ISBN 0 11 883299 9.

HS. Handbuch Storfalle (1983). *Technical report.* Umweltbundesamt, Bismarckplatz 1, 1000 Berlin 33. Eric Schmidt Verlag GmbH. ISBN 3 503 02385 2.

LEES, F.P. (1980). *Loss prevention in the process industries.* Number ISBN 0 408 10604. Butterworth Ltd.

LPB. List of incidents 1981-1986. *Loss Prevention Bulletin*, 1981 to 1986. The Institution of Chemical Engineers, Railway Terrace, Rugby, UK.

MARSHALL, V.C. (1977). How lethal are explosions and toxic escapes? *Chem. Eng. London,* 323, 573.

Chapter 5: Risk Tolerance, Communication and Policy Implications

Hazardous Goods Transportation: A Major UK Study*

Keith Cassidy
HM Superintending Specialist Inspector
Technology Division
Health and Safety Executive
United Kingdom

ABSTRACT

The UK Health and Safety Commission has recently published a major report on the risks associated with the transportation of dangerous goods, in major hazard quantities, within the United Kingdom. The preparation of this report has involved a novel approach to the investigation of the risks, and a development in the concept of risk 'tolerability' in the transportation context. This paper outlines the background to the report, describes the methodology derived to enable the assessment of risk levels and of tolerability to be carried out, and summarises the conclusions and recommendations contained in the report. The author of the paper was a member of the investigating body, and chairman of the various technical working parties overseeing the risk assessments.

1. INTRODUCTION

The Flixborough Disaster (HSE, 1974) led to the setting up in the United Kingdom of the Advisory Committee on Major Hazards (ACMH), whose three reports (HSC, 1976, 1979, 1984) have been seminal in the development of a UK strategy for the control of major risks from hazardous installations; and this strategy has now become the basis for control internationally (Cassidy, 1988). The

---

* The figures and tables in this paper have been previously published in the Health and Safety Commission publication "Major Hazard Aspects of the Transport of Dangerous Substances". This material is reproduced with the permission of the Controller of Her Majesty's Stationery Office.

ACMH Reports were concerned primarily with hazards and risks from fixed installations, but in its Third Report, the Committee identified a prima facie case for further investigation of the risks associated with the transport of hazardous substances, based on the assessed potential for harm, and the relative frequency of incidents which had occurred in the transportation mode. The Report particularly identified problems in:

a) water transportation - the very large volumes of dangerous goods carried, often in bulk;

b) rail transportation - lack of flexibility, closeness to population centres, concentrations of dangerous goods in, for example, marshalling yards, linear communication problems, and risks associated with long tunnels; and

c) road transportation - the omnipresence of the road network, and its reach into population centres, again concentrations of dangerous goods in for example, parking areas, and the very diverse responsibilities within the industry.

In response to these findings, the UK Health and Safety Commission (HSC) set up a subcommittee of the Advisory Committee on Dangerous Substances (ACDS), with terms of reference which required it to report to HSC - which has a role inter alia to advise ministers - on:

• the hazards and risks associated with the transport of large quantities of dangerous substances (excluding radioactive substances) by road, rail or water, which had the potential to present major accident hazards to employees or to the public from fire, explosion or toxic release

• its identification of any appropriate mandatory or voluntary control measures which seem to be desirable beyond present controls and

• any necessary additional action.

2. THE WORK AND APPROACH OF THE ACDS SUBCOMMITTEE

Initially, the subcommittee attempted to approach their task using as bases for their investigations current legislative standards and their engineering underpinning, set against the recent historical record. It soon became clear, however, that the latter was quite insufficient to provide a robust basis for prediction of high - N (i.e.

large numbers of casualties) incidents, or major events of low probability; and the use of wider data sources (e.g. international statistics) was fraught with difficulties of relevance and extrapolation. These and other prob-lems led to a decision to proceed on the basis of a systematic, quantified risk analysis (QRA) for each aspect of each mode of transport, to estimate existing levels of risk; and where these were not negligible (by whatever criterion), to consider measures which might reduce risks, and whether those measures were 'reasonably practicable'.

The decision to proceed using QRA approaches was taken only after extensive discussions on the advantages and disadvantages of such an approach, both within the subcommittee and in those organisations represented thereon; these discussions taking place in the context of a wider UK debate on the merits of using QRA in the decision making process (HSE, 1989). It was recognised that there would be problems in developing an effective methodology, in the short term, for assessing the issues. It was known to be an area where methodology was rapidly developing, but where great uncertainties still existed. It was accepted that the emerging results of current research (particularly into consequence modelling) could well affect outturns on a short term basis; and most importantly the relative immaturity of QRA methodology in the 'peopleware' context of activity (as opposed to 'hardware' assessment) was considered to be a major difficulty. Critically, it was acknowledged that the magnitude of uncertainty might well be greater than the freedom of action available to the decision maker whose judgements the analyses were intended to inform. But despite these considerations, the QRA route was felt to be the only route by which the terms of reference given to the subcommittee could be met.

3. THE STRUCTURE AND APPROACH OF THE QRA STUDIES

It was recognised that the QRA approach would necessarily involve lengthy, highly specialised, sophisticated technical analyses; and that for the credibility of the outcomes, the involvement and consensus of all affected organisations was essential. For these reasons, technical working parties were appointed to supervise in a continuing and interactive manner the assessment process. These working parties, chaired by a member of HSE, had memberships which were drawn from (or otherwise represented) all the interests involved, and played a major role not only in providing basic data available only from those interests, but also in providing peer review of the developing methodology and emerging results.

The working parties thus formed were:

a) the Technical Working Party on Hazard and Risk Analysis (TWP) which was charged with overseeing the development of an appropriate methodology for assessing road and rail risks (and its subsequent application);

b) The Marine Working Party (MWP) for developing and applying an existing methodology for marine risks; and

c) a criteria group (the 'ad-hoc' group) for considering and where appropriate recommending 'tolerability' criteria relevant to the transportation context.

The studies themselves were carried out:

a) for marine risks, by Technica Ltd, Consulting Scientists and Engineers (funded by HSE); and

b) for road and rail risks, jointly by the HSE's Technology Division (TD) and the Safety and Reliability Directorate of the UK Atomic Energy Authority (now known as SRD of AEA Technology) again funded by HSE.

Membership of the 'criteria' group was on the basis of personal reputation and knowledge.

All working groups reported, at appropriate intervals, to the ACDS subcommittee, which responded interactively as the studies proceeded; and the detailed findings of the groups form part of the subcommittee recommendations to ACDS, and thereafter to HSC. Interim findings were, on occasion, published (Canadine, Purdy 1989; Smith 1988).

4.   THE ACDS REPORT

The ACDS Report has now been published (HSC, 1991). Essentially based on the 'Canvey' format (HSC, 1978, 1981) this major study is in the form of a general report (essentially a non-technical overview containing a description of the background methodology, conclusions and recommendations of the study) and a number of (mainly technical) appendices in many of which the detailed methodology and conclusions of the working parties are described. Most of the appendices were prepared by the working parties themselves, but three appendices are personal contributions from members of the WPs or the subcommittee. The major appendices are:

- Membership of the subcommittee and the working groups
- Summary of selected major transportation accidents 1981-1989
- Summary of current movements and regulatory arrangements modes and consequences of failure of road and rail tankers (personal - Dr. V.C. Marshall)
- Individual and Societal risk criteria for transportation study
- Port Risks in Great Britain
- Rail Risks in Great Britain - non explosive substances
- Road Risks in Great Britain - non explosive substances
- Explosives - Road and Rail Risks
- Road and Rail Comparison - non explosive risks
- Tolerability, risk reduction and risk mitigation measures road and rail non explosive risks
- Transport Accident Emergency Planning (personal - Mr. W.D.C. Cooney)
- Management of Safety in Transport (personal - Dr. I.C. Canadine)

5. LIMITATIONS ON THE SCOPE OF THE STUDY

The terms of reference described in Section 1 contain some major general limitations to the study. Air transportation is excluded, as is transport damage to the environment per se (although not the environment as an indirect vector of harm to man). Nor were pipeline risks considered. The constraints of time (and the influences of developing initiatives elsewhere) necessitated some selectivity in the particular aspects of the risks assessed. The scale of the task necessitated some judgements as to reasonable limitations. For these and other reasons, therefore, the marine study does not address the risks from oxidizing substances and toxic liquids in ports, nor risks associated with packages, freight containers, and with break-bulk and Ro-Ro operations; similarly (but for a different reason) port risks from explosives are not considered by the report. For road and rail risks, some selectivity was also deemed essential and the study concentrated on those risks which were felt to demand the higher priority:

- those which seemed potentially the most hazardous (specifically liquefied, toxic and flammable gases with LPG, chlorine and ammonia as examples)

- those which were moved in greatest quantity (flammable liquids with motor spirit taken as an example)
- explosives (the study was extended to include explosive articles as well as substances). This part of the study was given further immediacy by the Peterborough explosion which occurred during the course of the work (HSE, 1990)

and thereby to be representative in providing a model on which to base guidelines for control. For similar reasons the Report only addresses loading and unloading risks in a relatively limited way in the non marine context, primarily because such road and rail operations are already subject to stringent controls as a result of legislation embodying European Community Directives (CEC, 1982), or UK national legislation (HMSO, 1982). Timetabling considerations precluded as full a consideration of mitigating measure cost effectiveness in a road/rail context as that carried out in the marine study.

6.   THE CONTEXT AND NATURE OF THE RISKS

6.1 The Context of the Risks

The risks assessed in the report are in the context of an existing legislative regime much of which is based in international convention and recommendation e.g. Rail (RID), Road (ADR) and Marine (IMDG), at the core of which lie unifying concepts centred in UN approaches. All such control leaves a residuum of risk, however, even with full observance: which is not always forthcoming, however, strict an enforcement regime, nor are the controls easily modified on a responsive basis, particularly at the international level; and such controls, based mainly on past experience, bring no firm assurance against the potentially very serious, very rare event.

6.2 The Nature of the Risks

Many of the risks associated with some elements of the transportation process are analogous with those presented by fixed installations - indeed loading, unloading and marshalling points, ports, stopover areas etc. form in essence a quasi-fixed installation in themselves. They are surrounded by an identifiable population and the classical methods of assessing and evaluating 'individual risk' can generally be applied. Furthermore, any 'societal risk' can be derived using similar, classical approaches. The dominant risks may well be a matter of political judgement. During the mobile phase of the transportation process individual risk is likely to be very (normally vanishingly) small, whether it is an 'on route' risk - that is to those persons living alongside the route which the hazardous substances pass - or the 'en route' risk - that is to those other persons also using the route concerned. The societal aspects of this combination of 'on route' and 'en route' risks may well be very significant and their modelling is extremely complex. The Report develops in detail the concept of 'route societal risk' to express this dimension. Risk aversion is recognised as a potentially important determinant in any discussion of tolerability in this societal risk context.

6.3 The Range of Events Considered

It was especially this 'high-N' factor which governed the choice of scenarios suggested by the assessments. Specific examples include:

a) puncture of motor spirit road tanker (pool fire, 20-30 en route casualties)

b) puncture of road or rail chlorine or ammonia tanker when passing through a town (up to 1000 casualties within a mile or so of release)

c) puncture followed by BLEVE of LPG road tanker (several hundred on route or en route casualties)

d) loss of LPG from a ship in collision or unloading (substantially bigger release than (c))

e) explosion of ammonium nitrate in a ship in harbour

f) high explosives incident following fire in road or rail vehicle.

Most of these events have already occurred somewhere; and all have had precursor events in the UK.

Only on very few occasions has the complete combination of 'worst case scenario' and 'worst case outcome' occurred; and the problems of extrapolating limited data has led to the need to synthesize probabilities for some such very rare events. Where this synthesis was necessary, sensitivity testing was also carried out, as well as judgement on confidence limits.

6.4 The Assessment Procedure Used

The procedures used followed a classical risk assessment architecture, involving:

* Review of (a) movements
           (b) historical accident record

* Identification of main or potential causes of classes of incidents

     Estimation of event frequencies (for 'specimen' substances)
     Estimation of appropriate 'source term(s)'

* Quantification of consequences

* Identification of   (a) individual risk levels
                      (b) societal risk levels

* Identification of remedial and mitigating measures

* Evaluation of remedial and mitigating measures

6.5 The Methodology Developed and Used

Several papers are to be presented to the Consensus Conference detailing aspects of the methodology used in the studies. These papers cover the marine assessment (Spouge, 1992), the non explosives road and rail assessment (Purdy, 1992) and the explosives road and rail assessment (Riley, 1992). A brief outline only is therefore presented in this paper of the major elements of this aspect of the studies.

6.5.1   Marine Study

Of the 42 ports in the UK handling significant quantities of dangerous substances, 3 were chosen (using judgement) as representative surrogates. The basis of this choice was:

a) a port handling very substantial quantities of a wide range of dangerous goods (Teeside)

b) a port handling a substantial traffic of one of the chosen substances (LPG) in a busy, multi-purpose harbour, with significant ferry and other general traffic (Felixstowe)

c) a small port with a varied trade, including in this case, ammonium nitrate (Shoreham)

A general methodology already existed for assessment of individual and societal risks from port operations and this was used along with statistical inputs from the industry and subject to interactive oversight by MWP to calculate individual and societal risks for port workers, ships crews, and members of the public who might be affected by incidents involving dangerous goods in the three port areas. Details of traffic and other characteristics of the remaining 39 ports were then combined with the outputs of the detailed assessments of the three ports and scaled to derive a national risk figure. Whilst this technique is not appropriate for estimating the risks from a port other than the three named ports (e.g. for evaluation of risk reduction measures) it is felt to provide a fair estimate of national port risks overall. The actual levels of risk are discussed in section 8 of this paper.

6.5.2 Road and Rail Study

After assembly of national traffic data provided by operators, handlers, and similar organisations, puncture frequencies were derived using historical evidence where available and consensus engineering judgement where such history was not available, or felt to be misleading. Frequencies of equipment failure and small leaks were similarly derived. The estimated emission or spill probabilities (expressed per wagon/kilometer, and therefore generic in nature) were combined with typical population densities, weather type and characteristics and vulnerability/effect models to estimate risks to those living adjacent to the road and rail routes, (the 'off road' and 'off rail' populations) having regard to population densities typical of urban, suburban and rural locations. Additional populations were also considered in particular for rail, other passengers en route, (or 'on rail'); and for road other road users, using typical road configurations.

This approach permits the national road or rail trade in that substance to be calculated, by scaling to total national trade, and to average route length. Clearly it is important to ensure that routes so chosen are truly characteristic, or carry a large proportion of the relevant trade.

7. PRINCIPLES OF RISK CONTROL

The general principles associated with risk control used in the study are described in a recent publication on the tolerability of risks from nuclear power stations (HSE, 1988) which developed more general arguments on risk and risk acceptability (Royal

Society, 1983) in the UK. The '3-zone' approach reproduced at Figure 1 embodies the concepts of:

a) an 'intolerable' level of risk at or above which immediate action to reduce is called for, irrespective of cost;

b) a 'negligible' level at or below which further risk reduction measures are not required (again irrespective of cost); and

c) a middle region where the 'ALARP' principle applies (ALARP being acronym for 'as low as reasonably practicable') where (briefly described) additional risk reduction measures are necessary until their overall cost becomes grossly disproportionate to their risk reduction effect.

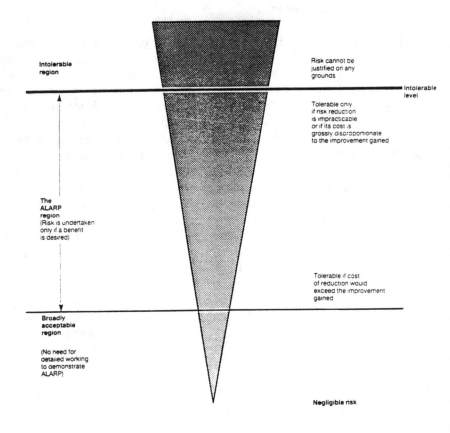

(Adapted from a diagram in "The tolerability of risk from nuclear power stations", HSE, 1988)

Figure 1: Levels of risk and ALARP.

Such concepts are applicable to both individual and societal risks, but numerical values for the zones have more commonly been applied to criteria for individual risk. For the purpose of this study, the report suggests an upper level of 1×10^{-4} yr^{-1} and a lower level of 1×10^{-6}yr^{-1}, except where the operations are more appropriately compared to those of a 'static' installation where other criteria may be more appropriate. The latter criteria suggested by (HSE, 1988) are 1×10^{-5}yr^{-1} and 1×10^{-6}yr^{-1} (3×10^{-7}yr^{-1} for very vulnerable populations) of receiving a 'dangerous dose'.

The Report discusses at length the complexities and difficulties in deriving tolerability criteria for aspects of societal risk; and in the light of all the problems involved suggests criteria for a 3-zone approach for ports and for road and rail risks expressed in one locality. Essentially, the 'intolerable' level is centered in the knowledge that the risks at Canvey (HSC, 1978, 1980) were judged just tolerable after an exhaustive analysis followed by explicit Ministerial and Parliamentary debate (the upper line on the F/N graph is based on a line of slope -1 through N=500 at a frequency of 0.0002 per year and the negligible line is three orders of magnitude (or decades) lower than this. Within these boundaries the ALARP principle applies. The disposition of the lines (at a slope of -1) is intended to imply 'neutral' aversion to events involving large numbers of deaths (although the Report accepts that some elements of aversion are implicit in such a slope); and it is recommended that where any extra aversion is appropriate it should be applied explicitly and transparently. These recommended criteria are shown at Figure 2. They apply to one locality or to one route or to the transportation of one substance in the context of a major national transportation activity. They may not be appropriate criteria for, for example, different types or sources of risk or judgements about single installations or for control of land use and development.

The Report suggests a further development of the 3 zone approach. The criteria proposed are derived from a decision about a situation which is felt to be reasonably comparable with the risks assessed by the study. But they apply essentially to locally concentrated risks to the people in that location. Higher criteria may apply to the total national port risk(s). A 'national scrutiny' level is proposed, therefore, suggesting a tolerability level based on a per ton rate of cargo shifted at Canvey scaled up to the national volume of trade.

This is a difficult concept, but it permits recognition of a situation in which the scale of overall risks relative to the quantity of dangerous goods handled has the potential to lead to an overall deterioration in

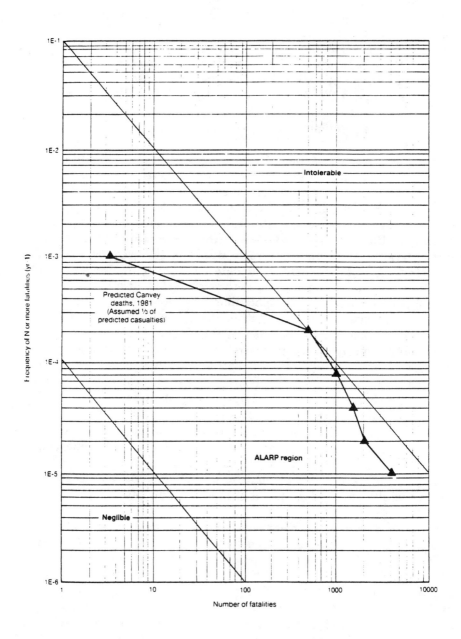

Figure 2: Proposed societal risk criteria for an identifiable community (e.g., living near a port).

the national risk. Such a national scrutiny level has a potential for prioritizing and focussing attention. A similar approach can be used to define a 'local scrutiny' level for smaller ports based on the Canvey risk level per ton of cargo scaled by the volume of trade at the port of interest. Risks at such a level although remaining tolerable, would signify a priority need for further analysis and possible risk reduction ALARP; and could also identify those elements of port trade which have a disproportionately high risk in relation to volume, or value of such trade. Local scrutiny levels are not appropriate where volume of trade exceeds the Canvey reference point.

Apart from the effects of any aversion, these criteria are also limited in the sense that they are 'death-only' and do not take account of injury, or property damage. Equally, such essentially localised criteria do not present the full picture for road and rail risks because of the 'en route' dimension. For this reason the overall road and rail risks were set against the national port scrutiny level (although this assumes a broad, if uncertain, equivalence between the economic importance of all port and road and rail traffic) and against local tolerability levels (a more certain test, since if this criterion is not reached the risks must be in the tolerable region).

8.   THE RESULTS OF THE QRA ANALYSIS

8.1 Port Risks

Detailed risk assessments were carried out for three very different ports and the national risks scaled therefrom. In terms of individual risk, all three ports fall below the intolerable line described above, but most are ALARP. Some of the risks would attract advice from HSE on land-use restriction. An example of these assessed risks is shown at Figure 3. The societal risks, set against intolerable and scrutiny levels described above both for local societal risks, and for national port risks are again in the ALARP region but very few are negligible. The levels of national societal risk are shown at Figure 4.

8.2 Rail Risks (non explosives)

Individual and societal risks incorporating both on route and en route populations were assessed. 'Hot Spots' aside, individual risks are in or very close to the 'negligible' region. Societal risks are 'tolerable', at both national and local level although some of them may be far from negligible. The ammonia traffic in particular appears to present the most prominent risk, but this conclusion is sensitive to the volume of trade, and toxicological and engineering assumptions. Figure 5 shows the assessed levels of societal risk.

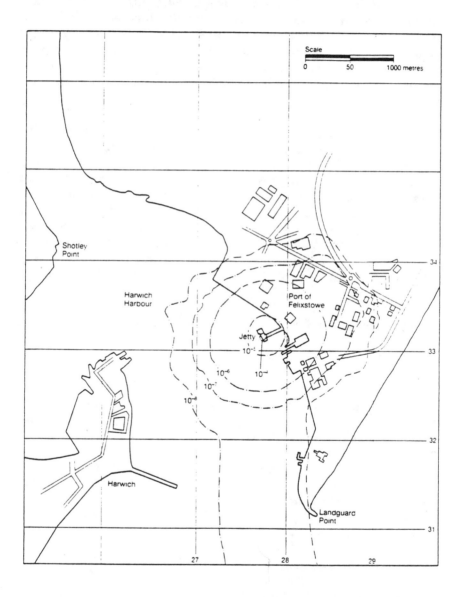

Figure 3: Felixstowe: Individual risk for gas carriers.

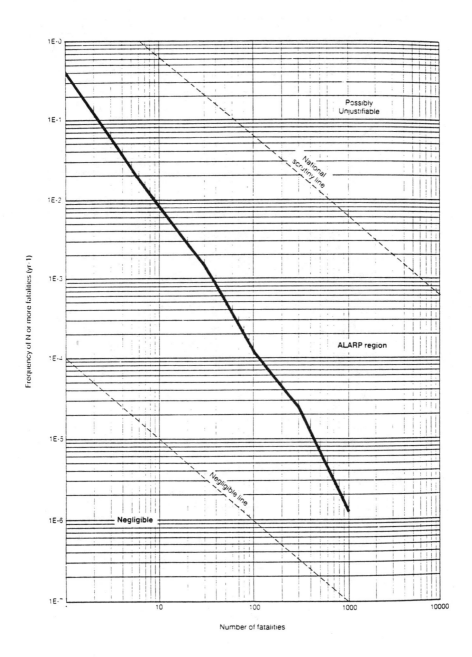

Figure 4: National societal risk tolerability for ports.

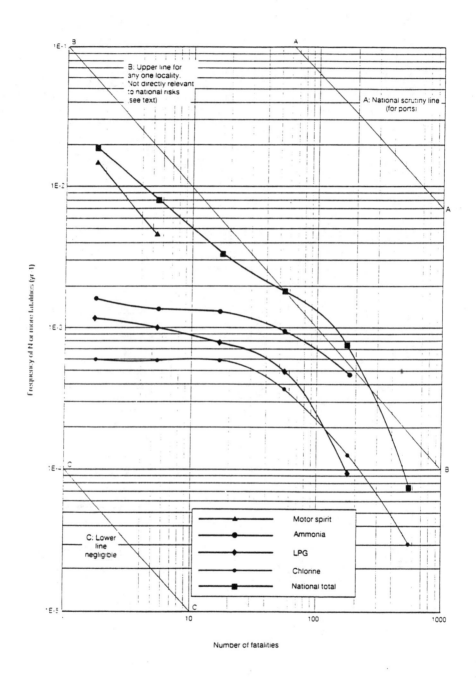

Figure 5: Total national societal risk en route - Transport of
substances by rail - Comparison with local risk criteria.

8.3 Road Risks (non-explosive)

Again, both on route and en route risks were assessed and in generic terms found to be negligible in terms of the recommended criteria, possible 'Hot Spots' apart. Societal risk is not negligible (but is only a small proportion of the risk from ordinary road accidents - which, of course rarely have capacity for high-N fatalities -) but is in the 'ALARP' region. Lorry stopover points (and to a much lesser extent, rail marshalling yards) may suggest some need for land use planning controls; and some aspects of ammonia and LPG unloading, particularly in the non-notifiable context, may need further investigation. Figure 6 shows the assessed levels of risk.

8.4 Road and Rail Risks (explosives)

Although the methodology involved was somewhat different in detail individual and societal risk figures for road and rail carriage of explosives were derived as shown in Figures 7 and 8. For rail, 'off rail' risks are the most significant component of an otherwise very low level; whereas for road, the risks to other road users are more significant but still tolerable (even for one locality). Any additional 'aversion' issues in the context of explosives are uncertain.

8.5 Prominent contributors to risk

Essentially, whilst all the risks assessed fall below the 'tolerable' line, most suggest an ALARP approach. The most prominent contributors in this ALARP region (it is emphasized within a very small total) are:

- in ports, cold rupture and ignition of consequential spill
- in ports, unloading point releases
- for roads, motor spirit (for low-N)
- for road, LPG (for high-N)
- for rail, aspects of ammonia transport
- for explosives, road and rail, initiation by fire

8.6 Risk Management

Clearly, in the course of the study various aspects of risk management were considered. The investigators were particularly interested, in the UK context, in the following issues:

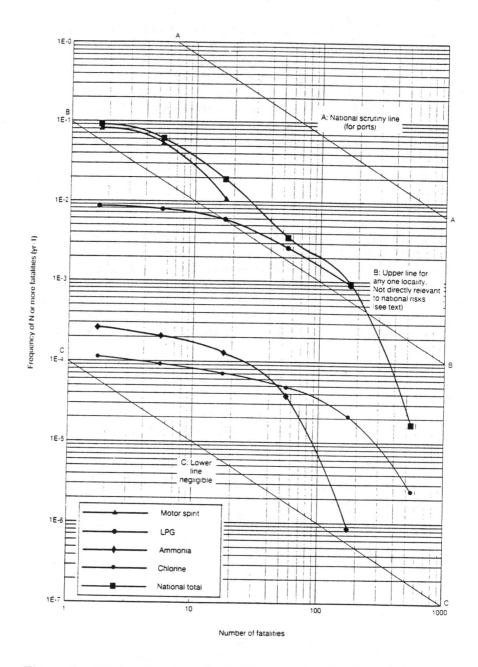

Figure 6: Total national societal risk en route - Transport of substances by road - Comparison with local risk criteria.

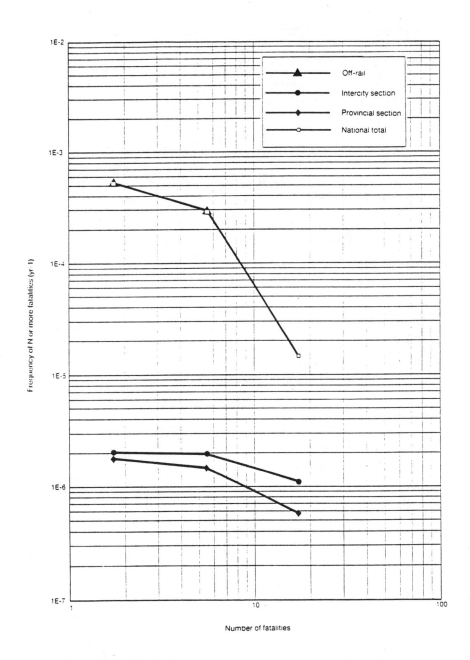

Figure 7: National societal risk - Transport of explosives by rail - Route B.

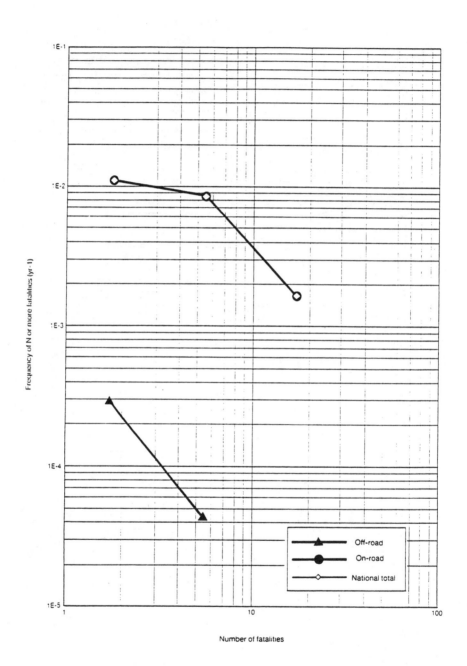

Figure 8: National societal risk - Transport of explosives by road.

a) risk removal, and in particular the possibility of on-site manufacture of hazardous goods. Of course, the offsetting factors associated with increased site risk need to be considered here

b) risk reduction (which is discussed in section 9 below)

c) risk mitigation, including aspects of emergency planning and the roles and duties of emergency services and local authorities; (some aspects of which are discussed further in Section 10 below)

d) land-use planning, and in particular the risks associated with

(i) 'subnotifiable' LPG and ammonia unloading operations

(ii) lorry stopover points (including motorway stations)

(iii) railway marshalling yards (particularly, but not exclusively for explosives)

e) cost effectiveness, and the debate (which was not resolved by those involved in the study) on whether or not cost effectiveness and/or cost benefit calculations should include any commercial gain aspects of risk mitigation measures

f) the relative merits (and associated uncertainties) in hardware versus 'peopleware' approaches

g) the value of prioritising risk reduction measures on the basis of calculated annual expectation values

h) the effects, if any, of types of 'aversion' and any differential aspects thereof

i) the calculation/suggestion of value of life

8.7 'Value of Life'

The assessments did not produce risk figures which, when crudely converted to annual expectation values, predicted more than 2 deaths per year; often in specific situations the prediction was much lower. In the context of a suggestion by the ad-hoc criteria group on the value of life at £2m (allowing to some extent for the 'gross disproportion' effect upon present values) the scope for improvements even where affected by this multiplier and possible further multipliers for non-fatal effects and for any aversion affects, is very limited. In particular:

a) very major capital expenditure seems to be ruled out

b) some restricted capital expenditure of restricted nature might be justified if phased over several years

c) low capital cost improvements (mainly in the 'peopleware' category) seemed more likely to be justifiable; but the 'on-costs' of such improvements might not be negligible and the uncertainties (in QRA terms, inter alia) of such changes might be problematical

It was in the context of these constraints that specific measures for risk reduction were assessed.

9.   ASSESSMENT OF SPECIFIC RISK REDUCTION MEASURES

9.1 Avoidance/Prevention

For a number of reasons a fuller analysis of some risk reduction measures than of others was undertaken. In particular, time constraints precluded all but a preliminary analysis of many potential road and rail measures. Notwithstanding this the Report contains many general and specific recommendations for further study of potential risk reduction measures for risks in the ALARP region. Before any implementation, all would require detailed analysis using QRA and CBA techniques, applied to the specific context of the analysis. It is unlikely that meaningful conclusions could be reached without a quantified risk analysis approach. These include:

General

- enforcement of or improved compliance with existing legislation management awareness and commitment
- technical devices (eg tyre pressure monitoring, middle distillate additives, engine overrun cut-offs, RPE provision) and training

For Road

- tanker design   - roll over protection/avoidance
                   - puncture protection/avoidance
- routeing   - mandatory/recommended
- improved radio communications   - in transport
                                  - on route
- medical checks for drivers
- passage scheduling

For Rail

- tanker design - ammonia tankers
- training and dedication of drivers

For Ports

- hard arms vs hoses
- handling of LPG near population centres - assessment
- land use planning
- water use planning
- improvements in terminal procedures
- improvements in training
- improvements in communications.

For Explosives (road and rail)

- Tyre fire prevention/protection
- cargo compartment protection
- routeing
- provision of radiotelephones
- placarding/marking of explosives vehicles
- segregation of mixed loads
- improved fire-fighter training
- fire resistant packaging
- phasing out nitroglycerine based explosives
- separation of H.D. 1.1 explosives
- risk assessment of marshalling yards
- double manning
- non-combustible packing/containers

Full details of all these isues are to be found in the Appendices to the Report.

9.2 The Contribution of Emergency Planning and Response

A further element of mitigation, as opposed to accident prevention or avoidance, is provided by preplanning for emergencies, and response to and management of an emergency in the event. The investigation identified and supported in principle, improvements in the following areas (although accepting that some aspects would need further investigation):

- Improvements in information systems
- Improvements in communications on route and en route
- Adequate and appropriate labelling and placarding
- Adequate provision of specialist assistance
- Adequate emergency planning and provision
  (especially 'novice' locations - that is to say, those locations lacking the expertise resulting from the existence of static hazardous installations therein)
- Adequate and appropriate funding of emergency planning

- Definition and delineation of 'primacy' in emergencies
- The benefits of an 'all hazards' approach

10. COMPARISON OF ROAD AND RAIL TRANSPORT RISKS

The Report accepts that modal choice is not always practicable or indeed possible. An island with a major international trading economy of necessity needs ports (and any effect, on this modal limitation, of the Channel Tunnel is far from being determined!) and even for internal transportation modal choice is not necessarily an option (for example remoteness of quarries may determine delivery mode of explosives; motor spirit delivery needs at the local level may preclude the use of rail). There will however be occasions when a 'real' modal choice is possible and it is important that such choice is made on a structured and objective basis. The Report recommends that any initiatives for modal control (such as the recent German Legislation) are so underpinned.

The UK analysis includes risk comparisons for various types of modal choice; these include general comparisons (on a tonne/kilometer basis) and a comparison of a specific (chlorine) route (Figure 9). A number of conclusions emerge. These include:

a) that for low-N, road risks are a factor of about 20 greater than those of rail (a function of the dominant effect of motor spirit releases and greater rail segregation in general)

b) that for N = approximately 100, road and rail risks are broadly equivalent; and

c) that for high-N, rail risks tend to be greater that those of road

Such considerations are of course very sensitive to realistic opportunity of modal choice, to volumes and mix of traffic, and to en route as opposed to on route victims.

On the basis of a tonne/kilometre comparison:

a) motor spirit would appear to be safer by rail than by road (but modal choice is only a limited option here)

b) LPG risks are broadly comparable overall (but rail passengers are less at risk than road users)

c) for chlorine and ammonia, road is safer than rail (particularly for on route victims (the en route comparison is inconclusive)

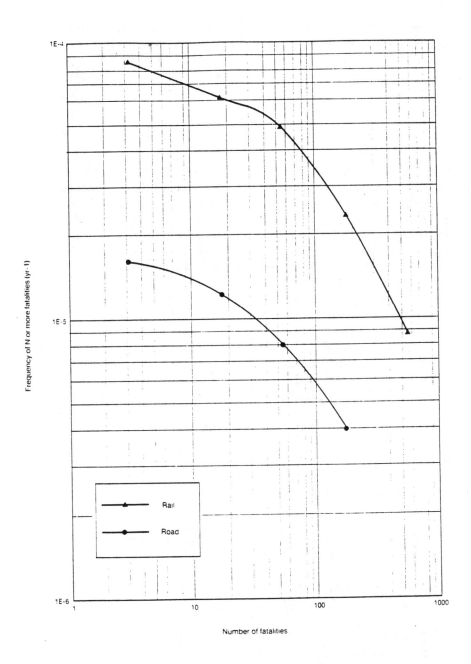

Figure 9: Transport of chlorine between Lostock and Fleetwood - Comparison of risks.

In the case of the specific route road risks for the transport of chlorine are some 5 times lower than for those or rail; but it must be emphasized that this is a substance specific/route specific conclusion and that it cannot lead to more general modal conclusions.

11. CONCLUSION

This Report will, it is hoped, lead to extensive public discussion on the conclusions and recommendations it contains. It is considered to be an important contribution to the current knowledge of the safety of hazardous industrial operations and the general approach is felt to provide a significant step forward in techniques of assessment of risks from the transportation of dangerous substances, and in the delineation of relevant criteria for tolerability of such risks.

12. REFERENCES

CANADINE, I., & PURDY G. (1989). The transport of chlorine by Road and Rail in Britain - a consideration of the risks. 6th International Symposium on Loss Prevention and Safety Promotion in the Process Industries, Oslo, June 19-22, 1989.

CASSIDY, K. (1988). The national and international framework of legislation for major industrial accident hazards. *I Chem. E. Symposium Series* 110 , Rugby.

CEC (1982). *Directive on the control of major industrial accident hazards*: 82/501/EEC (as amended 1988, 1990) OJEC, Brussels.

HMSO (1982). *The notification of installations handling hazardous substances regulations*. HMSO, London.

HSC Advisory Committee on Major Hazards. *First Report 1976, Second Report 1978, Third Report 1984*. HMSO, London.

HSC Canvey an Investigation (1978). *Canvey a second report* (1980). HMSO, London.

HSC (1991). *Major hazard aspects of the transport of dangerous substances*. HMSO, London.

HSE (1974). *Flixborough. Report of the public inquiry.* HMSO, London.

HSE (1988). *The tolerability of risk from nuclear power stations.* HMSO, London (Now revised September 1992).

HSE (1988). *Risk criteria for land use planning in the vicinity of major industrial hazards.* HMSO, London.

HSE (1989). *Quantified Risk Assessment: its input to decision making.* HMSO, London.

HSE (1990). *The Peterborough explosion.* HMSO, London.

PURDY, G. (1992). The measurement of risk from transporting dangerous goods by road and rail. Paper presented at the International Consensus Conference on the Risk of Transportation Dangerous Goods, April 6-8, 1992, Toronto.

RILEY, T.N.K. (1992). A novel risk assessment methodology for the transport of explosives. Paper presented at the International Consensus Conference on the Risk of Transportation Dangerous Goods, April 6-8, 1992, Toronto.

ROYAL SOCIETY (1983). *Risk assessment - A study group report.* Royal Society, London.

SMITH, L.M. (1988). *Major hazard aspect of transport.* Proceedings of a conference on the Transportation of Hazardous Materials. IBC Publications, London.

SPOUGE, J. (1992). Techniques for risk analysis of ships carrying hazardous cargo in port areas. Paper presented at the International Consensus Conference on the Risk of Transportation Dangerous Goods, April 6-8, 1992, Toronto.

A Comparison of Theoretical and Actual Consequences in Two Fatal Ammonia Incidents

Theodore S. Glickman
Resources for the Future
Washington, DC
U.S.A.

Phani K. Raj
Technology and Management Systems Inc.
Burlington, MA
U.S.A.

1. INTRODUCTION

Consequence analysis, as it relates to toxic gas incidents, is the process of generating fatality estimates from information on release conditions, cloud behavior, health effects and population exposure. Given the uncertain nature of such information, the usual practice, as noted by Beattie (1979), Lees (1980), and Purdy and Davies (1987), is to err on the high side when estimating fatalities. Unfortunately, this practice can lead to misperceptions of risk and inefficient allocations of risk management resources. Thus, we need to know how conservative (i.e., pessimistic) such estimates tend to be.

This paper takes a modest step in that direction by comparing theoretical fatality estimates to the actual number fatalities for each of two well-known transportation incidents involving anhydrous ammonia. The incidents in question are the 1976 truck accident in Houston, Texas, which resulted in 6 fatalities (5 due to the toxic gas) and 178 injuries, and the 1977 train derailment in Pensacola, Florida, which resulted in 2 fatalities and 46 injuries. In both cases, the release was so sudden that, despite the extraordinarily short initial response times (5 minutes in Houston and 7 minutes in Pensacola), everyone who eventually died from ammonia inhalation was exposed to a lethal dose before the emergency responders arrived. Many other potential fatalities were avoided because people took actions to protect themselves and others.

The findings of this paper, while far from conclusive, demonstrate the importance of taking into account the vulnerability of an area when estimating the consequences of toxic gas incidents. Vulnerability is a function of the number of people potentially exposed to lethal effects and their ability to avoid being exposed. The focus here is on the former aspect, not because it is more important but because it is more readily quantifiable. The major conclusion of this paper is that further research is needed to develop quantitative methods for dealing with vulnerability.

In Section 2, which follows, we provide background information on the transportation of ammonia, its physical properties and its health effects. Section 3 then summarizes the facts in the Houston and Pensacola incidents. The modelling of ammonia release behavior is discussed in Section 4. In Section 5 a heavy gas dispersion model is applied to each of the two incidents. The theoretical number of fatalities is estimated and compared to the actual number in Section 6. The conclusions are presented in Section 7.

2.   BACKGROUND ON AMMONIA

2.1 Transportation

Anhydrous ammonia ("ammonia") is a widely used chemical in agriculture, chemical processing, pulp and paper, mining and refining industries. It is transported in very large quantities on land and by water in the U.S. In the rail mode alone, ammonia ranked third in 1990 in number of carloads of hazardous materials moved in tank cars. Ammonia is classified by the U.S. Department of Transportation (DOT) both as a Division 2.3 material (poison gas) and a Division 2.2 material (nonflammable gas). The various identification numbers used by the transportation industry for ammonia include the following: UN number 1005 (anhydrous, liquefied), 2672 (solutions, 10%-35%) and 2073 (solutions, 35%- 40%); STCC number 4904210 (anhydrous), 4904220 (solutions of 44% and greater) and 4904221 (ammoniacal liquor); CAS number 7664-41-7; OHM-TADS number 7216584.

Anhydrous ammonia is transported in compressed liquefied state or as a cryogenic liquid. In tank cars and tank trucks it is shipped as a pressurized liquefied gas at ambient temperature. The tanks are not insulated. The largest tank cars allowed on American railroads carry about 75 tons, whereas tank trucks are limited to 25 tons. Ammonia is also transported on water as a cryogenic liquid at 240K in tankers and in insulated tanks on barges. Barge shipments of up to 3000 tons are common.

2.2 Physical Properties

Ammonia vapor is a colorless gas with a sharp, penetrating odor detectable by humans at very low concentrations in air. The chemical is in a gaseous state at normal ambient temperature (293K) and atmospheric pressure. The gas can be liquefied by pressure (about 8.5 bar at 293K) or by cooling to 239.8K at 1 bar pressure. When the compressed liquid is released to the atmosphere at ambient temperature it immediately flashes; a part of the released mass forms saturated vapor and the remainder is manifested as a cold liquid. Any of the liquid that impacts or collects on the ground will evaporate due to atmospheric heat transfer. Ammonia is miscible in water in infinite proportions. Dissolution of ammonia in water is an exothermic process.

Ammonia is generally nonflammable, but there are a number of documented cases in which ammonia-air mixtures have either burned or exploded under partial to full confinement. It forms a combustible mixture with air when the chemical concentration is in the range 15.8% to 25.7% by volume (Buckley and Husa, 1962). The magnitude of explosive pressure rise in a confined ammonia-air explosion can be three-quarters as high as that of common hydrocarbon fuels under similar conditions. Damages caused by ammonia fires or welding operations in ammonia plants have been documented by Gustin and Novacek (1978).

2.3 Health Effects

Ammonia is an irritant to skin, eyes and the respiratory tracts. Its odor can be discerned by "average" individuals at as low a concentration as 5 ppm; even at this level the vapors produce skin and eye irritation. Exposure of skin to 30,000 ppm causes blisters in a few minutes. At low concentrations, acute exposures can cause mild irritation of the respiratory tract. High concentrations, which produce severe symptoms involving laryngeal spasm and bronchopneumonia, can be fatal. The metabolic toxicity of ammonia in man, the toxicity symptoms, and related information have been documented by Kamin (1979).

Table 1 (Environment Canada, 1984) provides ammonia concentration values for different types of exposures and effects. Exposure standards differ widely: the IDLH value recommended by NIOSH in the U.S. is 500 ppm, whereas the LC50 (low) value is 30,000 ppm for 5 minutes of exposure.

Table 1: Health effects of ammonia.

Concentration (ppm)	Health Effects
20	first perceptible odor
40	a few individuals may suffer slight eye irritation
100	noticeable irritation of eyes and nasal passages, and of upper respiratory tract
400	severe irritation of the throat, nasal passages, and upper respiratory tract
700	severe eye irritation; no permanent effect if the exposure is limited to less than 1/2 hr.
1700	serious coughing, bronchial spasms; less than 1/2 hr. of exposure may be fatal
5000	serious edema, strangulation, asphyxia; fatal almost immediately.

DOT has recently adopted an index defined by the American Industrial Hygiene Association (AIHA, 1988) for use in evaluating potential hazard areas arising from transportation accidents involving a release of chemicals that are toxic by inhalation (DOT, 1990). In defining this index, called the Emergency Response Planning Guideline (ERPG), DOT recognized that, in general, chemical releases from transportation accidents can pose short-duration hazards to the public. The ERPG-2 value that DOT uses to estimate potential hazard areas is defined as "the maximum airborne concentration below which it is believed that nearly all individuals could be exposed for up to one hour without experiencing or developing irreversible or other serious health effects or symptoms which could impair an individual's ability to take protective action." This value is set at 200 ppm for anhydrous ammonia.

3.   SUMMARIES OF THE INCIDENTS

3.1 Houston, Texas (NTSB, 1977; NTSB, 1979)

At about 11:08 a.m. on May 11, 1976, a tractor-semitrailer tank transporting 7,509 gallons of anhydrous ammonia struck and penetrated a bridge rail on a ramp connecting I-610 with the Southwest Freeway (US 59) in Houston. It was a bright and sunny day, with temperatures in the low 80s. The tractor and trailer left the ramp, struck a support column of an overpass, and fell onto the freeway 15 ft. below. Anhydrous ammonia was released from the damaged tank, which was filled to 71.8% of capacity.

The tractor separated from the trailer and the front end of the trailer separated from the main body of the tank. This caused a sudden, rapid release of ammonia. In the process of falling over the rail and bouncing off columns, the tank was opened at both ends.

There were approximately 500 persons within 1/4 mi. of the release point. Six persons were killed and 178 injured. Five of the deaths and all the injuries were the result of ammonia inhalation. The truck driver" death was caused by the collision. Out of the 178 injuries, 100 were treated at the scene only and 78 were hospitalized. These 78 were within 1000 ft. of the estimated release point. Three automobile drivers and two of their passengers were killed; the injuries included 20 drivers and 12 passengers.

Emergency personnel responded promptly. The first ambulance was on the scene within 5 minutes and found the traffic moving with no significant backup. By 11:40 a.m., 14 ambulances and four pieces of fire equipment had responded. Firemen sprayed water to dissipate ammonia on the ground and wash down automobiles that contained ammonia fumes. Because all the fatalities were within 200 ft. of the release point, it was estimated by the NTSB that the ammonia concentration in this area exceeded 6,500 ppm for at least 2 minutes.

Wind pushed the vapor downwind 2,000 ft. until the effects were minimal. The cloud reached a height of 100 ft. before being carried by a 7 mph wind for half a mile. Within 3 minutes, the maximum width of the vapor cloud over the ground was 1,000 ft., diminishing in size downwind. After 5 minutes, most of the liquefied ammonia had boiled off and the cloud was completely dispersed.

3.2 Pensacola, Florida (NTSB, 1978)

The Pensacola accident occurred at 6:06 p.m. on November 9, 1977. Thirty-five cars of a train derailed and the adjacent tank heads of the 18th and 19th cars were punctured during the derailment, releasing the ammonia into the atmosphere. The 18th car was punctured on the right side, causing narrow cracks in the shell near the junction of the head and the shell of the tank. The puncture of the 19th car was a 3 in. wide by 38 in. long tear in the leading tank head. When the two tank cars were punctured, the pressure within each tank caused the anhydrous ammonia to vent and vaporize. Within 10 minutes, about 50% of the contents of the 19th car entered the atmosphere. The contents of the 18th car were released slowly over 12 hours.

An air traffic controller at Pensacola Airport first observed the cloud formation on radar at about 6:10 p.m. It appeared to be one mi. in diameter and about 125 ft. high, in the vicinity of Gull Point. The cloud moved northeast over Escambia Bay and into Santa Rosa County, traveling almost 15 mi. before it faded. At the time of the accident, it was dark and the temperature was about 68°F. The sky was overcast and a light rain was falling. The wind was from the southwest at 3.5 mph, and in the following 6 hours changed to southeasterly at about 12 mph.

There were 2 fatalities and 46 injuries. One man died several hours after the accident from severe pulmonary edema. His wife died 2 months after the accident from lung damage. They lived in a house adjacent to the railroad tracks. Eight of the injured required hospitalization.

A resident notified Pensacola police at about 6:07 p.m. Police and fire teams arrived at 6:13 p.m. The Pensacola fire chief ordered about 500 persons evacuated within a 3,500 foot radius. Escambia County and Santa Rosa County were alerted to the cloud's movement and evacuated about 500 more people living within the cloud path.

The escaping vapor was restrained by spraying the punctured areas of the two tank cars with a fog-type water spray. Water was also pumped through the dome of the 18th car to dilute the ammonia. By the following morning only a 30% solution remained in the car, with minimal escaping vapors.

4.   MODELLING THE RELEASE BEHAVIOR OF AMMONIA

The behavior of ammonia released into the atmosphere depends on the state of the chemical in the tank before release and the substrate onto which the chemical is released. The density of saturated anhydrous ammonia vapor at atmospheric pressure is lower than that of ambient air. Hence, pure vapor released into the atmosphere would rise and pose considerably less threat to people at or close to ground level. On the other hand, if a release occurs in which the vapor is mixed in with liquid aerosols, the combined density of the vapor-aerosol will be higher than that of ambient air, resulting in a ground-hugging dispersion of the vapor cloud. Such a release, therefore, poses a considerably higher toxic threat in a populated area.

When pressurized liquefied gas is released continuously from a hole in the wall of a container, a high-velocity liquid jet results which subsequently and very close to the hole flashes to form a cold, heavy, two-phase jet containing liquid aerosols and vapor. This two-phase jet begins to entrain air. The mixing with air results in vaporization of the aerosols, reduction in the volumetric concentration of ammonia and reduction in the velocity of the jet. Also, because of the relatively high density the jet expands laterally due to the gravity-induced expansion. The mixing of air with the ammonia also results in the condensation of atmospheric moisture and the dissolution of ammonia in the condensed water, forming ammonium hydroxide aerosols. The heat of reaction/dissolution released in the water-ammonia reaction affects the temperature, density and liquid concentration in the vapor-aerosol cloud. When the jet velocity is nearly equal to the wind speed, the dispersion of the vapor plume (and any unevaporated aerosol) is influenced by the atmospheric wind and meteorological conditions rather than by the release conditions.

During this second phase of dispersion the vapor plume moves downwind at or near the wind speed and expands both laterally due to the higher-than-air density of the plume, and vertically due to entrainment of air. The dispersion process is dominated by the turbulence in the atmosphere when the density of the plume is very close to that of air. During this third dispersion phase, the concentration of vapor in the plume can be calculated by using neutral density gas dispersion models.

In the case of an instantaneous release of ammonia, e.g., from a ruptured tank car, there is no initial jet; instead a massive cloud is formed which then moves downwind and is diluted by the prevailing wind. The physical phenomena of air entrainment, chemical reaction, and lateral and downwind movement of the cloud are modeled in much the same way as in the case of a plume.

This sequence of dispersion phenomena has been incorporated into a heavy gas dispersion model called ADAM (Raj et al., 1987; Raj and Morris, 1988) which includes both plume dispersion and the puff or cloud dispersion. This model calculates the rate of release for specified tank size and thermodynamic conditions of the chemical in the tank, and size of the hole. The model further determines the flashing fraction and the characteristics of the jet formed (velocity, density, temperature, cross-section size and chemical concentration). Wind effects such as entrainment of air into the jet and drag (interfacial and ground) are considered. Chemical reaction effects are calculated using an equilibrium thermodynamic model for the mixing of anhydrous ammonia and humid air. The parameters of the model are determined at each downwind spatial-step position. These calculated values form the input values for determining the plume spread and plume conditions for the subsequent step. The transition from the jet phase of dispersion to the second phase (heavy gas dispersion) occurs when the jet velocity is within 5% of the wind speed. The final phase of dispersion occurs when the density of the vapor-air mixture in the plume is very close to that of air.

The centerline total concentration of ammonia and the concentration in the crosswind position are calculated at each of the downwind spatial positions. The crosswind width to the contour of the specified hazard concentration is then determined. Finally the contour for this concentration is plotted and the area within the contour is calculated. The model calculates the time-averaged concentration values only and does not provide any results on the concentration variations with time at any given location. Also, in the calculation of the downwind concentration hazards arising from a long-term release, the release rate is assumed to be constant with time. This provides a conservative estimate of the hazard area.

The model results have been found to agree reasonably well with values of concentrations and other plume/cloud parameters reported in the literature from several field-scale tests conducted with different chemical releases (Raj & Morris, 1988).

5.   APPLICATION OF THE MODEL

For each incident, the heavy gas dispersion model was run for three cases: reported meteorological conditions, average meteorological conditions, and worst case (very stable) meteorological conditions. In each case, we used the wind direction, type of release (instantaneous or continuous) and release quantities that were actually reported in each of the two incidents. Under average conditions, we used the mean temperature, relative humidity and wind speed from the preceding year, which is what a risk analyst might have used as inputs prior to the incidents. Under worst case conditions, we used the most stable combination of temperature, relative humidity and wind speed for any day in the preceding year.

A number of assumptions had to be made to convert the accident data into the appropriate form for model input. Listed below are the assumptions common to both accidents and unique to each accident.

5.1 Common Assumptions

a) The ammonia was assumed to be transported at atmospheric temperature with the tank pressure equal to the saturation pressure at that temperature.

b) Neutral atmospheric conditions (D stability) were assumed for both reported and average meteorological conditions. Extremely stable conditions (F stability) were assumed for worst case release conditions, as was a wind speed of 1 meter per second.

5.2 Houston Assumptions

a) The release of the entire contents of the tank was assumed to have occurred instantaneously (no venting) due to the separation of the tank shell and head.

b) Accumulation of any liquid on the ground was assumed to be negligible.

5.3 Pensacola Assumptions

a) The release of anhydrous ammonia venting from the 18th car over a 12- hour period was assumed to be insignificant compared to the continuous venting of half of the contents of the 19th car over a period of 10 minutes (the model is not capable of superimposing two separate release events).

b) The release was assumed to emanate from a circular hole at the bottom of the tank shell. Based on the reported tank diameter, it was assumed that the liquid head within the tank was at a height of 1.51 meters (approximately half the diameter of the tank). Flow rate calculations (as a function of initial liquid head and tank pressure) indicated that the hole would have to be 0.066 meters in diameter in order for 16,750 gallons of ammonia to vent from the 33,500 gallon tank within 10 minutes.

c) The flow rate of material through the hole in the tank wall was assumed to be the same as the initial flow rate.

d) All the liquid formed after the flash process was assumed to be entrained into the dispersing plume.

Figures 1 and 2 show the cloud contours for the reported case at concentrations of 400 ppm (minimal physiological effects) and 6500 ppm (lethal effects). The latter concentration was noted by NTSB (1977) for Houston and is within the fatal concentration range reported by Raj (1982). No plots are shown for the average case or worst case. Table 2 presents the output for each case, which consists of the maximum downwind distance of the cloud for each of the two selected concentration levels and the ground area covered by it.

6.   ESTIMATION AND COMPARISON OF CONSEQUENCES

The number of persons potentially exposed to the lethal effects of a toxic gas incident is commonly estimated by multiplying the size of the lethal hazard area by the density of the residential population within it. This simple approach assumes that all the residents of the lethal area are there when the incident occurs and that no one else is involved.

Table 3 shows the results applying this approach to Houston and Pensacola. The lethal areas that went into these calculations were the 6500 ppm entries in Table 2. Population density in each location was determined by dividing the total population for the appropriate year by the total area of the census tracts in question, where the total population was estimated by interpolating between the 1970 and 1980 population counts.

Experience shows that the number of motorists in the lethal area also has to be taken into account, especially in the case of releases caused by highway accidents. Table 4 shows the factors that enter into estimating this number for the reported case in Houston, where two

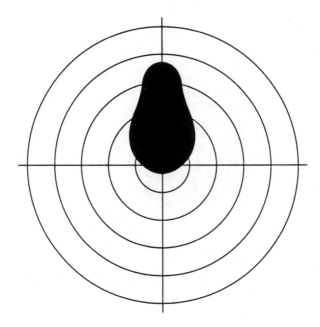

(a) Houston (grid size = 500m)

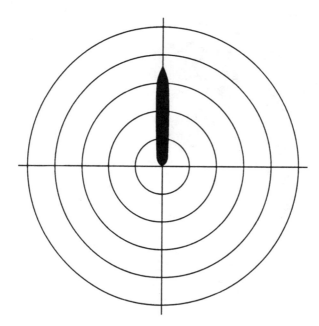

(b) Pensacola (grid = 2000m)

Figure 1: Cloud contours for 400 ppm.

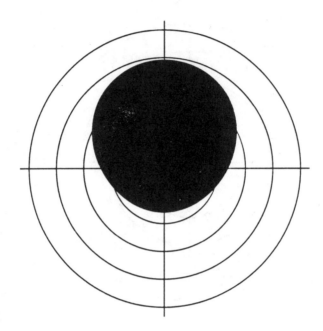

(a) Houston (grid size = 100m)

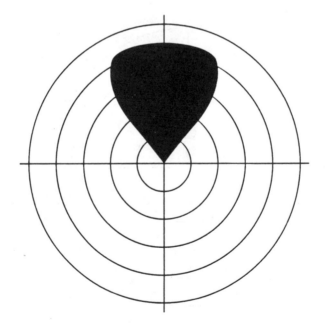

(b) Pensacola (grid size = 100m)

Figure 2: Cloud contours for 6500 ppm.

Table 2: Model output for the two incidents.

a) Maximum Downwind Distance (mi.)

	Reported Case		Average Case		Worst Case	
	400 ppm	6500 ppm	400 ppm	6500 ppm	400 ppm	6500 ppm
Houston	1.14	0.24	1.21	0.24	5.53	0.27
Pensacola	4.42	0.24	1.80	0.22	22.11	0.11

b) Ground Area Covered (sq. mi.)

	Reported Case		Average Case		Worst Case	
	400 ppm	6500 ppm	400 ppm	6500 ppm	400 ppm	6500 ppm
Houston	0.62	0.08	0.59	0.08	1.54	0.15
Pensacola	1.88	0.03	0.63	0.02	24.10	0.04

Table 3: Estimated number of residents potentially exposed to 6500 ppm.

	Reported Case	Average Case	Worst Case
Houston	266	257	491
Pensacola	134	90	218

Table 4: Estimated number of motorists exposed in Houston (reported case).

	US 59	I-610
Traffic Flow (vehicle/hour)	9720	9640
Traffic Density (vehicle/mile)	177	175
Length of Road Exposed (mile)	0.34	0.32
Number of Motorists	70	65

major roads were involved. The traffic flows, in vehicles per hour, are actual hourly traffic counts provided by the Texas Department of Transportation. The corresponding traffic densities, in vehicles per mile, are the flows divided by the assumed speed of 55 mph. The distance exposed along each road was estimated by overlaying the contours of the lethal areas onto local road maps. The number of motorists exposed on each road is then the product of the density, the exposed distance and the vehicle occupancy rate of 1.16 persons per vehicle.

No such calculation was made for Pensacola because the cloud contour did not overlap the local roads. Even if it had, the effects would have been much smaller than in Houston because of the relatively low traffic volumes.

Table 5 compares the number of residents and motorists potentially exposed to the actual number of fatalities in each location. The estimates far exceed the actual number of inhalation fatalities in each of the two incidents. The dashes in the "others" category indicate that we did not attempt to estimate the number of non-residents in the lethal area, other than any non-resident motorists. Any such attempt would no doubt have widened the gap even further for the Houston incident, which happened in mid-morning in a commercial area, at which time the daytime population would have exceeded the residential population.

Table 5: Estimated potential exposure (reported case) vs. actual fatalities.

	Houston			Pensacola		
	No. of Persons Exposed	Adjusted No. of Persons	Actual Fatalities	No. of Persons Exposed	Adjusted No. of Persons	Actual Fatalities
Residents	266	8	0	134	2	2
Motorists	135	–	5	0	0	0
Others	–	–	0	–	–	0

The vulnerability of the areas where the accidents occurred depends not only on the number of persons potentially exposed but other factors, too, such as the number of persons outdoors vs. indoors, the presence of "special" populations in the area (e.g., in schools and hospitals), the effectiveness of emergency response and the success of individual survival actions. Unfortunately, the literature only proved to be helpful for adjusting the estimated number of people potentially exposed according to the time of day and the fraction outdoors. Yet, these factors alone reduce the observed differences by at least one order of magnitude, as the following calculations show.

The Canvey Island study (HSE, 1978) and the Rijnmond study (Rijnmond Public Authority, 1982), both of which dealt with industrial complexes, estimated the ratio of daytime population to nighttime population in the vicinities of concern to be 42% and 45%, respectively, or about 44% on the average. This is the value we use for the time-of-day adjustment factor for the Houston accident, which happened in the middle of the day at about 11 am. For the Pensacola accident, which happened in the early evening at about 6 pm, we use a value of 90% instead; this is the value estimated by Glickman (1986) for the percentage of the nighttime population at home in a residential area at this hour.

Petts et al. (1987) estimated the average proportion of the population outdoors to be about 7% by day and 1% by night, without distinguishing by day of the week or time of the year. The 7% value applies directly to Houston, whereas for Pensacola we weight the 1% night-time factor by 90% and the 7% daytime factor by 10% to get an average value of 1.6%. For Houston, the combined adjustment factor for time of day and fraction outdoors is thus 44% x 7% = 3.1%, and for Pensacola it is 90% x 1.6% = 1.4%. Therefore, the expected number of residents potentially exposed drops from 266 persons to about 8 in Houston and from 134 to about 2 in Pensacola, as displayed in Table 5.

No similar adjustments were made to the number of motorists potentially exposed because we found nothing helpful in the research literature. Obviously, the potential number of motorists exposed depends on the volume of traffic, which varies by time of day. Such information is readily available and should be used, but as is the case with non-motorists, we also need to be able to evaluate the ability of an individual to avoid or escape from a toxic cloud. The opportunity for a vehicle occupant to avoid being exposed depends on whether the vehicle can be driven away safely or whether the occupant can flee on foot. The escape option, once the vehicle is caught in the cloud, depends on the driver's visibility and his access to an escape route, but in some cases, as has happened with chlorine, the chemical vapor may choke off the engine. In this event, too, the vehicle occupants might succeed in getting away from the cloud, but might be better off staying put and taking shelter in the vehicle. In Houston, some motorists whose cars were disabled by the accident or who attempted to flee by leaving their cars were fatally injured by the toxic fumes. NTSB (1979) reported that some of the Houston victims "would have avoided the aerosolized cloud which presented the greatest threat" if they had shut off their motors and rolled up their windows. In the future, a better understanding of the complexities of motorist exposure in the event of toxic release incidents would not only improve the process of consequence analysis, but would also contribute to better emergency guidance.

7. CONCLUSIONS

These results demonstrate the importance of having realistic estimates of population exposure when analyzing the consequences of toxic gas incidents. Residential census statistics alone do not reflect outdoor exposure levels in the daytime and do not take motorist exposure into account. Further research is needed to develop realistic approaches to consequence analysis that take account of these factors and other determinants of vulnerability, such as the effectiveness of self-protective measures (e.g., sheltering) and actions taken by emergency responders. We believe that, at this time, such research would be more beneficial than tinkering with dispersion models to make them more accurate.

Monte Carlo simulation might prove to be a useful methodology because the number of fatalities in a toxic gas incident depends on the dynamic interactions of a number of random variables. Figure 3 illustrates in simplified, schematic form how a simulation approach might work, assuming only three macroscopic variables: (1) release behavior, (2) population exposure and (3) emergency response.

The "values" of the release behavior and population exposure variables in the first increment of time t_1, during which there has not yet been any emergency response, influence the number of fatalities caused in t_1. These values also influence what the emergency response will be in t_2, the second increment of time. This phase of the emergency response then influences the release behavior and the population exposure in $t_2 + t_3$, which in turn influence the number of fatalities caused in $t_2 + t_3$. The process continues iteratively in this fashion until the emergency is over.

In practice, the variables would have to be far more specific and the relationships among them would have to be made explicit, but the basic logic of Figure 3 would still apply.

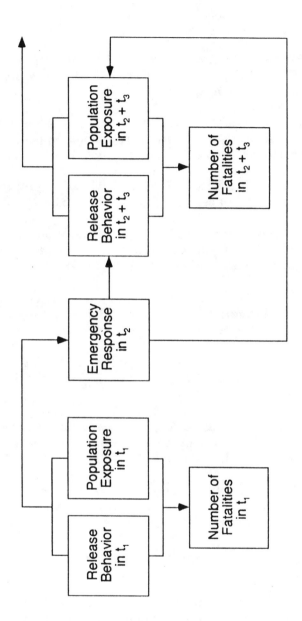

Figure 3: A simulation approach to consequence analysis.

Acknowledgement The authors are grateful to Clayton Turner of TMS, Inc., for his contributions to the model application.

8. REFERENCES

AIHA (1988). Emergency Response Planning Guidelines. American Industrial Hygiene Association Emergency Response Planning Guideline Committee, Akron, OH.

BEATTIE, J.R. (1978). A Quantitative Study of Factors Tending to Reduce the Hazards from an Airborne Toxic Cloud. Appendix to the Canvey Report, Her Majesty's Stationery Office, London.

BUCKLEY, W.L. & HUSA, H.W. (1962). Combustion Properties of Ammonia. Chemical Engineering Progress, 58, 2, 81-84.

DOT (1990). Guidebook for First Response to Hazardous Materials Incidents. U.S. Department of Transportation, Research and Special Programs Administration, Washington, DC.

ENVIRONMENT CANADA (1984). Ammonia: Environmental and Technical Information for Problem Spills, Environmental Protection Service, Ottawa, Ontario, July.

GLICKMAN, T.S. (1986). A Methodology for Estimating Time-of-Day Variations in the Size of a Population Exposed to Risk. Risk Analysis, 6, 3, 317-324.

GUSTIN, R.E. & NOVACEK, D.A. (1978). Ammonia Storage Vent Accident. Paper 43f, Symposium on Safety in Ammonia Plants & Related Facilities, AIChE, Miami Beach, FL, November.

HSE (1978). Canvey: An Investigation of Potential Hazards from Operations in the Canvey/Thurrock Area, Health and Safety Executive, Her Majesty's Stationery Office, London.

LEES, F.P. (1980). Loss Prevention in the Process Industries, Butterworths, London.

NTSB (1977). Highway Accident Report--Semitrailer (Tank) Collision with Bridge Column and Sudden Dispersal of Anhydrous Ammonia Cargo, Houston, Texas. National Transportation Safety Board Report No. NTSB-HAR-77-1, Washington, DC.

NTSB (1978). Railroad Accident Report--Louisville & Nashville Railroad Company Freight Train Derailment and Puncture of Anhydrous Ammonia Tank Cars at Pensacola, Florida. National Transportation Safety Board Report No. NTSB-RAR-78-4, Washington, DC.

PETTS, J.I., WITHERS, R.M.J. & LEES, F.P. (1987). The Assessment of Major Hazards: The Density and Other Characteristics of the Exposed Population Around a Hazard Source. Journal of Hazardous Materials, 14, 337-363.

PURDY, G. & DAVIES, P.C. (1987). Toxic Gas Incidents--Some Important Considerations for Emergency Planning. Loss Prevention Bulletin 062, Institute of Chemical Engineers.

RAJ, P.K. (1982). Ammonia. In Hazardous Materials Handbook, G.F. Bennett, I. Wilder and G. Feakes (eds.), McGraw Hill Publication, New York.

RAJ, P.K., & MORRIS, J.A. (1988). Source Characterization and Heavy Gas Dispersion Models for Reactive Chemicals (NTIS No. AD A-200-121). TMS Report No. G-85-42 to the U.S. Air Force, Air Force Systems Command, Hanscom AFB, January.

RAJ, P.K., MORRIS, J. & REID, R.C. (1987). A Hybrid Box-Gaussian Dispersion Model Including Considerations of Chemical Reactions and Liquid Aerosol Effects. Presented at the International Conference on Vapor Cloud Modeling, Cambridge, MA, November.

RIJNMOND PUBLIC AUTHORITY (1982). Risk Analysis of Six Potentially Hazardous Industrial Objects in the Rijnmond Area. A Pilot Study, D. Reidel, Dordrecht, Holland.

Using a Delphi Technique to Develop High-Risk Scenarios and Countermeasures

E.R. Russell, Sr.
Department of Civil Engineering
Kansas State University
Seaton Hall
Manhattan, Kansas
66506-2905, U.S.A.

ABSTRACT

Kansas State University (KSU) conducted a comprehensive study that included development of a set of prioritized, extreme-risk or catastrophic scenarios related to incidents occurring during the transport of hazardous materials on highways, and the development of a set of feasible, practical and implementable, physical protective systems to mitigate these potentially catastrophic incidents. Developing catastrophic scenarios that were meaningful to the states, and concomitant, feasible, practical protective systems that would be accepted by states' personnel was an early stumbling block. It was decided to form a panel of state personnel and include them in the development process. Although not generally used in technical studies, Delphi techniques were used to develop the scenarios, sets of protective systems tied to each scenario and finally provide an estimate of their effectiveness. This paper details the use of the Delphi process to accomplish these objectives. The technique was successful in getting a consensus from the states' panel on the ranked scenarios, the protective systems and estimates of their effectiveness, thus providing assurance that the scenarios developed were real concerns of the states and that the protective systems were credible with states' personnel.

1. INTRODUCTION

Kansas State University (KSU) conducted a comprehensive study of the development of a set of prioritized, extreme-risk scenarios, development of a set of feasible, practical and implementable protective systems, and development of a manual to provide guidelines on the use and implementation of these protective systems to mitigate potential, extreme-risk situations that could occur during the transport of hazardous materials (hazmat) on our highway system. (This study will be referred to as the Guideline Study.)

The specific objectives of this research study were:

1. identification of potential, extreme-risk situations which develop when hazmat are spilled on the highway systems,

2. identification of effective, practical, feasible, and implementable protective systems, and

3. development of guidelines for implementing protective systems.

The research study was limited to materials spilled within the highway system. It focused on potential risks which would result in severe, long-term, permanent, irreparable or catastrophic consequences, and existing technology and state-of-the-art knowledge for development of protective systems to mitigate these consequences. The protective systems within the scope of this study are systems that can be constructed or physically incorporated into the highway system or modifications thereto. Measures such as routing, response procedures, regulations, and prohibition of shipments and other regulatory and policy approaches to control hazardous materials (hazmat) shipments, were not within the scope of the study.

2. RESEARCH APPROACH

Early in the project it was concluded that that concept of designing protective systems into highway systems specifically to prevent or mitigate hazardous materials spills was a new and unique concept. No U.S. literature existed that directly addressed this concept. Literature on systems that could be adapted to the concept, such as drainage containment systems, high-strength barrier rail, etc., was available, but first the catastrophic nature of spills had to be defined.

By its nature, a catastrophic occurrence resulting from the highway transportation of hazmat is a rare event. Not everyone agrees to a universal definition of "catastrophic occurrence." Catastrophic can mean many things to different people. Developing catastrophic scenarios that would be meaningful to the states and lead to concomitant, feasible, practical protective systems that would be useful to the states and have credibility with states' personnel, was an early stumbling block. All sorts of catastrophic scenarios could be dreamed up, but would they be realistic or meaningful?

It was decided to contact all states through official Federal Highway Administration (FHWA) channels and form a project advisory panel of personnel from those states wishing to participate in the research. This panel was then used to develop potentially catastrophic scenarios. With this approach, there was assurance that the scenarios developed were real state concerns and not merely creations of overactive imaginations of the researchers. Eleven ranked catastrophic scenarios were developed.

The next task was to develop protective systems keyed to the scenarios that had been developed. These protective systems were to be feasible, practical, and implementable. In this age of modern technology, almost any system imaginable is feasible and implementable--given unlimited resources. Obviously states do not have unlimited resources; therefore, what is "practical" in this sense will ultimately be a state policy decision based on some sort of cost-effectiveness analysis. Once again, to come up with a range or a set of possible protective systems that would have credibility with states' personnel, the states' panel was used to develop such a set. This ensured that the set should not be too far outside the range of what an individual state decision-maker would consider practical or at least worth investigation, i.e., determine its cost-effectiveness.

The approach used to develop both the ranked scenarios, and related protective systems, was the Delphi approach. It was also used to get a first estimate of the effectiveness of the protective systems.

3.   THE DELPHI TECHNIQUE

The Delphi Technique is a consensus technique using a series of questionnaires interspersed with information feedback in the form of written summaries. The classic form of Delphi restricts all interactions to written form although it has been used successfully in combination with group discussions.

The Delphi process is a way of guiding the interactions among members of a group. It has several advantages:

- It overcomes the weakness of a one-shot group approach and allows for feedback of group opinion before a final determination is made.
- Repeating the process allows participants to benefit from group opinion and composite results.
- Opinions or preferences can be expressed in total anonymity.
- The vagaries of group discussion can be minimized.
- Ideas can be judged only on their merits, with no influence by the group or individuals.

The Delphi process is especially useful where:

- A relatively small number of people are involved in the process.
- The level of knowledge, influence and communications skills varies widely within the group.
- There is a strong possibility that one or more group members may dominate the outcome.
- The issues to be addressed are highly judgmental or subjective and ideas may not be fully developed.

The Delphi is a proven method of formulating goals and objectives within groups of varying backgrounds. It is not normally thought of as a technique to obtain results in a technical study. Such a technique definitely would not be used if good data were available or if results could be calculated or analyzed in the traditional scientific or engineering approach.

4.   CATASTROPHIC SCENARIO DEVELOPMENT

4.1 General Procedure

In regard to catastrophic scenario development, several rounds of mailings were undertaken, along the lines of the following general steps:

1. All panel members were asked to suggest as many scenarios, real or hypothetical, as they felt appropriate.

2. All scenarios from the responses were collected, organized, combined to eliminate redundancy, and rewritten to have a clear single style.

3. The scenarios were sent out to the entire panel for comment, rating and ranking.

4. The returned responses and comments were again collected, analyzed and edited. Edited, ranked and rated lists of scenarios were returned to the panelists.

5. This process went on through seven rounds of mailings and resulted in a ranked list of eleven, generalized catastrophic scenarios that the entire panel was generally in agreement with.

A more detailed description of the process is described in the following sections.

Some critics of the use of the technique on this project were concerned that initially there were wide ranges or differences of opinion on what was being proposed by individual panel members. The concepts were new and no previous information existed on which to base an evaluation. These are the conditions where the Delphi technique can be valuable.

4.2 Developing the Ranked List

The first two rounds were to acquaint the states' contacts with the project objectives and the procedures that the author was going to use.

In Round 2, a questionnaire asked each respondent to list five real or hypothetical scenarios with catastrophic potential that had occurred or could occur in their state. It was decided that this was the best way to get a set that reflected the real concerns of the states. Response was good. Out of approximately 60 scenarios returned in the responses, 44 scenarios were determined as being unique and appropriate to the study.

The 44 scenarios were sent out as Round 3. The panel was asked to rank these on a scale of one to seven, as shown in Table 1. In number 45 they were also asked to list any scenario they felt should have been included but hadn't been covered in the original 44. Question 46 asked them to list in ranked order the top five as would concern their State.

Table 1: Key to incident scale values.

Scale Value*	Key
1	Very minor incident; of little or no consequence under normal conditions.
2	Minor incident; little chance of escalation, little danger to life or serious or long-term environmental damage (aquifer, reservoir, or water supply) unless grossly mismanaged.
3	Potentially dangerous incident; but not likely catastrophic, danger to life or environment (aquifer, reservoir, or water supply) only if not handled properly.
4	Neutral; no clear catastrophic potential yet hard to predict.
5	Definitely dangerous incident; could be catastrophic under certain conditions of traffic, weather, or inadequate response. Could easily escalate to catastrophic situation.
6	Very dangerous incident; high catastrophic potential, high probability of loss of life, serious injury, or long-term damage to environment (particularly aquifer, reservoir, or water supply).
7	Definitely catastrophic incident; loss of life, serious injury, serious damage to environment (particularly aquifer, reservoir, or water supply) is certain to be avoidable only with extreme good luck.

* In general terms, where all replies are averaged, a mean value greater than 4 was interpreted to mean the scenario is catastrophic or has catastrophic potential.

The responses were excellent and several new scenarios were added through number 45. It appeared that some respondents had thought of other catastrophic scenarios or were getting a better feel for what was wanted. It was decided that another round should be sent out after reducing the list of 44 to those with "high" scores, plus new candidates from question 45 of the previous round.

A new list of 25 was determined by using the highest mean ranking scores, plus new candidates from the previous round, question 45.

These 25 scenarios were again sent out as Round 6. Another "top 5" list was requested. The 31 responses to Round 6 were used to rank and evaluate the 25 scenarios. Three separate rankings were made; one based on mean score, one based on the count of combined 6 & 7 score, and one based on the count of number of times the scenario appeared in the respondents' "top 5" list.

It should be noted that in the Delphi Technique how the scenarios were rated and ranked was not important. What is important is that there is agreement on the final result.

A total score was obtained by adding the respective ranks of each scenario. Ties were resolved by the 6 & 7 score rank, and if still tied, by the top 5 list rank; and if still tied, by the mean score rank. A summary of the ranking is shown in Table 2. The ranked list of 25 scenarios can be seen, as ranked, in Table 3.

It was decided that 25 scenarios was still too many. The first thought was that another round would be needed to further reduce the list to a set of about 10 ranked scenarios that would be appropriate for developing guidelines. However, while analyzing the list, it became clear that several scenarios were similar and could be combined in a generalized scenario that was still specific and detailed enough to lead to an appropriate guideline "category." This grouping can be clearly seen in Table 3.

4.2.1   The Final Ranked List

The combining and subsequent reranking was done simultaneously. It was found that 11 general statements could be written to cover all 25 of the more specific statements in Table 2. (In fact all 60 of the original statements would fit into one of the 11 categories.) These are shown in Table 4. Thus, development was not a case of picking out the top 11 and throwing out 49 of the original 60, but a process of collecting and rewriting statements that were broad enough to include all or most of the original statements.

5.   ENVIRONMENTAL SCENARIO

When the original scenarios were returned by the participating state representatives, it was noted that there were only a very few that related to environmental problems. Some experts believe that the public's greatest fear is contamination of water supplies. A separate round of questions was sent to the panel dealing only with environmental issues. This set of scenarios and the results can be seen in Table 5.

Also, advisory panel contacts in New York, Minnesota, and Rhode Island noted that the potential for contamination of reservoirs or aquifers is a major concern in these states. It was appropriate that an environmental scenario be among the top set which will direct the guidelines.

Table 2: Ranked scenarios from Round 6.

Rank	Scenario No.*	Scenario
1	2	Peak hour urban freeway accident involving toxic fumes/flammable or explosive gas, with commuter vehicles downwind trapped between interchanges.
2	45d	Chemical spills (poisonous or explosive) that could enter underground METRO stations or transit tunnels through sidewalk fresh air vents, etc.
3	37	Flammable liquid crashes and burns in a congested tunnel.
4	45b	Gasoline or LNG exiting an elevated highway going through rail and crashing into a high-rise building.
5	34	Gasoline tank truck on elevated freeway ruptures and spills onto buildings, people and vehicles below.
6	45a	Amtrak train hits an anhydrous ammonia tanker in an urban area with a mental institution downwind.
7	41	Poisonous gas release, such as phosgene, hydrogen fluoride etc. on any urban freeway.
8	16	Toxic cloud, such as Nitric Acid, Anhydrous Ammonia, Chlorine, leaks from container in heavily populated area.
9	18	Hazardous materials, such as chlorine accident and release in elevated (or depressed) area of a downtown freeway.
10	45e	Tank truck carrying propane crashes through median fence and hits passenger train running in median strip of highway.
11	36	Truck loaded with dynamite explodes on an urban freeway.
12	17	Chlorine tank truck release in front of a large hotel near highway.
13	19	Chlorine tanker accident near a hospital complex.
14	1	At-grade, urban railroad crossing accident between truck carrying propane or gasoline with hazmat cargo.
15	26	Interstate adjacent to school or hospital, a load of chlorine crashes through rail into the yard.
16	28	Propane tank truck fires & explosions in urban areas.
17	6	A truckload of herbicide rolled over, spilled and contaminated a reservoir.
18	42	Elevated urban freeway with gasoline tanker going through barrier rail and exploding on street below.

Table 2: Cont'd.

Rank	Scenario No.*	Scenario
19	38	Brake failure on 16% grade in mountains causes gasoline tanker to crash in town at the bottom of the hill.
20	4	Accident under a major freeway bridge or viaduct involving materials that could damage structure, such as, explosives, gasoline or propane.
21	45c	Spills of nuclear waste products in populated urban areas.
22	3	Accident on bridge over waterway causing major spill of poisonous or dangerous biological agents.
23	40	Propane tanker rollover and rupture on bridge abutment under freeway overpass over urban freeway.
24	15	Toxic material leaking from container during transit in heavily congested area.
25	5	Accident on a major freeway or viaduct involving materials that could damage structure such as gasoline or propane.

*Numbered according to Round 6.

Any of the 11 scenarios could be subdivided into more specific cases, such as the round 5 results subdivided as the environmental category, i.e., the results shown in Table 5 are a more detailed breakdown of generalized scenario number 8. A state may wish to detail other scenarios in this manner in the process of making the generalized scenario list state-specific.

6. RELATING EXTREME RISK TO HIGHWAY FACILITY TYPE

During the last rounds of determining the 11 ranked scenarios, additional forms were presented to the panelists to rate and rank the extreme risk potential of highway facility types, e.g., elevated highways, depressed highways, ramps, weaving sections, etc. The result of the top 48 are shown in Table 6. A more meaningful, generalized summary is shown in Table 7. This table follows a statement of one of the panel members that gives good insight into the perceived risk related to facilities: "All spilled material has to go down, up or laterally." As can be seen in Table 8, the danger is perceived in that order.

Table 3: Ranked scenarios grouped subjectively by general
consequence/type.

Rank	Scenario No.	Scenario
1	2	Peak hour urban freeway accident involving toxic fumes/flammable or explosive gas, with commuter vehicles downwind trapped between interchanges.
2	45d	Chemical spills of poisonous or explosive materials that could enter underground METRO stations or transit tunnels through sidewalk
3	37	Flammable liquid carrier crashes and burns in a congested tunnel.
4	45b	Gasoline or LNG exiting an elevated highway going through rail and crashing into a high-rise building.
	34	Gasoline tank truck on elevated freeway ruptures and spills onto buildings, people and vehicles below.
	42	Elevated urban freeway with gasoline tanker going through barrier rail and exploding on street below.
5	41	Poisonous gas release, such as phosgene, hydrogen fluoride, etc. on an urban freeway.
	16	Toxic cloud, such as Nitric Acid, Anhydrous Ammonia, Chlorine, from container in heavily populated area.
	18	Hazardous materials, such as chlorine accident and release in depressed (or elevated) area of a downtown freeway.
	17	Chlorine tank truck release in front of large hotel near highway.
	19	Chlorine tanker accident near a hospital complex.
	26	Interstate adjacent to school or hospital, and a load of chlorine crashes through rail into the yard.
6	45a	Amtrak train hits an anhydrous ammonia tanker in an urban area with a mental institution downwind.
	45e	Tank truck carrying propane crashes through median fence and hits passenger train running in median strip of highway.
	1	At grade, urban railroad crossing accident--propane or gasoline.
7	36	Truck loaded with dynamite explodes on an urban freeway.
	28	Propane tank truck burns and explodes in an urban area.
	4	Accident under a major freeway bridge or viaduct involving materials that could damage structure, such as gasoline or propane.
	40	Propane tanker rollover and rupture on bridge abutment under freeway overpass or over urban freeway.
	5	Accident on a major freeway bridge or viaduct involving materials that could damage structure, such as gasoline or propane.
8	6	A truckload of herbicide rolled over, spilled and contaminated a reservoir.
	3	Accident on bridge over waterway causing major spill or poisonous or dangerous biological agents.
9	38	Brake failure on 6% grade in mountains causes gasoline tanker to crash in town at the bottom of the hill.
10	45c	Spills of nuclear waste products in populated urban areas.
11	15	Toxic material leaking from container during transit in heavily congested area.

Table 4: Ranked, generalized scenarios.

	General Scenario Description
1	Poisonous, toxic flammable or explosive material endangers large numbers of trapped motorists, e.g., between interchanges, in cut section or in traffic jam downwind of poisonous or toxic gas release.
2	Chemical spills of poisonous or explosive materials that could enter underground "METRO" stations or transit tunnels through sidewalk vents, etc. (Includes entry of lighter-than-air toxic or poisonous gases into adjacent or overhead transit stations.)
3	Hazardous materials accidents causing release of toxic, flammable or explosive materials in tunnels.
4	Gasoline, LNG, propane (flammables, explosive gases), etc., accidents and releases on elevated facilities, including ramps thereto, with people at risk below or in adjacent buildings.
5	Release of poisonous toxic or explosive gases in populated areas in general and/or in locations and situations where special populations and/or institutions such as schools, hospitals, hotels, nursing homes, apartment complexes, etc., are at risk.
6	Releases from accidents between hazardous materials containers on highways and passenger trains or trains carrying hazardous cargo either at rail-highway crossings at grade or in situations with shared rights-of-way, such as freeways with transit in the median.
7	Explosive materials on facilities in populated areas and particularly in situations and areas where catastrophic consequences could occur to highway structures or apartments--adjacent or on air rights. Includes situation with adjacent petro-chemical plant that could result in conflagration.
8	Sufficient quantities of poisonous materials, such as herbicides, or dangerous biological/agents (or any material causing long-term or permanent damage) being released into a potable water supply, particularly reservoirs and susceptible aquifers and/or watersheds.
9	Rural, hilly or mountainous areas with cities or towns at bottom of long or steep grades where brake failure of hazardous materials carriers could cause catastrophic consequences to the populated area.
10	Spills of nuclear wastes or other nuclear materials, particularly in populated areas, areas affecting water supply, or areas particularly difficult to respond to and/or clean up.
11	Carriers of toxic flammable or explosive materials leaking material during transit in heavily populated or congested areas.

Table 5: Environmental scenario questionnaire and summary of results.

Rating*	
Mean	
	1. Direct spill into potable water supply
5.78	(a) Reservoir-direct spill
5.59	(b) Aquifer - little or no soil cover
4.31	(c) Aquifer - soil cover > 25 ft. (7.6 m)
4.13	(d) Area of wells; within 1 mi. (1.6 km)
	2. Spill into waterbed or stream within 1 mi. (1.6 km)
4.88	(a) Reservoir
4.69	(b) Aquifer
	3. River
5.00	(a) Immediately upstream of urban area
4.44	(b) Rural
	4. Stream
4.31	(a) Rural
4.47	(b) Urban
4.00	5. Crop land
3.68	6. Open ground, agricultural
	7. Open ground, non-agricultural
4.00	(a) High runoff
4.03	(b) High permeability
4.22	(c) Sinkhole area
3.96	8. Ecosystem flora, fauna
	9. Sewage drainage system
4.00	(a) Rural
4.25	(b) Urban
	10. Storm water
4.13	(a) Rural
4.28	(b) Urban

* Based on 1-7 scale explained in the text.

Table 6: Top 48 ranked facility descriptors.

Rank	Components
1	elevated basic segment over shopping center
2	elevated weaving area (non-ramp over shopping center)
3	elevated ramp/ramp junction/accel.-decel. lanes over shopping center
4	depressed basic segment with air-rights development
5	depressed weaving area with air-rights development
6	elevated, at-grade depressed (nothing over or under) basic segment within 1 block of nursing home or hospital
7	depressed ramp/ramp junction with air-rights development
8	elevated, at-grade or depressed (nothing over or under) weave section (within one block of) nursing home or hospital
9	elevated, at-grade or depressed (nothing over or under) ramp/ramp junction/accel.-decel./lanes within 1 block of nursing home or hospital
10	elevated, at-grade or depressed (nothing over or under) basic segment (within one block of) school
11	elevated, at-grade or depressed (nothing over or under) weave section (within one block of) school
12	elevated, at-grade or depressed (nothing over or under) ramp/ramp junction/accel.-decel./lanes within 1 block of school
13	elevated, at-grade or depressed (nothing over or under) basic segment (within one block of) apartments
14	elevated basic segment over parking
15	elevated ramp/ramp junction/accel.-decel. lanes over parking
16	elevated weaving area (non-ramp) over parking
17	elevated, at-grade or depressed (nothing over or under) basic segment (within one block of) shopping center
18	elevated, at-grade or depressed (nothing over or under) basic segment (within one block of) factory
19	elevated, at-grade or depressed (nothing over or under) basic segment (within one block of) apartments
20	elevated, at-grade or depressed (nothing over or under) weave section (within one block of) hotel
21	elevated, at-grade or depressed (nothing over or under) basic segment (within one block of) office building
22	elevated, at grade or depressed (nothing over or under) ramp/ramp junction/accel.-decel./lanes within 1 block of apartments
23	elevated, at-grade or depressed (nothing over or under) ramp/ramp junction/accel.-decel./lanes within 1 block of factory
24	elevated, at-grade or depressed (nothing over or under) weave section (within one block of) factory

Table 6: cont'd

25	elevated, at-grade or depressed (nothing over or under) weave section (within one block of) shopping center
26	depressed drainage into storm sewer
27	elevated, at-grade or depressed (nothing over or under) ramp/ramp junction/accel.-decel./lanes within 1 block of shopping center
28	depressed drainage into combined sewer
29	elevated, at-grade or depressed (nothing over or under) weave section (within one block of) hotel
30	elevated, at-grade or depressed (nothing over or under) ramp/ramp junction/accel.-decel./lanes within 1 block of office building
31	elevated drainage (from el.) to storm sewer
32	elevated drainage (from el.) to combined sewer
33	at-grade drainage into storm sewer
34	at-grade or depressed (nothing over under) ramp/ramp junction/accel.-decel./lanes within 1 block of hotel
35	elevated, at-grade or depressed (nothing over or under) weave section (within one block of) office building
36	at-grade drainage into combined sewer
37	at-grade basic segment
38	at-grade weaving area (non-ramp)
39	at-grade ramp/ramp junction/accel.-decel. lanes
40	depressed ramp/ramp junction without air-rights development
41	depressed weaving area without air-rights development
42	depressed basic segment without air-rights development
43	elevated, at-grade or depressed (nothing over or under) basic segment (within one block of) storage or hazardous mat.
44	elevated, at-grade or depressed (nothing over or under) weave section (within one block of) storage of hazardous mat.
45	elevated, at-grade or depressed (nothing over or under) ramp/ramp junction/accel.-decel./lanes within 1 block of storage of hazardous mat.
46	elevated basic segment no development under
47	elevated weaving area (non-ramp) no development under
48	elevated ramp/ramp junction/accel.-decel. lanes no development under

Table 7: Draft summary of the ranked facility descriptions.

Rank	Approx. Avg. Mean Score	Generalized Highway Facility
1	5.6+	Elevated facilities with development below
2	5.5	Depressed facilities with development over
3	5.0 to 5.4	Any facility adjacent to vulnerable population in order of: a) nursing home or hospital b) schools c) apartments d) shopping center e) hotel f) factory g) hazmat storage facilities
4	4.0	Runoff into drainage system

Table 8: Result of statistical test of significant difference of mean response scores between materials ("t" - test on mean of all possible pairs; Table 2 score values).

	Propane	Chlorine	Anhydrous Ammonia	Gasoline	Nitric Acid	Phosphorous
Propane	---	YES	NO	NO	NO	NO
Chlorine	YES	---	YES	YES	YES	YES
Anhydrous Ammonia	NO	YES	---	NO	NO	NO
Gasoline	NO	YES	NO	---	NO	NO
Nitric Acid	NO	YES	NO	NO	---	NO
Phosphorous	NO	YES	NO	NO	NO	---

6.1 Relating Extreme Risk to Highway Facility Type and Material Type

Throughout this project, six materials were used to represent the types of extreme risk consequences that could occur from all materials. (This was also determined by the panel on one of the early rounds.) These six were: propane (pro), chlorine (CHL), anhydrous ammonia (AA), nitric acid (NA) and phosphorous (PH).

To develop the information found in Table 6, forms with about 50 geometric components were sent out for each material to determine if there were differences in how these six materials were rated and ranked according to material. A t-test of mean response ratings of all possible pairs were performed on the results. It was found that only chlorine was determined to be more dangerous. This can be seen in Table 8.

It should be noted that throughout the early rounds, when scenarios were being developed, scenarios specifically mentioning chlorine dominated the responses. From these results, it could be concluded that there is more "fear" of chlorine incidents than any other hazardous material. (These were responses not limited to the six discussed in this section).

7.   USING THE PANEL TO DEVELOP PROTECTIVE SYSTEMS

7.1 Introduction

After developing the ranked list of 11 prioritized extreme-risk scenarios, the next task was to develop "feasible, implementable, and practical" protective systems keyed to the 11 scenarios.

Deciding what were feasible, implementable, and practical protective systems proved to be very difficult. It was concluded that the practical aspect of the protective systems was the key criterion, and this could only be decided by an individual state considering its risk versus the cost/benefits of the protective system, within the context of overall state priorities and resources--requiring a management decision. Ideally, benefits should be measured in terms of risk reduction, but this would be very elusive indeed because the data necessary to do a meaningful risk reduction analysis on previously untried protective systems is just not available. Accident reduction values, for example, would initially have to be based on the judgment of traffic engineers with expertise in accident causation.

7.2 Panel Survey

The states' advisory panel, formed for developing and ranking scenarios, was again utilized.

The original protective system ideas had come from this panel, and, after these were stored and organized, the panel was surveyed to evaluate 98 protective system ideas that had been generated. The last round of the scenario prioritization process asked the panelists to present ideas on protective systems keyed to each of the eleven scenarios. Again, the objective was to send out as many rounds as necessary to get a panel consensus on protective systems.

The response was very good. Although some panelists responded to some scenarios with comments such as "no hope" and left some blank, several good ideas were returned for all scenarios. These ideas for scenarios were sorted, edited, and returned to the panel. Editing of the responses was kept very minimal to keep the ideas essentially as the panelists had presented them.

Several responses focused on regulatory type solutions, which were outside the scope of the project. However, the panelists had been informed of this many times. Since many still felt that these sorts of solutions were best, this was considered to be significant information. Thus, it was decided to leave these in the list to be evaluated to see how they were rated and ranked, both individually and as a group.

The key to the "1 to 7" rating scale used to rate suggested scenarios is presented in Table 9. Thirty-two responses were analyzed and the mean, standard deviation, maximum and minimum scores of each protective system were calculated.

Table 9: Key to scale values--systems.

	Scale Value*	Key Guidelines to Assist Raters
Bad (Worst)	1	Nearly impossible to implement, not at all practical, will serve no useful purpose
	2	Very difficult to implement; little value
	3	Difficult to implement; some value possible but probably not worth the effort or cost
"Neutral"	4	Hard to judge; not clearly a "good" or "bad" idea
	5	Possible merit as practical and implementable protective system; worth further thought or development
	6	Clear cut merit as practical and implementable protective system
Excellent (Best)	7	Highly feasible, very practical, useful and efficient, excellent and very desirable

* In general terms, where all replies are averaged, a value less than four would suggest that the idea would be highly difficult to design/construct and install or would not be very useful/desirable--inappropriate.

One category of protective systems (communication and detection systems) was mentioned in the responses to almost all scenarios. Exploratory work was done on these types of systems, and specific examples were sent back to the panel in a separate section.

7.2.1 Analyzing Results

The initial responses were so varied that it was not immediately clear how to interpret the results. It was decided that, as in the case of the rating and ranking of the scenarios, these fluctuations represented real world differences of opinions in a new area, highway protective systems, where varied and/or limited knowledge and experience on which to base evaluations is prevalent. Thus, it was concluded that the fluctuations did not have any clear meaning applicable to the results. The decision was made to use a mean that reflected the various biases of the responders versus using other individuals' or groups' opinions or using weighted opinions.

The next decision was to determine what mean value should be considered "high" and what mean value should be considered "low." Because there was no rational way to determine this, an arbitrary mean rating of 4.0 was chosen as the cut-off point. The protective systems were listed from highest to lowest, and a reasonable number of them were picked from the top that could be handled well with the available project resources.

The protective systems that had a mean rating of 4.0 or greater were categorized by the 11 scenarios. The last group of protective systems in the table is a summary of communication and detection systems that were suggested and highly ranked for many of the scenarios. An example is presented for scenario number one in Table 10. Again, the panel generally agreed with the results which is a key requirement of the Delphi techniques.

7.2.2 Summary of Protective System Selection

Table 11 presents a concise summary of the relevant, physical protective systems that were proposed. They are broken down into two main groups: I. Mitigating and II. Preventive.

Table 10: Protective system rating results all proposed protective systems rated 4.0 or greater.

SCENARIO 1 -- Poisonous, toxic flammable or explosive material endangers large numbers of trapped motorists; e.g., between interchanges, in cut section or in traffic jam downwind in poisonous or toxic gas release.		
Protective type solutions		
Rank	**Mean**	
1	5.1	Traversable medians
2	5.0	Emergency phone call boxes on all hazardous cargo routes
3	4.7	Crossovers
3	4.7	Median openings
4	4.6	Highway exits designed for traffic entrance (response team)
	X̄ = 4.8	from opposite direction
Regulatory type solutions		
Rank	**Mean**	
R1	5.2	Routing restrictions
R2	5.0	Prohibition on hours (curfews)
R2	5.0	Prohibit large trucks through congested areas (routing)
	R̄ = 5.1	

Mitigation are further categorized into:

- detection and warning
- systems to facilitate escape and response
- system to mitigate fire/explosion consequences
- system to mitigate spill consequences
- specialized situation

Preventative are categorized into:

- containment
- control

7.2.3 Conclusions on Protective Systems Development

General. It can be concluded from this phase of the study, based on the responses of the large panel representing a broad cross-section of states' concerns, that regulatory type preventative measures dominate suggested solutions. Conversely, it can be concluded that the physical, protective system concept is not applicable as a general preventive or mitigating approach. It is limited to a few site-specific, high-risk situations where the protective system approach is clearly effective and the risk is deemed high enough to offset the cost. This is a policy decision of each state, and this decision is the heart of the practicality criteria.

Table 11: Categorization of proposed physical, protective systems for highways.

I. MITIGATING	
Category	**System**
A. Detection and Warning	Built-in PA systems Emergency call boxes Gas detectors/alarms Monitoring for quick response Communication and detection systems
B. Systems to Facilitate Escape and Response	Crossovers Traversable medians Median openings Highway exit/entrance redesign for emergency response vehicles Emergency exits with heavy doors (tunnels) Arrows pointing to nearest exit (tunnels)
C. System to Mitigate Fire/Explosion Consequences	Foam blanketing systems Large sprinkler systems Effective vent systems
D. Systems to Mitigate Spill Consequences Robust drainage with holding	Pea-style vents to trap gases Effective vent systems (closed areas) reservoirs Avoid use of open rails on structures Large sumps Grease trap sedimentation basins Floating surface barriers Drainage gutters directed toward collection points Retention basins that automatically close Clay blankets or barrier membranes
E. Specialized Situations	Fresh air vents at elevated levels (METRO) Coamings over street-level intake vents (METRO) Air intake away from roads (tunnels, METRO) Massive barriers with energy absorbing material (runaway trucks)
II. PREVENTATIVE	
A. Containment	High performance barrier systems
B. Control	Truck escape ramps Upgrade truck runoffs Wide shoulders

8.   USING THE PANEL TO GET A FIRST ESTIMATE OF PROTECTIVE SYSTEM EFFECTIVENESS

8.1 General Approach

The protective system ideas obtained through the states panel is a new concept in hazardous materials research. Most of these systems have never been implemented in highway situations, and where they have data on their effectiveness is not readily available, and thus their effectiveness is not known. It was decided that the best alternative was to provide a first approximation of the value of their effectiveness until better values are available. This was done by using Delphi techniques using the same states panel. Forms were provided for the user to input his/her own values of effectiveness. The panel responses were summarized and a mean value and standard deviation calculated. These values were returned to the panel members to determine if they could accept the mean value, i.e., a consensus that it was a reasonable value.

8.2 Results

The survey was carried out in stages. Physical protective system ideas were sent out first. The results obtained were compiled and the mean response sent back to the panel for their consensus along with the regulatory protective system ideas. On the second found the majority of panel members said they could accept the mean values so the exercise was not continued past two rounds.

Table 12 shows the summary of results of the survey. As emphasized previously, this is a very recent concept and hence there is no verifiable data to compare with these results. However, the author contends that the numbers provide a starting point that is reasonable and valid for making initial decisions about the effectiveness of the protective systems, particularly relative effectiveness.

What was of importance is that the panel members came to a consensus that they accepted the mean values. Until factual data is available from field experience, those numbers should be a good first estimate.

Table 12: Protective system rating and effectiveness results for scenario number 1.

Scenario 1 -- Poisonous, toxic flammable of explosive material endangers large numbers of trapped motorists; e.g., between interchanges, in cut section or in traffic jam downwind in poisonous or toxic gas release.		Effectiveness % Reduction of Incidents*
Protective type solutions		
Rank		Mean
1	Traversable medians	50
2	Emergency phone call boxes on all hazardous cargo routes	25
3	Crossover	30
3	Median openings	30
5	Highway exits designed for traffic entrance (response team) from opposite direction	30
$\overline{X_1}$ = 4.8		
Regulator type solutions		
Rank		
R1	Routing restrictions	65
R2	Prohibition on hours (curfews)	45
R2	Prohibition large through congested areas (routing)	55
$\overline{R_1}$ = 5.1		

* Rounded to nearest 5% increment

9. CONCLUSIONS

Based on responses of a large panel representing a broad cross-section of states' concerns, regulatory-type preventative measures dominated suggested solutions. Conversely, it can be concluded that the physical, protective system concept is not applicable as a general preventative or mitigating approach. It is limited to a few site-specific, high-risk situations where the protective system approach is clearly effective and the risk is deemed high enough to offset the cost. This is a policy decision of each individual state and is the heart of the "practicality" criteria. The effectiveness values, albeit far from perfect, could be used in a cost/benefit or cost/effectiveness approach to determining if they are worth considering, particularly when considering relative benefits of alternative strategies.

Engineering studies of this nature must be made on the best data available, as they have been made for centuries.

In regard to the Delphi technique, it was well suited to obtain necessary information that was the best available for this project.

Acknowledgements The research reported in this paper was conducted under the partial sponsorship of the USDOT Federal Highway Administration, and the partial sponsorship of the State of Pennsylvania (through Virginia Tech). However, the findings and conclusions in this paper are those of the author and do not necessarily represent the views of the Federal Highway Administration or the State of Pennsylvania or Virginia Tech. Also, the author is grateful to those panel members who patiently filled out many long forms.

10. REFERENCES

RUSSELL, E.R., SR. (1991). Developing High-Risk Scenarios and Countermeasure Ideas for Mitigation of Hazardous Materials Incidents, State and Local Issues in Transportation of Hazardous Materials: Toward a National Strategy, Proceedings of the National Conference on Hazardous Materials Transportation, St. Louis, MO, May 13-16, 1990, ASCE, New York.

RUSSELL, E.R., SR. (1990). Protective Systems for Spills of Hazardous Materials, Volume I: Final Report, Federal Highway Administration, FHWA-RD-89-173 (Unpublished).

RUSSELL, E.R., SR. (1990). Protective Systems for Spills of Hazardous Materials, Volume II: Guidelines, Federal Highway Administration, FHWA-RD-89-174 (Unpublished).

Risk Management for the Movement of Dangerous Goods

J.H. Shortreed
F.F. Saccomanno
K.W. Hipel
Institute for Risk Research
University of Waterloo
Waterloo, Ontario
Canada N2L 3G1

ABSTRACT

Improvements in the risk management of dangerous goods are now possible due to the development of new techniques for risk estimation, risk communication, conflict analysis and the development of criteria for the "public interest". The paper develops a framework for utilizing these new techniques and uses the Toronto area rail system to illustrate the concepts.

Public interest is defined as the maximization of individuals' quality of life. The method of accounting for risk reduction and management costs in terms of quality of life is described. Techniques for risk communication are presented. Their relationship to risk estimation and the decision criterion are developed. Moreover, the technique of conflict analysis is presented as a possible method to improve decisions about appropriate risk control measures by improving the decision makers understanding of the possible impact of alternative decisions.

The paper argues that improvements in risk estimation, while important, are only one part of the improvement in risk management and that improved risk estimates must be complemented by improved criteria for society decision making, better risk communication and the use of decision theory models.

1. INTRODUCTION

San Carlos, Mississauga, and other transport accidents provide
tragic evidence of the risks of transporting dangerous goods
(Glickman, 1992). The hazards are significant and the risks are
real. Moreover, each year there are increased numbers of
dangerous goods being transported.

Fortunately, the risk management of dangerous goods has been
successful in reducing risks and limiting the hazard potential of
the dangerous goods being transported. For example, in Canada
there are no recorded deaths due to dangerous goods transport by
rail. Risk management has three objectives: to keep the risks low, to
maintain the benefits of the transport of goods and to minimise the
costs of risk management.

If there is a hazard, such as a dangerous good, there is always risk -
zero risk is not possible. The objective of effective risk management
is to balance risks, costs and benefits so that the net benefit to all
individuals is maximised (Hipel and Shortreed, 1992).

The objective of this paper is to present a comprehensive approach to
the management of transporting dangerous goods. This approach
uses a number of techniques and methods that have been developed
in recent years by members of the Institute for Risk Research (IRR)
and others. Collectively these methods provide a practical means of
managing the risks of transporting dangerous goods. The
movement of dangerous goods in the Toronto area is used as an
illustrative example. The paper uses the perspective of a regulatory
authority.

A general model of risk management is illustrated in Figure 1.
Individuals in society undertake activities (e.g. work, play,
shopping, games, etc.). The activities are provided through
facilities and operations (e.g. factories, sports facilities, recreation
programmes, production of goods, transport, etc.). The activities of
individuals and the operation of facilities result in issues, (e.g.
risks of sports, air pollution from factory operations, costs of taxes
for health care, benefits of medical services, etc.). The issues are
resolved through a variety of societal decisions, which result in
"controls" in the form of laws, regulation, emergency services,
standards, inspections, and so on.

In Figure 1 it is possible to identify traditional areas of study that form the basis of practical risk management. System analysis (Miser and Quade, 1985, 1988) provides for the analysis and optimisation of operations and facilities in response to a number of objectives and constraints, including control by society. The methods of system analysis allow for the testing of "what if" questions through the extensive modelling of the systems operation. Risk analysis and risk estimation are methods of system analysis.

Group decision theory is concerned with the methods and procedures that are used to develop controls that reflect the collective values of individuals in the society. The objective is a search for decisions that result in "acceptable" or "tolerable" or "good" controls. Conflict Analysis is a method of modelling the society decision process in terms of possible control measures, the interaction between individuals and groups involved with the decision, and the preference structure of individuals and groups (Fraser and Hipel, 1984; Hipel, 1990).

The importance of criteria for the public interest illustrated in Figure 1 is perhaps most easily seen in decision processes where the public interest is not explicitly defined and the process is observed to wander, have considerable controversy, be subject to reversal of decisions, cause loss of public credibility, etc.

The Netherlands is in the process of defining criteria for the public interest for a number of risk situations. Their objective is to deal consistently with risk, no matter what its source (Directorate General, 1989).

The last stage in the risk management process identified in Figure 1 is communication. Communication is clearly a key determinant in the explanation of issues, in the expression of individual values and preferences, in the debate to define the public interest, in the effectiveness of implementing control measures, in the acceptability of the decision process, and so forth. In particular the methods of risk communication and their role in the risk management of dangerous goods will be described.

Figure 2 illustrates the transportation of dangerous goods in terms of the general model of risk management outlined above. The issues to be resolved by the decision process include; risk, benefits of transport, costs of operations, equity issues between those at risk near the transport and those benefiting from the transport, credibility of public authorities and the uncertainty about risks.

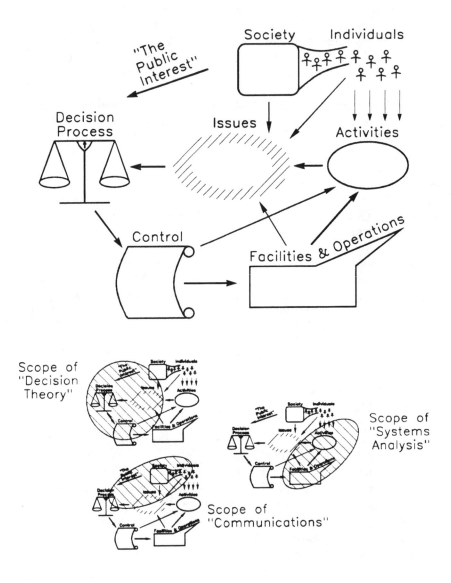

Figure 1: A general model of risk management.

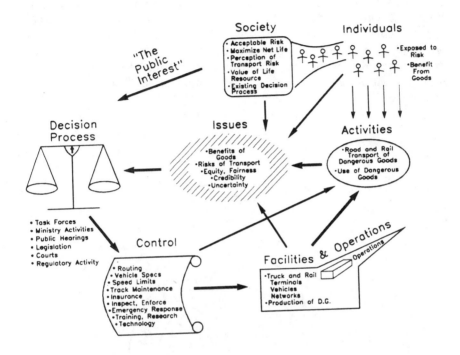

Figure 2: Risk management for the transport of dangerous goods.

The "public interest" is characterised by levels of acceptable risk to individuals adjacent to the transportation activity as well as a criterion of maximising net life values, a criterion which will be outlined below. The "public interest" is the aggregation of 1) society's perception of risks, benefits and costs, 2) the values society places on these life factors, and 3) the existing decision process and its inherent values, attitudes and precedence.

Figure 2 also lists the common risk management control options that are applied to the transport of dangerous goods. Each of these control measures has an extent of application, an associated cost of implementation, impacts on the benefits of transportation and estimated reductions in risk.

The management of the risks of transporting dangerous goods will be approached, in this paper, as a series of layers of analysis, understanding, and social evaluation, surrounding the actual transport of dangerous goods. The general scheme is illustrated in Figure 3. The structure is one of the traditional onion, where each layer incorporates and encloses all the underlying layers. The whole onion represents the risk management process while the core of the onion represents the actual transportation of dangerous goods and the implemented risk control measures. The situation is dynamic.

Figure 3: An inclusive framework for risk management techniques.

The hypothesis of the paper is that risk management of dangerous goods can be improved although it still will be problematic and difficult. Improvement will result from the structured application of techniques of: risk estimation, risk communication, conflict analysis and the development of public interest criteria. These techniques should be considered within the conceptual structure and relationships illustrated in Figure 3. For instance, the use of conflict analysis to assist in the decision process incorporates the results of risk communication and risk estimation.

The paper first presents an overview of the risk management of the movement of dangerous goods in the Toronto area as an example for illustrating the methods. Then the concepts of the public interest criterion are presented. The risk estimation process is discussed next, followed by risk communication. Finally, the method of conflict analysis is presented.

2.   TRANSPORTING DANGEROUS GOODS IN THE TORONTO AREA

In Canada each year there are about 3-5 persons who die as a result of the transport of dangerous goods. To date all have been on the road (Matthews, 1984). The typical situation is an accident involving a truck transporting gasoline, there may be a pool fire and people using the roadway can be involved in the accident and suffer injury or death.

There are also a number of road and rail accidents and incidents involving dangerous goods each year which lead to closures of the transport network, delays to traffic and media coverage of the resulting clean up of the spilled dangerous goods. The level of risk is accurately reflected in the media coverage within one mode (IRR, 1988), but there is variation in the level of coverage by mode of transport with differences in excess of 1000% between the rail and auto modes (Shortreed and Wallace, 1989).

The control of the risks of dangerous goods has been on the public agenda in Canada for many years due to the risks and the reporting of these risks in the media. In particular the Mississauga rail derailment, the subsequent fire and explosion of LPG and release of chlorine, followed by the evacuation of about 250,000 people focused considerable attention on these risks. It should be noted that there was no loss of life in the accident and only a few minor injuries. The accident also demonstrated the excellent performance capability of the existing emergency response system.

Canada passed federal and provincial legislation to control the movement of dangerous goods in 1985 (Canada, 1985). The act places the onus on the shipper to ensure that the movement complies with packaging requirements, moves in approved vehicles and that there is in place an emergency response plan in the event of an accident. The act defines dangerous goods according to class and for each class has specific controls. The act is enforced through an inspection and enforcement system. Currently the act is being updated.

Mississauga raised the issue of risk and the adequacy of the risk management controls. A royal commission into the event recommended a number of control measures including: improved vehicle equipment, better detection of vehicle failures that would lead to accidents and release of product, speed limits in populated areas, and so forth. In addition, the public concerns resulted in the formation of a public interest group Metro-Toronto Residents Action Committee (M-TRAC). M-TRAC proposed the control of risks through rail relocation, expanded speed limit controls, extension of controls to "empty" vehicles, and so forth.

In response to the issues the "Toronto Area Rail Task Force" was created to estimate the risk of fatalities per year, assess the risks and propose appropriate risk control measures. There was no formal risk communication method used or any decision model. However, the stakeholder decision method was invoked with all interested parties being represented on the Task Force as a means of rationalizing conflicts.

The Task Force was addressing the risks of existing rail lines carrying in excess of 50,000 cars per year of classified dangerous goods through a heavily populated area of Toronto. Eventually the Task Force focussed on three alternative routes illustrated in Figure 4 (Toronto Area Rail Task Force, 1988).

a) Technical improvements to the existing route (Plan A2).

b) Rerouting of the traffic to nearby routes that had lower population numbers but were longer (Parkway Belt alternatives).

c) Construction of a new even longer rail line in relatively rural areas (Northern Corridor).

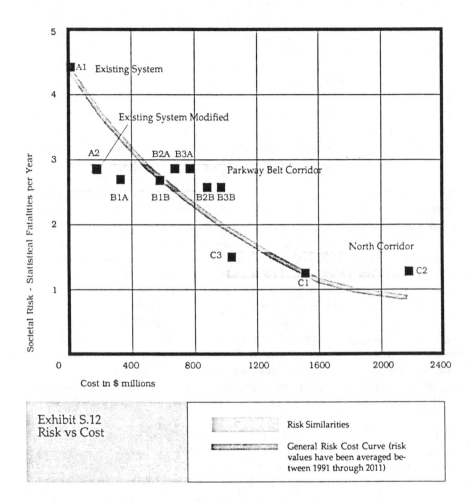

Figure 4: Risk versus cost from Toronto Area Rail Task Force
(1988)

Figure 4 taken from the Task Force Report illustrates the results of
the analysis in terms of the costs and the expected annual risks. The
Task Force recommended that the rail lines not be moved. The
results in Figure 4 raised the issue that the risk estimates were
overestimated since there have been no recorded rail deaths in all of
Canada for all the years of record. Saccomanno, et al. (1990) discuss
this issue in some depth.

The Task Force issued a Final Report recommending some technical safety measures but no rerouting of traffic (Toronto Area Rail Task Force, 1988). The M-TRAC organisation who were members of the Task Force, issued a minority report proposing more controls. Subsequently some of the recommendations of the Task Force and minority report were implemented by the government. There was considerable concern about the controls that were implemented. This was due to the lack of knowledge and the high levels of uncertainty concerning the risk estimates associated with the implemented controls, in particular the issue of the impact of train speeds on risk. The government subsequently embarked on a research study to clarify some of the risk estimates.

As a result of the Task Force there was improved communication through the media of the risks and the costs of controls for the transport of dangerous goods. There was also the development of a comprehensive risk estimation model (Concord, 1988). Research was stimulated into the uncertainty surrounding the risk estimates (Saccomanno et al, 1990). There was also concern about the large expenditure of money on the decision process and the selection of controls to be implemented given the very low level of residual risk associated with the movement of dangerous goods, the number of controls already in place and the uncertainty about the effectiveness of the controls in reducing risk. In particular there was concern that the Task Force was established as a stakeholder decision process which produced consensus recommendations that were not followed, perhaps due to public pressure by opponents of the recommendations.

The Toronto Task Force illustrates the need for improved methods of managing risks in terms of the techniques identified in Figure 3. There were issues of risk analysis, especially on the impacts of speed on risks, which could not be satisfactorily resolved. The Toronto Area Task Force used standard methods of communication. For example, they received 183 submissions, they issued many reports, held public hearings, etc. However, at that time special techniques for risk communication (NRC, 1988, W. Leiss, 1989) were not available.

The Task Force utilized the stakeholder approach in seeking a consensus but in the end there was a conflict of opinions and the issue was resolved by a ministerial choice which produced "winners" and "losers" without the benefit of a model of the decision choices such as conflict analysis to understand the consequences of the decisions to the various groups. It is proposed that conflict analysis may have been of considerable assistance to decision makers in clarifying the decision.

Finally, the Task Force did not have available to it any criterion to define the "public interest". They used a cost-effectiveness technique, as illustrated in Figure 4, to assist them in reaching a recommendation. It is proposed that their job would have been easier and their recommendations more acceptable if there was general agreement on balancing risks, costs, and benefits in the public interest.

3. WARNING: TECHNICAL METHODS MAY BE MISLEADING

Brunk et al. (1991) documented in convincing detail how the values of scientists modify their technical conclusions, rather than being based entirely on facts, and accepted concepts. While their results are not unexpected, it questions the validity of the usual risk management paradigm whereby risk estimation first establishes the facts; then risk evaluation assess these facts. There is a need for technique of checks and balances similar to legal cross examination methods. These should be built into the risk management process so that the risk estimates will not be biased by the values of the risk estimator. Some public interest groups, such as Greenpeace, argue that risk analysis is simply a presentation of a set of values and has no factual or objective content at all. (Greenpeace Newsletter, 1991). While disagreeing with this extreme position it is clear that caution is required to ensure that risk estimates and other components of the risk management process are as free as possible from the bias of the risk managers. For example, the use of ranges of possible values or distributions of values to express uncertainty are techniques that may minimize the effects of value bias.

4. THE PUBLIC INTEREST

Lind et al. (1991) has proposed a criterion for managing risk in the public interest. The criterion is to maximise the net quality of life. All risks, benefits and costs are considered in terms of their impact both on the length of life and the quality of life, for individuals. Lind proposed a "Life Product Index" to measure the length and quality of life and has demonstrated how the index can be used to evaluate the production and use of chemicals as well as its use in setting standards for risk levels for radiation (Lind, 1992).

The central argument is that the life of the individual is the highest possible value in a democratic society and the objective of maximising both length of life and quality of life defines the "public interest" of the society of individuals. This approach is currently being applied in the modified form in health care with the use of cost per Quality Adjusted Life Year as a basis for setting priorities. The 1992 US budget document included a table of cost per life saved as one indicator of the value of regulatory risk control measures.

In the proposed criterion all risks, benefits and costs are expressed in life values as the product of (length of life) times (quality of life) weighted by the respective probability of occurrence. Quality of life is equally weighted for each individual in the "one person, one vote" democratic tradition.

Risk is a low to very low probability condition that involves the loss of life, health, property or the environment. The use of Loss of Life Expectancy (LLE) as a measure of monetary risk is a well established technique which assumes a quality of life of 1.0 for each individual in the population. Quality Adjusted Life (Quebec, 1986) extends this method to include adjustments for lifetime in poor health. Interestingly, differences between male and female life expectancies is reduced when quality of life is considered. The use of drugs to alleviate pain is measured by the change in quality of life (Drummond, 1988).

Risk of property loss is measured in life terms in the same fashion as cost. Thoreau in Walden (1852) observed that "the cost of a thing is the amount of what I will call life which is required to be exchanged for it, immediately or in the long run." The basic concept is that all costs are due to human effort or the expenditure of human life. The time spent working is, for most people, not as enjoyable as non working time. Thus, work represents a reduction in the quality of life, and the cost of labour can be expressed in terms of person years of work or other measures effecting the length of life and the quality of life. Thus, the life equivalent of a "cost" can be expressed in the same terms as risk, namely Quality Adjusted Life Years (QALY's) or alternatively as Days of Life Lived (DLL). Also capital costs for roads, hospitals etc. are investments of life that are expected to pay dividends in benefits or risk reduction throughout their life.

Benefits also can be measured as the product of the length of life and the improved quality of life. For example, the benefits of goods transport can be measured in the "value added" to peoples lives. The life value of benefits is often measurable in people's "willingness to pay" or the purchased price of goods. This monetary measure can then be translated into QALY's.

The criterion for the public interest is to maximize the discounted net quality of life, summed over all individuals in society. The changes in risks, costs and benefits should be converted into the Days of Life Lived or QALY's. So long as it is clear that it is life that is being optimized then it may be convenient to do the analysis in monetary terms and use the cost of saving a life as a measure of efficiency.

A direct consequence of a life based decision criterion is that no risk management control should be implemented where there is, on balance, a loss of QALY's. This happens if more than 1 work year of cost produces less than 1 Quality Adjusted Life Year of risk reduction. Unfortunately due to the existing decision procedures, many such "life negative" investments take place. The removal of asbestos from buildings is one example. Given the very low observed residual risks for the transportation of dangerous goods, it is likely that many of the existing and proposed risk control measures for dangerous goods could also have a negative impact on life value.

If each person in society was allocated a monetary value of $50,000 per year (Canada's GDP/capita is now about $25,000) as their share of society's wealth and as a return on their labour, and if the average death represented a discounted loss value of 20 QALY's (i.e., about a 4% discount rate for the average age for transport accident victims), then the maximum life value that should be allocated to save a statistical life would be one million dollars in life terms. This can be compared to the cost of saving a life values used for transport analysis in Canada of $1.5 million. It is possible that we are investing too many lives in transport safety.

It should be noted that in the decision process other considerations might be introduced besides the proposed criterion for the public interest. For example, some jurisdictions place extra weight on multiple deaths (Directorate General, 1989). The authors' view is that it is appropriate for decision makers to consider these factors in their decision, but that it should not be calculated in the technical analysis. Otherwise, these considerations tend to be double counted, once in the analysis and once more in the decision process.

In the case of the Toronto Area Rail Task Force for the A2 alternative in Figure 4 (technical safety improvements in the existing system) the QALY benefit/QALY cost ratio is in the range of 0.1 to 0.2 and is not in the public interest according to the proposed criterion.

5. RISK ESTIMATES

The estimate of risk and the uncertainty associated with the risk estimate is, as shown in Figure 2, central to the risk management task. Uncertainty is of two types; a) unknowable and b) knowable or estimation errors. The latter uncertainty exists because of resource allocation limits. For instance, a recent study of plume dispersion in Alberta was done at a cost of about 2 million dollars and improvements were made in the estimation of toxic gas concentrations down wind from a H_2S release. (ERCB, Concord, 1990). However, there are still many unknowns associated with release direction, temperature, effects of topography, wind speeds, etc. which would require additional experiments. The level of investment in removing this type of uncertainty can be estimated in terms of the ratio of the QALY benefits of improved estimation divided by the QALY cost of the experiments.

Unknowable uncertainty is typically associated with random processes. Many of the large risks associated with transporting dangerous goods have annual probabilities of occurrence of 1 in 10,000 or less. While the expected value of the risk is the most likely risk that will be observed, there is a very high probability of not observing any of the large risks during the lifetime of the transport facility and movement of the goods (Saccomanno et al. 1990). For example, Figure 4 suggests an "expected" risk of 4 fatalities per year for rail only, in the Toronto area. Yet there have been no fatalities recorded for all of Canada for the 25+ years that records have been available. The uncertainty about the occurrence or non occurrence of risk events is simply not known and is unknowable. It is thought that much of the risky behaviours of individuals (i.e., boaters who don't wear life jackets) is associated with unknowable random risks using the belief that the risks are so small that they won't happen in our life time.

In the Toronto Area Task Force the need for improved risk estimation was one of the major concerns - "we found that databases, from which critical decisions are made, are inadequate and incomplete... a great deal of railway research and development ... is uncoordinated". The Task Force recommended more research into the relationship between speed and risk and other aspects of risk estimates for dangerous goods.

Extensive coverage of risk estimation methods is given in other papers in this document.

6.   RISK COMMUNICATION

As indicated in Figure 2, once an acceptable risk estimate is available then there follows a number of risk communication tasks which are illustrated in Figure 5. The risk estimates must be communicated to individuals, and to decision makers such that:

> *"Risk communication is successful to the extent that it raises the level of understanding of relevant issues or actions and satisfies those involved that they are adequately informed within the limits of available knowledge"* (NRC, 1990).

Covello has identified 19 characteristics of risks that must be considered in the risk message if there is to be sufficient information for people to be able to evaluate the risks (Covello, 1989). Those factors of importance for the movement of dangerous goods are:

1.	Catastrophic potential	Fatalities and injuries grouped in space
2.	Understanding	Mechanisms or process not understood
3.	Uncertainty	Risks scientifically unknown or uncertain
4.	Controllability	Uncontrollable by individuals
5.	Voluntariness of exposure	Involuntary
6.	Victim identity	Identifiable victims near transport route
7.	Media attention	Much media attention
8.	Accident history	Major and sometimes minor accidents
9.	Equity	Inequitable distribution of risks and benefits
10.	Benefits	Unclear benefits
11.	Reversibility	Effects irreversible
12.	Personal stake	Individual personally at risk
13.	Origin	Caused by human actions or failures

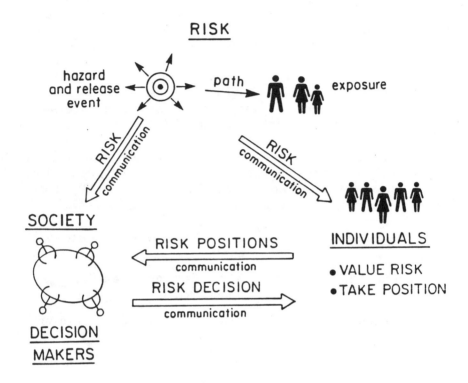

Figure 5: Four risk communication tasks (Shortreed, 1989).

For decision makers the risk communication must include the estimated range of the changes in risks, costs and benefits due to the proposed risk control measure. The combined quality of life measure should also be included as a range.

These techniques for risk communication are well known and have been studied extensively (Leiss, 1989). The recent book by the National Research Council (NRC, 1989) provides a comprehensive set of guidelines for risk communications which should be considered.

As indicated by Figure 5 there must also be a process to facilitate the communication of values from people to the decision makers and from decision makers to people. This communication link is essential if the public interest is to be served.

It is clearly to the advantage of individuals to reduce their risks, maintain or increase their benefits provided others pay the cost. There is almost always a call for "zero risk" using the argument "there is a risk which can be reduced by spending $x we can't afford not to control the risk". One effective means of dealing with "zero risk" is to expand the risk communication to include all life effects (risks, costs and benefits) to who so ever they impact. The "zero risk is our only choice" approach cannot be maintained when the life cost of control expenditures enters the discussion.

Improving communications will not reach consensus. People will still claim their "rights" to safety and firms and governments will claim their "rights" to provide goods and services to the public. Conflict will exist whenever there are distributional effects for risks, benefits and costs and when there are precedence for positions of "rights".

The role of risk communication is two fold, firstly to ensure that the risks are adequately understood, secondly that the position of the participants in terms of gains and losses and ranges of gains and losses are clear to all involved. The selection of risk control measures is then up to the decision makers. However, the playing field is level.

The Toronto Area Task Force concluded that "the railways should pay more attention to community information exchange and cooperate in order to provide greater assurance...." and "Governments at all levels, and the railway, promote and encourage an increased level of knowledge and communication within the general public...".

These recommendations were based on the findings that "the Canadian railway system is one of the safest in the world and at the same time the finding that the public "does not feel confident that this is the case, despite the facts that there has never been a fatality due to a dangerous goods rail accident in Canada." Thus the Task Force study clearly identified the need for improved risk communication.

However, the Task Force also used the expression "to err on the side of safety" to describe a desirable approach to rail safety. This expression is one of the main tenants of the zero risk position and ignores the life cost of expenditures. Erring on the side of safety inevitably leads to a loss of quality of life as the life cost of risk reduction is in excess of the life savings from the lower risk. The desirable level of risk management is where the marginal life cost of management equals the marginal life value of reduced risk.

It should be noted that since the rerouting alternative in the Toronto study involved residential areas that would have reduced risks and residential areas with increased risk, there was a controversy with media appeal and extensive and lively coverage. This particular circumstance led to a good communication and understanding of the relative risks.

For example, a public opinion survey found that "Toronto area residents (are) willing to live with dangerous goods being transported on rail lines... found little support for shifting risks to someone else by rerouting rail lines to less populated areas" (Globe and Mail, December 8, 1987).

7.   CONFLICT ANALYSIS

Given an acceptable risk analysis, adequate risk communication and agreed criteria for the "public interest" it is proposed that additional information can be provided to decision makers to assess the validity of their decisions by providing information on the stability of alternative decisions. Conflict analysis is a technique developed to provide decisions makers with a clear understanding of the strategic implications of possible decision outcomes, states, or scenarios.

In any decision process there are a number of decision makers or "actors" who are both involved in the decision and who can take action to influence the outcome of the decision. For illustrative purposes only, consider the simplified situation shown in Table 1 with the actors and possible actions for the Toronto rail example. In this table, the actors having real decision making power in the conflict are listed in the left column. The Task Force does not appear as an actor since it is a part of the decision process and has no effective action that it can take to effect the decision outcome. The possible courses of action or options under the control of each actor are written immediately to the right of the actor in Table 1. Notice, for example, that the basic actions under the control of the Federal Government, as represented by the Minister of Transportation, are rerouting the rail line (action A) or putting controls on existing service (action B).

Table 1: Conflict model for the Toronto rail dispute.

Actors	Actions	Example Outcome
1. Federal Government (Minister)	(A) Reroute rail line	No
	(B) Controls on existing service	Yes
2. Railways	(C) Modify service to improve safety	Yes
3. M-TRAC (residents impacted by existing risks)	(D) Support rerouting (A) in media, lobby, etc.	Yes
4. Municipalities near proposed rerouting of line	(E) Oppose rerouting (A) in media, lobby, etc.	No

Conflict analysis provides a convenient medium for representing the possible decision outcomes that can take place in the real world. For the conflict shown in Table 1, each actor has the choice of "yes" taking an action under its control or "no" not selecting the action. After all of the actors have decided which options to select, the overall result is a decision outcome. An example of a decision outcome is shown in the right hand column of Table 1. In this outcome, the federal government does not reroute the rail line but places controls on existing transportation service, the railways implement a community emergency response program, M-TRAC continues to support rerouting of the lines, and the municipalities in the rerouted area do nothing. In fact, this decision outcome represents the current situation in the Toronto rail dispute. At this point in time conflict analyses could be used to model the existing situation and the existing possible actions.

There is a decision outcome for each feasible set of actions. For each of the actors, the possible decision outcomes are ordered by preference for the outcome. For example M-TRAC would clearly prefer all outcomes which include (A) Yes compared to those with (A) No.

Given the actors preferences and the actions they control, actors will attempt to move the decision outcome to a more preferred outcome. For example, the Task Force recommendation (in terms of the illustrative simplified example) was the outcome (A) No Rerouting; (B) No controls on existing service; (C) No continuation of voluntary speed controls (D) Support of rerouting (E) No comment. Writing horizontally in text, this is outcome (No, No, No, Yes, No).

This outcome was unstable because the effective actions of the M-TRAC in generating public support for some improved safety action and the preference structure of the federal government meant that this outcome was not at "equilibrium" and would therefore revert to the existing outcome with the change of (B) from No to Yes (controls being placed on existing services).

Conflict analysis determines the preference structure of each actor through examination of position statements in interviews with the actors or through published documents. These same techniques are used to identify the actors' possible actions and also infeasible decision outcomes. The methods used to determine the equilibrium and stability of different decision outcomes is beyond the scope of this paper. However, the technique has been used extensively and is supported by an implemented software system (Fraser and Hipel, 1990). Consequently, the technique is well past the research and initial development stage. Comprehensive references for the theoretical development of conflict analysis as well as documented case studies of a wide variety of disputes can be found in publications such as the book by Fraser and Hipel (1984) and the paper by Hipel (1990). For papers regarding recent developments in conflict analysis, the reader may wish to refer to special issues on conflict analysis appearing in the *Journal of Environmental Management* (Vol. 27, No. 2, pp. 129-228, 1988) and *Information and Decision Technologies* (Vol. 16, Nos. 3 and 4, pp. 183-371, 1990). Further contributions to conflict analysis are contained in papers published in proceedings for special sessions on conflict analysis held at conferences in France (Singh and Travé-Massuyès, 1991) and the United States (IEEE, 1991). Finally, as part of the activities of the Institute for Risk Research at the University of Waterloo, conflict analysis has been applied to environmental (Hipel and Fraser, 1982), hydroelectrical (Yin et al., 1987) and nuclear power (Fang et al., 1989) disputes, all of which involve various types of risk.

The results of a conflict analysis are a set of "equilibrium" decision outcomes. The actual outcome is clearly at equilibrium. However there are other outcomes which are also stable and at equilibrium. This set of outcomes provides the decision process with a small number of well defined possible outcomes that constitute possible compromise resolutions, one of which eventually takes place. Information on these outcomes is very helpful to the decision maker.

Conflict analysis can be extended to:

1) analyse the expected series of events in the decision process that would be required to reach a given decision;

2) identify for any particular actor what changes in preferences of other actors would be required to change the decision outcome;

3) estimate the likelihood of each possible outcome; and

4) to generally explore the intricacies of the decision process.

For decision makers, conflict analyses allows them to rise above the complex issues, facts, opinions, and so on in order to get a "birds eye" view of the overall decision process and decision outcomes.

In the Toronto area example, it could be hypothesised that if the railway had done a conflict analysis of the situation, it might have predicted the outcome and taken a different approach for dealing with the Task Force. In fact, the railway participated actively in the work of the Task Force and supported the recommendations. However, subsequent to the Task Force, actions, which might have been predictable by conflict analysis, have led to an outcome which is negative to rail transportation relative to their position with respect to trucks.

Similarly, the public interest does not appear to be properly served by the final outcome. Since the risk analysis indicated no safety gain for the implemented controls it follows that the net impact on quality of life would be negative. Conflict analysis could be used to identify what strategic initiatives would be necessary to achieve a decision process which would better reflect the public interest.

8.   CONCLUSIONS

The transportation of dangerous goods have three overriding characteristics:

1) The hazards are very high. Large quantities of potentially very dangerous substances are moved past millions of people every day.

2) The risks from the movement of dangerous goods are very low, based on both the historical record and risk estimates.

3) There is considerable concern and media attention about the safety of the movement of dangerous goods.

These three characteristics define a situation which requires effective risk management to provide appropriate control measures to ensure that while the hazards are high the risks are low. The characteristics also mean that the decision process will be difficult and there is a great risk of over control to the extent that the public interest is not served and the net quality of life is diminished.

The recent past has seen the rapid development of a number of techniques essential to the risk management of dangerous goods. This conference marks a milestone in the rapid development of complex risk estimate models to produce improved estimates of the risks. The field of risk communication and conflict analysis have been developed, documented, and effectively applied using the available step by step procedures for their use.

The establishment of criterion for risk management in the public interest is in the early stages of development but many of the inherent concepts have been usefully applied in system analyses, cost benefit and cost-effectiveness applications. The major advance is the clear identification of monetary cost as a life cost. This allows for the measurement of all risks, costs and benefits in a common and acceptable measurement scale - life.

This paper proposes a hierarchic framework for incorporating risk analysis into an effective risk management framework. Elements of this framework were illustrated by the Toronto Area rail situation. Clearly, the proposed approach requires either extensive preparation or sufficient time for implementation.

It is recommended that this framework be considered as the next step in the ongoing improvement in the techniques and processes for risk management of dangerous goods.

9.   REFERENCES

BRUNK, G.G., HAWORTH, L., & LEE, B. (1991). Value Assumptions in Risk Assessment: A Case Study of the Alachlor Controversy. Wilfrid Laurier University Press, Waterloo, Canada.

CANADA (1985). Transportation of Dangerous Goods Act.

CONCORD SCIENTIFIC (1988). Report on Risk Analysis Model, The Toronto Area Rail Transportation of Dangerous Goods Task Force. Supply and Services Canada, Ottawa.

COVELLO, V.T. (1989). Informing People About Risks from Chemicals, Radiation and Other Toxic Substances: A Review of Obstacles to Public Understanding and Effective Risk Communication. In W. Leiss (Ed.), Prospects and Problems in Risk Communication, University of Waterloo Press.

DIRECTORATE GENERAL FOR ENVIRONMENTAL PROTECTION AT THE MINISTRY OF HOUSING (1989). Premises for Risk Management: Risk Limits in the Context of Environmental Policy, (Annex to the Dutch National Environmental Policy Plan, Second Chamber of States General, 1988-89), session 21137, No. 5.

DRUMMOND, M., TEELING SMITH, G., & WELLS, N. (1988). Economic Evaluation in the Development of Medicines. Office of Health Economics, London.

ENERGY RESOURCES CONSERVATION BOARD (ERCB). Concord Environmental (1990), Summary - ERCB Field Measurement Program. Calgary Alberta.

FANG, L., HIPEL, K.W., & KILGOUR, D.M. (1989, May). Conflict Analysis of the Darlington Nuclear Power Dispute, Paper No. 14. Institute for Risk Research, University of Waterloo, Waterloo, Ontario, Canada.

FRASER, N.M., & HIPEL, K.W. (1984). Conflict Analysis: Models and Resolutions. North-Holland, New York.

FRASER, N.M., & HIPEL, K.W. (1990). DecisionMaker: The Conflict Analysis Program, copyright owned by N.M. Fraser & K.W. Hipel. Distributed by Waterloo Engineering Software, 22 King St. S., Ste. 302, Waterloo, ON, Canada N2J 1N8, (519) 885-2450.

GLICKMAN, T.S., GOLDING, D., & SILVERMAN, E.D. (1992). Acts of God and Acts of Man. Paper CRM92-02 Resources for the Future, Washington, DC.

GUTIN, J. (1991, Mar./Apr.) At Our Peril. The False Promise of Risk Assessment. Greenpeace Magazine. Washington, DC.

HIPEL, K.W., & FRASER, N.M. (1982). Socio-Political Implications of Risk. In N.C. Lind (Ed.) Technological Risk: Proceedings of a Symposium on Risk in New Technologies (pp. 41-75). University of Waterloo Press, Waterloo, Ontario, Canada.

HIPEL, K.W. (1990). Decision Technologies for Conflict Analysis. Information and Decision Technologies, 16, 3, pp. 185-214.

HIPEL, K.W., & SHORTREED, J. (1992). Risk Analysis. Invited paper in the McGraw-Hill Encyclopedia of Science and Technology, published by McGraw-Hill, New York.

INSTITUTE FOR RISK RESEARCH (IRR) (1988). Assessing the Risks of Transporting Dangerous Goods by Truck and Rail. University of Waterloo.

INSTITUTE OF ELECTRONICS AND ELECTRICAL ENGINEERS (IEEE) (1991). Conference Proceedings of the 1991 IEEE International Conference on Systems, Man and Cybernetics, held in Charlottesville, Virginia, Oct. 13-16, 1991, Volume 3, Sessions TA8-Conflict Analysis I and TB8-Conflict Analysis II, organized by K.W. Hipel, pp. 1978-2022.

LEISS, W. (Ed.) (1989). Prospects and Problems in Risk Communication. University of Waterloo Press, Canada.

LIND, N.C., NATHWANI, J.S., & SIDDALL, E. (1991). Managing Risks in the Public Interest. Institute for Risk Research, University of Waterloo, Canada.

LIND, N.C. (1992) Critères d'efficaceté des soins de santé à portir d'indicateurs sociaux. Presented at Management of Risks, Benefits and Costs of Pharmaceuticles. Canadian Public Health Association, Montreal.

MATTHEWS, M.K. (1984). An Overview - Support Programs for Risk Assessment in the Transport of Dangerous Goods. In S. Yagar (Ed.) Transport Risk Assessment. University of Waterloo Press.

MISER, J.H., & QUADE, E.S. (Eds) (1985). Handbook of Systems Analysis: Overview of Uses, Procedures, Applications and Practice. North-Holland, New York.

MISER, H.J., & QUADE, E.S. (Eds) (1988). Handbook of Systems Analysis: Craft Issues and Procedural Choices. North-Holland, New York.

NATIONAL RESEARCH COUNCIL (NRC) (1989). Improving Risk Communication. National Academy Press, Washington, DC.

QUEBEC (PROVINCE) (1986). Objective: A Health Concept in Quebec. Canadian Hospital Association, Ottawa.

SACCOMANNO, F. F., SHORTREED, J.H., & MEHTA, R. (1990). Fatality Risk Curves for Transporting Chlorine and Liquefied Petroleum Gas by Truck and Rail. Transportation Research Record 1264, Transportation Research Board, Washington.

SHORTREED, J.H., & WALLACE R.S. (1989). Assessment of Aviation Safety in Canada. Prepared for the Ontario Ministry of Transportation. Institute for Risk Research, Waterloo.

SHORTREED, J.H. (1989). Response to Liston. In W. Leiss (Ed.) Prospects and Problems in Risk Communication, University of Waterloo Press, Waterloo, Ontario.

SINGH, M.G., & TRAVE-MASSUYES, L. (Eds) (1991). Decision Support Systems and Qualitative Reasoning, Proceedings of the IMACS International Workshop on Decision Support Systems and Qualitative Reasoning held in Toulouse, France, March 13-15, 1991, Section II.2-Conflict Analysis organized by K.W. Hipel, North-Holland, Amsterdam, pp. 101-137.

TORONTO AREA RAIL TRANSPORTATION OF DANGEROUS GOODS TASK FORCE (1988). Final Report. Ministry of Supply and Services, Ottawa, ISBN 0-660-13034-3.

YIN, X., HIPEL, K.W., & LIND, N.C. (1987). Conflict Analysis of Technological Risk. Paper No. 10. Institute for Risk Research, University of Waterloo, Waterloo, Ontario.

QRA-Aided Risk Management of Dutch Inland Waterway Transport

H.L. Stipdonk
R.J. Houben
Ministry of Transport and Public Works
P.O. Box 1031 3000 BA
Rotterdam, The Netherlands

ABSTRACT

The Dutch ministry of transport and public works is concerned with the risks caused by transport, mainly of hazardous material, along the Dutch inland waterways. Therefore, a project called "Safety in inland waterway transport" was set up. This project is aimed at a Risk Effect Model, with which it should be possible to not just calculate inland waterway transport risks, but also to predict the effect of safety measures. This would enable a cost-benefit analysis of measures to be taken. At this moment, the project is not yet finished. However, the many intermediate results yield sufficient interesting information to be presented at the symposium.

1. INTRODUCTION

When in 1987 six-barge pushtow vessels were admitted to some of the most important Dutch inland waterways, the then minister of transport stated that the current safety level would not be affected by six-barge traffic. This was based upon a thorough investigation (Brolsma, 1988), during the six-barge trial year 1986. Nevertheless, the question arose as to what this current safety level really was. How are safety levels to be expressed, how are they measured, and if the level exceeds a limit, how can it be brought down to an accepted level? And finally, is it possible to estimate the effect of safety measures to enable cost-benefit comparison and/or analysis for different measures?

To answer these questions, the minister decided to start a five year project. This project, called "Safety in Inland Waterway Transport" was started in 1988 and is now more than halfway completed, and many intermediate results have already been achieved. In this paper we discuss results so far as well as what has to be done.

In section 2 the general ideas of Dutch risk management are evaluated. Section 3 goes into the problem of defining risk-dimensions and limits, especially those for transport routes. In section 4 the currently used and planned risk assessment methods are outlined, and section 5 deals with estimation of risk effects of safety measures.

2.   RISK MANAGEMENT

Setting limits for acceptable risk, assessing existing risk levels and predicting the costs and benefits of proposed safety measures are all meant to assist the risk management process. A clear picture of this process, the parties involved and the types of choices to be made is indispensable when drawing up the Risk Effect Model. Since risk management comes down to deciding on and implementing safety influencing measures, the categorization and characterization of safety measures (cf. table 1) was a logical starting activity of the project.

2.1 Safety measures and their characteristics

Risks: Accidents on waterways pose the following risks: damage to the health of crew and passengers; damage to the health of people in the vicinity; damage to the natural environment; material loss (ships, cargo, bridges, locks).

Parties: "Safety is a joint responsibility of industry and government" is a principle that is not only widely acknowledged but also put into practice in the Netherlands and the other states bordering the Rhine. Which of these two parties takes the lead and often pays for it, depends on the nature of the safety measure under consideration.

Scale: The scale on which a safety measure has its effect can differ considerably from one case to another. In the European situation, all requirements regarding the vessels and their equipment must be agreed upon internationally in Strasbourg (France), where the

Central Committee for the Navigation on the Rhine meets. This means that, though negotiations may be tough, the results can be far-reaching. At the other end of the scale are e.g. local improvements of the waterway, which will only locally enhance safety.

Term: Structural improvements to vessels will normally only be feasible for newly built ships.

In Table 1, safety measures are grouped into nine categories and characterized according to risks affected, executing party, scale and term of effect, as mentioned above.

Table 1 : Categories of safety measures and their characteristics.

#	Measures concerning:	Risks	Party	Scale	Term
1	Vessels, equipment	all	i	i	l,s
2	Crew: training, labour times	all	i	n/i	l,s
3	Traffic: VTS, regulations	all	g	l/r	l,s
4	Infrastructure	all	g	l	l
5	Cargo: packing, stowing	pv/ne	i	i	s
6	Transport: distance, amount	pv/ne	i	r/n	l
7	Disaster management	pv/ne	g	l/r	s
8	Routing	pv	i	r	s
9	Land use restrictions	pv	g	l	s

Risks: pv = damage to the health of people in the vicinity
ne = damage to the natural environment
Party: i = industry
g = government
Scale: l = local
r = regional   n = national   i = international
Term: l = long
s = short
# 1, 2 and 3: first item long term, second item short term

2.2 Applicability of safety measures for external safety

Focussing for the moment on one of the risks, that of damage to the health of people in the vicinity, one should realise that the level of risk is determined by three factors:

- the nature and amount of transported hazardous material;
- the local frequency and severity of accidents;
- the distance between vulnerable sites and the traffic lane.

Since a safety problem is typically associated with a point on the map, people tend to start thinking of solutions on a local level and then gradually move up through the regional and national levels to the international level.

If local factors have led to a relatively high accident frequency, improvements to the infrastructure (waterway, bridges, locks) or the local traffic situation is the logical way to look for a solution. The accident frequency elsewhere gives a clue as to what measure of reduction may be expected from these changes. If this reduction is enough, a local problem is solved locally.

There may be reasons however why a local problem may deserve a more generic solution. The first reason is that local improvements may not suffice: a reduction in accident frequency of say 20% by infrastructural improvements or traffic control is quite an achievement but sometimes not enough to reduce risks that increase on a logarithmic scale. The second reason is that those measures may not be the most cost-effective. The third is that one generic solution can be a good alternative to many local solutions aimed at solving many local problems.

The routing (Table 1, #8) of hazardous material transportation along waterways that do not run through densely populated areas could be the next option to consider. According to European standards, the Dutch waterway infrastructure is relatively dense. However, cases where an inland waterway alternative exists without just shifting the problem to another part of the country, are rare. In the case of transport between two seaports, transport by sea may be feasible. But a rise in operational cost could force the cargo into trucks or trains, which may not improve safety.

Trying to influence the amount or distance of hazardous material transportation (Table 1, #6) is in principle contrary to the Dutch Government's policy of free enterprise. But if the transportation risk can be removed completely by bringing production and use of hazardous semi-products together, this option should certainly be considered.

Because mostly, none of these regional measures are feasible, and for reasons mentioned earlier, one will have to resort to measures on the national or international level (Table 1, #1, 2 and 5) as well. From current developments it can be anticipated that in future all hazardous (to people or nature) cargo will be transported in double-hulled ships. Furthermore the improvement of human and organisational factors will play a significant part.

So far this section deals with possible solutions to a given external safety problem. Since it is better to avoid problems than having to solve them, land use restriction (Table 1, #9) should be given serious consideration wherever possible. For some substances, 15 m extra distance between accident and residence quarter already halves the risk. However, in urban areas, costs of land use restriction are high.

2.3 The role of risk analysis

The way in which risk analysis can help the decision making process differs from case to case. Sometimes a comparison of calculated risks to - preferably legally based - risk limits is needed while in other cases a comparison of alternative measures may suffice. Also, different risk dimensions may be needed.

Whereas individual risk (cf. paragraph 3.2) is very practicable for determining land use restriction zones, it is by definition unsuitable for expressing the benefits of emergency planning. But societal risk figures may help in setting priorities for disaster management schemes. Prioritizing substances with regard to the obligation of shipment in double-hulled ships requires a comparative approach, using standard release scenarios.

Although quantification of safety benefits is always desirable, safety measures, the benefits of which cannot be expressed quantitatively, should not be left out of the risk management process for that sole reason. If that were the case, quantitative risk assessment could even become counter-productive with respect to safety.

3.   RISK DIMENSIONS AND RISK LIMITS FOR TRANSPORT ROUTES

3.1   Scope

Why do we need risk dimensions and risk limits to express lack of safety? Why is it insufficient to just count accidents, or fatalities and measure safety that way? The answer is not difficult. As long as we are interested in activities like car driving, where the number of accidents is large enough to enable statistical analysis, there is no immediate need for more sophisticated risk dimensions. But as soon as comparison between different activities is involved, a dimension like "death rate per unit of time spent" (with this dimension mountaineering appears to be more dangerous than home cooking) or like "death rate per unit of distance travelled" (which proves motorcycle-driving to be more dangerous than car-driving) becomes necessary, even for activities with a large number of incidents.

When the number of accidents is low, things are different. When the mean number of fatalities per accident is also low, who cares? But if we deal with rare accidents with more than 10 fatalities each, the activities giving rise to the risk are of interest to risk assessors. Transport of hazardous material is an example of such an activity. Although up until now no Dutch inland waterway accident led to fatalities on the banks of the waterway, there is a small probability of an accident of such a severity, and it is conceivable that an accident with an LPG tanker might kill several hundred persons.

The Dutch government plans to make an inventory of these risks. To describe the risks of plants, shunting-yards, and polders (risk of floods) several risk dimensions and risk limits are already in use. For transport risks to be of optimal utility, it is desirable that transport risk dimensions be as comparable as possible to those for plants etc.

Apart from appropriate risk dimensions for transport, risk limits are to be defined as well. In the project "Safety in Inland Waterway Transport" the main concern is to develop dimensions and limits which ensure safe sailing. But a transport risk limit that is only used for ships is by far not as powerful as one that is applicable for all modes of transport. Thus another project was started by the ministry of transport, in cooperation with the ministry of environment, with the purpose of defining and establishing uniform transport risk limits for all transport modes. Once these limits are set, it is possible to guarantee some generally accepted safety level in any residential area, not only with respect to all sorts of transport, but also due to plants etc. Also, because the transport risk limits hold for any mode, when the limit is exceeded for one mode, transfer of transport to another mode will not be accepted if the limit for that mode is exceeded.

3.2 State of the art

The current risk limits for plants are such that the additional probability of death due to a specific activity should be no more than one percent of the natural death rate of a healthy, eighteen year old person. The dimension in which this limit is expressed is the "individual risk". Individual risk is calculated for a specific position. It is the probability of death for someone who is continuously present (and unprotected) at this position during a year.

In addition to individual risk, there is another risk dimension currently in use, namely "societal risk". Societal risk is a cumulative probability function, describing the probability of N fatalities all at once, caused by a single accident. This dimension is meant to cope with the fact that there is a difference between many fatalities all at once, and the same number of fatalities due to (equally) many accidents. The former attracts far more attention, and is therefore politically more important, and besides hundreds of deaths in one residential quarter can cause severe social disruption. The Dutch ministry of environment uses both individual risk and societal risk for licensing of (e.g. chemical) plants and shunting-yards.

3.3 Future developments

For transport-routes, individual risk can be a perfectly suitable dimension. It is to be expected that the risk limits for immovable risk sources will be applied to transportation as well. Unofficially, they sometimes are already in use. However, for societal risk the parallel is a lot more difficult to draw. There is a conceptual problem in calculating societal risk of a (part of a) transport route, and this is caused by the fact that the probability of an accident is generally proportional to the length of the route. For individual risk this yields no additional difficulty, as individual risk is calculated for a single point. Societal risk results on the other hand will be augmented with increasing length of the route considered.

There are two ways to handle this conceptual problem. One is to adapt the definition of societal risk by choosing an arbitrary route-length as equivalent to an immovable risk source, or by defining a ship as this equivalent (or even both). The other way is to admit that the problem really is conceptual (and therefore unsolvable) and to define a new risk dimension which is the most appropriate for transport. Both have pros and cons.

Adapting the societal risk dimension by defining a certain representative route length leads to a difficult decision to be made. Should this length be equal for road, waterway, rail and pipeline? Should it depend on the maximum distance over which a severe accident could cause deaths? Many varieties are possible and all have far reaching consequences for risk management. And what about the risk limits? Should they be equal for all modes (which may be realistic if the structural differences between them is cleverly accounted for in the definition of societal risk) or is it better to have different limits? How are the risks to be compared to the limits if e.g. a small town (say 200 m of length) near a waterway is entirely under influence of the risks caused by waterway transport, while the route length of concern (200 m) is much shorter than the standard length (say 1000 m)? Is the risk limit to be scaled to this length, or is the risk still to be calculated for the standard route length? There may be an important difference between the two.

The other possibility of adapting the societal risk dimension, namely by connecting societal risk to its source rather than its destination, is interesting but insufficient. Indeed societal risk per ship would allow for suppressing the risk caused by very dangerous ships, but for risk management of vulnerable locations it is of no value.

The authors of this paper would argue strongly for the development of a special transport societal risk definition. Instead of connecting societal risk to either risk source or risk destination, it may rather be dependent on the actual activity. Transport is movement of mass along a distance, expressed in terms of (metric) ton km (or ton mile). Nothing is more fundamental than to express societal risk in the same way: societal risk per ton km of transport. There are lots of pros but there is also one con: it is unconventional. Those who regard a transport route as an immobile risk source cannot accept that societal risk may increase beyond the limit (for immobile plants) due to transport growth over the years. This is indeed a minor problem, that fortunately can be dealt with by regular redefinition of the risk limits. This could be done with a cumulative national societal risk for all activities of which the accepted level is known (and invariable), and conversion of this total risk level to each contributing field.

In the Netherlands risk analysis experts expect that they will be able to solve the problem before our project is finished. It is not only expected, it is actually necessary, as one needs a risk dimension to express risk.

4. RISK ASSESSMENT

4.1 Scope

The core of the risk assessment model is the calculation of the probability of death (POD) at a specific point. The POD depends on many parameters, of which we will first mention the nature of the effects of the incident. Four possible scenarios are considered: release of a toxic cloud, of an explosive cloud, heat radiation from a burning pool, and heat radiation from a BLEVE (Boiling Liquid Expanding Vapor Explosion).

For an accident where a toxic gas (e.g. NH_3) is released, calculation of the concentration c as a function of space and time, $c(r,t)$ combined with the poisonousness of the toxic gas enables estimation of the POD as a function of space, $POD(r)$. For explosive clouds (e.g. LPG) a similar approach is followed. Given an ignition of the explosive cloud (the probability of which is a function of $c(r,t)$), and the location of the cloud at the moment of ignition, the $POD(r)$ can be calculated. In this case a $POD \neq 0$ may occur outside the cloud, hence $POD(r) \neq f(c(r))$ as with toxic clouds.

The third contribution to the POD consists of heat radiation from a burning pool of inflammable liquid, floating on the waterway. If this pool ignites, the POD inside the pool is 1, and outside the pool, on the banks of the waterway, it is a function of the heat of radiation. The heat of radiation is calculated from the size of the pool, and the geometry (and many other properties) of the flame.

Heat radiation from a BLEVE leads to a similar POD-calculation. Inside the fireball of the BLEVE the $POD = 1$, and outside it is determined by heat radiation as with a burning pool.

The method of risk assessment is as follows. The accident rate per containment (a ship-cargo combination) and the amount of transport per containment are combined to give a probability of an accident with that containment. Conditioned by the containment (e.g. single vs double-hulled) a probability of loss of containment is defined. Meteorological parameters, ignition rates, temperature parameters (air, water and cargo) and also physical properties of the hazardous cargo (density, boiling temperature, specific heat and so on) determine which of the four above mentioned scenarios is appropriate, and how the cargo spreads and diffuses in water and air. A more detailed description of the method is described by Veraart, Janssen and Stipdonk, 1992.

As to the results achieved so far, we will only remark that the individual risks appeared to be limited to 10^{-5} anywhere on the banks of any Dutch waterway. However, for most locations even the 10^{-6} limit is not exceeded (Figures 1 and 2).

4.2 State of the art

A detailed description of all physics involved in risk assessment is far beyond the scope of this paper. We will therefore only describe some of the aspects that we consider to be specifically of interest for waterway transport risk. Those aspects are the estimation of accident rate from accident histories, and the distribution of accident rate over the width of the waterway.

Accident casuistry is influenced by chance. When one just takes the number of reported accidents to be the measure of accident-probability, an unnecessary error is made. Firstly, parts of the waterway with zero accidents do not really have zero accident rate, and secondly, parts of the waterway with a high number of registered accidents may have been struck with "bad luck". It is sometimes seen that this problem is overcome by using the mean accident rate for most parts of the route, and using the actual number of accidents for those parts that have a significantly higher number of accidents. This approach is not too bad, but there is a better one, based on a technique developed by Hauer (Hauer, 1986). Hauer explains that for a large number of seemingly equal locations (parts of a transport route with equal length or, as in Hauer's paper, road intersections) the hypothesis that accidents are poisson-distributed over these locations should be tested. The test he proposes gives a readily applicable estimate of the accident rate T for a location with n reported accidents, T(n). The estimator takes into account that a reported number of accidents may be occasionally high, while the underlying accident rate is actually low, or the other way around.

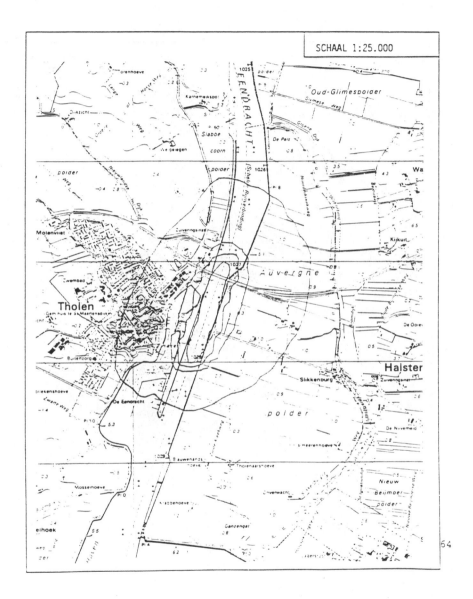

Figure 1: Contours of individual risk, ranging from 10^{-5} to 10^{-8}, on the Scheldt-Rhine canal, near the city of Tholen, due to ship collisions.

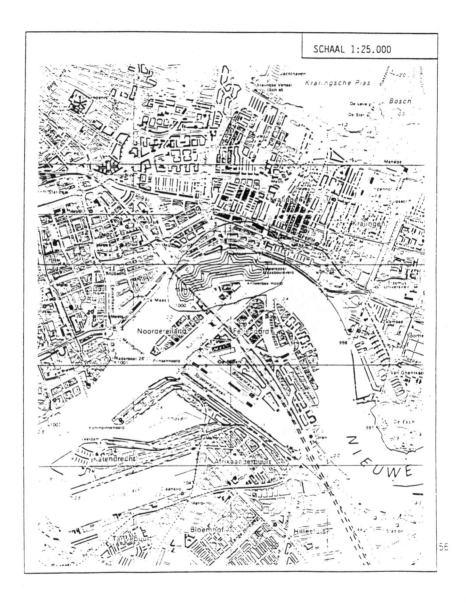

Figure 2: Contours of individual risk, ranging from 10^{-4} to 10^{-8} in the city of Rotterdam, due to ship collisions on the "Nieuwe Maas".

The estimation of accident rate as function of the location within the waterway is especially important because of the highly nonlinear behavior of the POD as function of r (for relatively small r). For transport risk assessment, up until this year the accident locations were assumed to be at the centre of the waterway, though in real life the locations are spread over the entire width. In order to allow for this modification, it is necessary to adjust the models describing the spreading of the pool. A choice also has to be made for the form of the probability function. On enquiring about this subject, and it turned out that a best first guess should be a uniform distribution. Even for locations where the accident rate is strongly influenced by nonsymmetric circumstances (as in the case of a bridge pile in the middle of the waterway) the accidents are reasonably uniformly distributed. This is probably due to the fact that while the cause of the accidents may be strongly located, a collision may actually take place some distance beyond this location, on either side. If, for example, a ship is approaching a waterway junction, it may decide to cross the waterway some distance away from the junction. This way the risk, due to the junction is partly replaced by risks further away, due to waterway crossings.

Another example of such a location is a bridge near Dordrecht. Many accidents, in which the cause was shown to be this bridge, actually take place up to 200 metres away from the bridge (Figure 3).

4.3 Future developments

Accurate calculation of risk involves incorporation of many varieties of parameter settings. For risk assessment of plants, the number of settings is large but manageable. For transport risk assessment, where the risk source location is a two dimensional, continuous variable, where the (variable) width of the waterway is of influence, and where many types of ships and cargo sail, it may take weeks of computer time to calculate the individual risk for one region. Even then, some simplifications are to be made: a fixed air temperature, not too many wind speed, wind direction and stability classes, and coarse discretization of space. To reduce computer time without loss of quality, there are two solutions: use a large supercomputer, or apply Monte Carlo Integration.

The possibilities of Monte Carlo Integration (MCI) in risk assessment are already known in the field of structural reliability. It is however not yet widely used in the field of external safety, possibly because it was not really a necessary tool, unless the variances of risk assessment results were needed.

Figure 3: Location of accidents near the bridges across the "Oude Maas" in the city of Dordrecht.

Generally, MCI should only be used when there is no other integration method available. The disadvantage of MCI is that the convergence of the result is relatively slow, so many evaluations are needed to establish a specific accuracy level. So why should we apply MCI to risk assessment? There are two special properties of risk assessment that make MCI a very attractive method. The first is that the function to be integrated over the variables, the POD, is limited to 1. The other is that there are very many random variables involved. All possible combinations (these are numerous) may be important, but for each variable the individual probability distribution is to be taken into account.

Each combination of random variables can at most generate a POD of 1. So when the probability of any combination is 10^{-10}, the maximum contribution to the individual risk is also 10^{-10}. If one is interested in risk levels down to only 10^{-9}, it is no longer necessary to check all possible combinations of random variables. In fact, the desired accuracy defines the number of combinations to be randomly generated. In the case of waterway transport risk assessment, it seems that MCI will enable us to add as many stochastic variables (most of which are usually taken as fixed) without increasing the runtime of the calculations.

Apart from this integration method, of which a detailed description is given by (Houben and Wessels, 1992), other aspects of the risk assessment are in development. We will only briefly mention just one of them, concerning accident rate. There is one aspect that Hauer's T(n) does not take into account, and that is the correlation of accident rates of near-by locations. Actually, the expected accident rate should not only be a function of the number of accidents at some location, but also of the number of accidents at nearby locations. As yet we have no technique that gives such an estimate.

5.   SAFETY MEASURES AND THEIR EFFECTS

5.1 Scope

Once risks can be calculated and expressed in appropriate terms, the technical part of the problem seems to be solved, but it is not. If the calculated risk exceeds a limit, some measure should be taken, if possible the most cost-effective one. And even if the risks are not too high, there is nothing against taking cost-effective safety measures. But although it is almost never difficult to think of safety-measures, it is hard to prove they are cost-effective, because this means that the effects should outweigh the costs. So the costs should be known, the effects should be known, and the weight of one with respect to the other should be known.

The last subject is of course the most controversial. We will leave it open, and postpone a discussion to a later date. Both other subjects, cost and effects of safety-measures, leave enough to be discussed. For if one wants to build a model that can estimate costs and effects of measures, these measures are to be known in advance. Economic aspects involved with each measure are to be taken into account. The enhancement of safety, but sometimes also any induced reduction of safety due to side effects, is to be estimated.

Even if the monetary value of safety is not known, one can at least compare different measures, once costs and benefits of measures are estimated. In this procedure, estimation of the effect of safety measures is the most difficult part, especially if more than one measure is to be taken at the same time. Some measures have effects that are mutually independent, but measures may also exclude each other or, on the other hand, enlarge each others effect.

5.2 State of the art

In 1989, an inventory was made of the possible effective measures experts could think of. A list of more than 350 measures was generated, but not all of them were to be considered really important. A survey of expert opinion was carried out, among many different inland waterway experts, in which each expert could comment on those measures in the list that he could be expected to give a valuable opinion on. The result of this project gave a list to select from. Finally a list of about 70 likely measures was generated, some of which are appropriate for a quantitative treatment in the Risk Effect Model, and others more appropriate for a qualitative treatment.

The measures in Table 1 were considered, with special attention being given to inspection (related to # 1, vessels and equipment), communication (related to # 2, crew, and # 3, traffic) and operation of bridges and locks (related to # 4, infrastructure). Within this paper we will not go into a detailed discussion of all of these. However, some of them are worth mentioning. One has to do with the anchors of ships, that usually stick out of the ship's bow. If the anchors were smoothed away, the potential damage after a collision, especially when the striking ship hits the struck ship on its side, would be much less. This measure is not expensive (about $ 25,000 for each new ship) and deserves to be carried out. The difficulty with this measure is that it has to be prescribed internationally to avoid problems with the competitive position of sailing companies. Nevertheless, the measure is considered very effective. Exactly how effective can be easily estimated from registered accident information. It has so far been done only for one of the Dutch inland waterways, where it turned out that only in three out of 200 registered accidents had a ship been seriously damaged by an anchor. On the other hand, in only 48 of those accidents, a ship had been damaged to the extent of a leak. So actually about 6% (3 out of 48) of the damage would have been less serious if anchors had been smoothed away. Of course this is only an example of how the effect of this measure could be estimated. Not just one waterway, but a representative amount is to be taken into account.

From this same study it was concluded that VTS (Vessel Traffic Services) is more cost-effective than widening the canal at a narrow passage or in a sharp curve to cope with two points with an increased accident rate. Other measures which were concluded to be among the most cost-effective were:

- improved adaptation to fog condition (i.e. no navigation without RADAR)
- more re-education of navigators
- more frequent check-up of vital installations

To reduce societal risk, transport of toxic gases under pressure should be discouraged. Cooled tankers, preferably with pressure resistant tanks, are much safer.

Another interesting measure is readjustment of the prevailing laws about single and double hulls of ships. An obligation to transport not only the very toxic, but also the less toxic fluids (which are now transported in single-hulled ships) in double-hulled ships, turned out to be the only measure to significantly reduce calculated individual risk for people living close to the waterway. In Europe, benzene is one of the chemicals that will be transported in double-hulled (inland waterway) ships in the near future. Double-hulled ships are much safer than single-hulled ships, but exactly how much safer is not really known. Until recently the probability of fatal damage (with loss of containment) for double-hulled inland waterway ships was only guessed. In the last two years some numerical simulations of ship collisions were carried out, which resulted in more insight in double-hulled ship strength (Lenselink, Thung, Stipdonk and Van der Weide, 1992). This insight has not yet led to definite probabilities for loss of containment, because it was first necessary to establish the accuracy of the numerical finite element technique. In order to assess this, a series of full scale ship collision tests have been carried out in the Netherlands, of which an extensive report will be given this year (TUD 1992).

5.3 Future developments

In paragraph 5.2 some examples of measures and the estimation of their safety effects were given. In this section we will explain how most of the other measures will be treated in the Risk Effect Model, particularly with regard to measures concerning traffic and waterway. We will describe a model with which we are able to relate accident influencing factors to accident rate. The estimation of the safety effects of measures with the use of this model is explained.

Accidents are caused by several kinds of failures. Perrow (1984) distinguishes six main causes: Design, Equipment, Procedures, Operators, Supplies and materials, and Environment (DEPOSE-components). Some of our measures involve improvements in such components, but estimation of their effect is difficult to model. Many other measures do not apply directly to these components, but rather to circumstances in which they can lead to an accident. Circumstances like a narrow bridge passage, ferries and quays (all leading to manoeuvres), curves in the waterway, and the amount of traffic, or meteorological circumstances like fog or wind, are themselves not causes for accidents, but they augment the probability that a potential cause "takes its chance". These circumstantial influences can be modeled, and many measures like improved curves, speed measures, interdiction of being at anchor, of overtaking, and so on, involve these circumstances.

The model is based on the assumption that accident rates can well be described with a function of only a few parameters, namely (mean) traffic volume, waterway curvature, waterway width (static circumstances), and visibility, windspeed and water current (dynamic circumstances). As the number of different waterway sections with constant traffic volume, width and curvature is large, and the number of registered accidents for each waterway section is not very high, it is not possible to generate one single model in which not only static, but also dynamic circumstances are embodied. But it is possible to generate two models, one of which describes the effect of wind, visibility and water current on accident rate, and the other describing the influence of waterway width and curvature on the accident rate per elementary traffic situation. An elementary traffic situation is defined as a specific manoeuvre like overtaking, encountering, but also like passing a length of quay, or crossing (at intersections) and so on.

The method of modeling is straightforward. For a total of 274 sections of waterway the static parameters (width, quay length etc.) was gathered. Then the model describing accident rate as a function of those parameters was fitted to the number of accidents for each section. In fact not one but five models were made for different kinds of elementary traffic situations. The result showed that for more than 95% of the waterway sections, the accident rate could be described with the model. For the remaining 5% the difference between measured accident rate and model accident rate was too large to be attributed to a stochastic effect. The discrepancy could in fact be understood from special circumstances for these sections, such as a bridge, or unusually high water current. In a future version of the model these aspects will be taken into account as well.

An example of the results so far is the following. If a curve in the waterway which is 90 m wide and with a radius of 1000 m is widened to 115 m and a radius of 1150 m, a reduction of the accident probability per elementary traffic situation by 18% can be expected. This example illustrates that as long as one is interested in a relative reduction of accident rate, the model does not have to give absolute rates, as long as the estimated reduction of accident-rate is sufficiently correct.

For the model for the dynamic parameters (weather and current) a different approach was followed. For these parameters all accident information for the different sections was put together, and classed into 36 groups. There were 3 classes for wind speed, 3 for visibility and 4 for water current. The accident rate for each combination of parameters was then defined as the number of accidents with the same combination, divided by the frequency of occurrence of this combination. For this model the difficulty is to establish these frequencies. Information about visibility or about wind speeds is readily available, but the Dutch meteorological institute can not easily give frequencies of combinations of both.

Roeleven, Kok, De Vries and Stipdonk (1992) describe the model in more detail, and also give results. Additionally, in their paper the probability of a severe accident in relation to an arbitrary accident is also considered.

6. CONCLUSION

Risk assessment is a valuable tool in the management of transport risk. It should be well tuned to the specific needs of decision making. Uncertainty in, or inability of, quantification should not be used as a pretext for taking no decision at all.

Calculated individual risks nowhere exceed 10^{-5} on the banks of a waterway.

Developments in the modeling of costs and benefits of safety measures are promising in some parts of the field. The results will enable policy makers in both industry and government to serve safety more cost-effectively.

Legally based risk limits are essential to guarantee safety in potential clashes with other interests. However, their use should not prohibit the enhancement of safety ... beyond the limit.

7. REFERENCES

BROLSMA, et al. (1988). PIANC AIPCN Bulletin, 62, 1-64.

HAUER (1986). Accid. Anal. & Prev. 18, 1-12.

HOUBEN & WESSELS (1993). Manuscript submitted for publication.

LENSELINK, THUNG, STIPDONK & VAN DER WEIDE (1992). Proc. of the Second International Offshore and Polar Engineering Conference.

PERROW (1984). Normal Accidents, Living With High-Risk Technologies. New York: Basic Books, Inc.

ROELEVEN, KOK, DE VRIES & STIPDONK (1993). Manuscript submitted for publication.

TUD (1992). Marine Safety and Environment and Ship Production Symposium, 1 June 1992, Delft, The Netherlands.

VERAART, JANSSEN & STIPDONK (in press). PIANC AIPCN Bulletin.

Transport Canada Dangerous Goods Accident Costing Study and Model[*]

J. Wright
Transport Dangerous Goods Directorate
Transport Canada
344 Slater Street, 14th Floor
Ottawa, Ontario
Canada K1A 0N5

ABSTRACT

Transportation accidents involving dangerous goods result in unanticipated losses for industry, government agencies and members of the general public. Knowing the dollar value of historical losses can help predict the costs of future accidents, as well as aid in assessing the financial benefits of safety measures and funds needed to recover from loss events. Until now, however, there has been little published information on costs associated with dangerous occurrences and no guidelines on how to predict future losses.

To address these problems, the Transport Dangerous Goods (TDG) Directorate of Transport Canada initiated a national research program and database on accident costs. This study collected and analyzed cost data for 70 dangerous goods incidents that occurred in Canada over the last ten years. The study also developed and tested a prototype evaluation method using commercially available software to predict future losses from dangerous goods incidents.

---

[*] This paper contains excerpts from the final report "Accident Costing Study for the Transportation of Dangerous Goods in Canada"

1. SUMMARY

The main purposes of this study were to 1) identify the types of losses various members of Canadian society are exposed to, and 2) provide detailed cost data on damages to people, property and the environment. Loss categories included costs related to injuries and fatalities, evacuation, property damage, business interruption, pollution and environmental clean-up, and emergency response, among others.

A micro-computer database and program were developed to manage the large amount of loss data collected during the research tasks. The database and evaluation software is contained in an information management system called the Risk Management Information System (RMIS), residing in the Transport Dangerous Goods Directorate offices in Ottawa.

The total loss associated with all 70 incidents amounts to more than $41.6 million, reported in 1990 dollars. An average per incident loss for the study group approximates $594,300. Of the total loss, 54 percent were attributable to the presence of dangerous goods. The dangerous goods portion of all general public losses amounted to about 74 percent. Much of this loss accounted for fatalities, road closures, evacuations and legal fees connected with dangerous goods incidents. Industry losses attributable directly to dangerous goods mounted to 33 percent of all industry losses, reflecting the relatively high cost of vehicle damage that can occur even without hazardous materials on board. Municipal, provincial and federal costs were largely assigned to the presence of dangerous goods where they provided response capability.

The results of our research are also reported in terms of loss types and dollar losses for the 1) General public, 2) Industry, 3) Municipalities, 4) Provinces and 5) Federal government. Industry suffered the greatest amount of direct loss among the five major groups of entities affected by dangerous goods incidents, according to the study data. Industrial firms involved in shipping, transporting and receiving hazardous materials reported losses averaging more than $300,000 per incident.

Public losses were second with an average loss nearly equalling industry losses, at $269,000 per incident. Public losses include fatalities, injuries and interruption of commercial businesses due to dangerous goods occurrences, and far exceed reported costs of municipal, provincial or federal agencies. This distinction likely reflects the fact that public and industrial interests must pay to recover from direct injuries and property damage, whereas government agencies largely expend funds to respond to incidents.

Several samples of possible analyses are included within the report as to demonstrate the types of data and evaluation possible with the Risk Management Information System program and database. Readers interested in accessing the database are encouraged to contact the TDG Directorate in Ottawa.

2. INTRODUCTION

Safety serves a number of important objectives for Canadian organizations involved with the transportation of dangerous goods and with the protection of the public. These objectives usually reflect the legal and humanitarian responsibilities for protecting the public from possible harm associated with dangerous goods transport accidents and include, but should not be limited to, reducing injuries and deaths associated with these accidents.

It is a rare event; however, where the dangerous goods are the cause of injuries and an even rarer event where death results from exposure to the cargo. Transport Canada data show that, on average, fewer than three people per year die as a result of dangerous goods accidents. This compares to some 4,000 traffic fatalities per year on Canadian highways. A comparison of the numbers of injuries shows similar results.

Nevertheless, dangerous goods accidents do extract a heavy toll on Canadian society in terms of property, environmental, net income and liability losses. Preliminary estimates indicate that total losses to society associated with dangerous goods accidents could approach one billion dollars annually.

We have been finding throughout our investigations that most Canadian companies have never had a serious dangerous goods accident, although many appreciate that accidents can happen to any organization regardless of their loss prevention efforts. Further, we find that very few companies appreciate the magnitude of costs associated with even the smallest of spills. In short, and most companies readily admit this, they do not have full knowledge of their exposure to risk.

Decision-makers in both public and private organizations expect to see hard evidence that the dollar benefits of risk control safety measures equal or exceed the cost of their implementation. And if safety can be shown to pay, they wish to know by how much. To demonstrate the benefits and costs of safety, three pieces of information are needed:

- Costs associated with dangerous goods occurrences

- Reduction in these losses due to safety measures

- Costs of safety measures

While the last item is often estimated with some confidence, the first two have been, to date, very difficult to quantify.

One answer to these problems is a system which allows for the identification of possible exposures to loss that might be associated with a transport accident involving dangerous goods, and a system which allows for the expression of these losses in financial terms, namely dollars. This provides for the risk of dangerous goods transportation to be communicated in a convenient set of units, and permits industry and government alike to:

- Predict costs of dangerous goods occurrences

- Predict benefits of risk reduction

- Compare dissimilar options

- Select cost-effective safety measures

- Examine insurance provisions

- Set aside funds to pay for losses and compensation

3. STUDY OBJECTIVES

The main purpose of this study is to develop such a system which will allow a risk manager to identify, assess and predict costs associated with dangerous goods accidents. This project follows a pilot study, completed in 1989, that tested available sources of information and developed the basic analytical structure. The purpose of Phase II of the study is to:

1. Identify the types of losses, associated with dangerous goods transport accidents, that various members of Canadian society might be exposed to. These data are needed to facilitate planning for and preventing where possible, future losses.

2. Investigate detailed cost data for 70 dangerous goods accidents, including: data on damages to people, property and the environment; social costs including business interruption costs; costs of pollution; clean-up costs; emergency response costs; costs associated with evacuation and a host of others. These data are used to estimate possible exposures to accidental loss and to identify feasible risk control alternatives.

3. Develop and test a computer model to be used in analyzing the dollar losses of past dangerous goods accidents and for predicting future losses. These data are used in setting risk management priorities and analysing the benefits and costs of risk control alternatives. The data are also used in setting the Transport Dangerous Goods Directorate's priorities.

The study contributes a comprehensive database of details, including costs, associated with 70 dangerous goods occurrences, with estimations of dollar losses to various entities within the general public, industry, municipalities, provinces and the federal government. These are linked with operational details included in accident reports supplied to the Directorate. The computer model will forecast the likelihood, type and magnitude of losses that might be associated with future incidents.

3.1 Why Investigate Costs?

This study expresses dangerous goods incident losses in economic terms. Using dollars to measure risk offers several advantages.

First, expressing risk in financial terms puts dangerous goods safety on an equal footing with other decisions affecting an organization. People who authorize safety actions in any organization typically hold the purse strings. They face daily requests for funds, and usually make their decisions based on economic analysis. They deal with dollars. The managers who make safety happen would benefit from a method that spoke their language.

Second, safety must compete with other objectives of an organization, and safety decisions that demonstrate economic sense have definite advantages over those that do not. Tangible dollar returns on a safety investment serve the objectives of any enterprise, while also serving the public interest. Both public and private agents seek to maximize the present value of net cash flows in meeting their basic objectives; they want to get the most for their money.

Third, the ultimate objective of risk analysis is to influence safety. Risk is often expressed in terms of fatalities per year, or number of people exposed to a danger. But the number of lives saved by a certain safety measure may do little to motivate those who must pay for the technique. One life saved per 150 years does not help the general manager of a trucking or rail firm justify speed controls or training programs.

Because fatalities are rare, other factors must be considered in evaluating risk. The PCB warehouse fire in St. Basil le Grande in 1989 injured no one, yet led to losses approaching $60 million. The results of any risk analysis should motivate decision-makers to pursue a course of action with confidence. Dollars can motivate people.

Expressing consequences as dollars will not solve every concern. Many decision-makers feel uncomfortable, for example, when a human life is reduced to a dollar value. But even if other humanitarian criteria are considered, regulators, executive officers and emergency responders must know the projected costs and returns of optional safety measures.

3.2 Why Develop a Cost Model?

Organizations across the Canadian social spectrum have expressed a need for information on the types of losses and costs associated with transport accidents involving dangerous goods. Few companies have the resources required to conduct a comprehensive survey of dangerous goods transport accidents, yet they must make economic decisions on risk control and insurance. Those who invest in such research have a limited scope of interest, and are not likely to share results with their competitors. An organization's own loss record is unlikely to be large enough to develop reliable statistics, and data on the combined experience of others are often not sufficiently specific enough or current enough, if they are available at all.

Municipalities are often left "holding the bag" following an incident. They respond to protect the public, and collect taxes to do so. Yet, they know little about the types of losses they are exposed to or how much they are spending each year on the risk of dangerous goods transportation. Some cost information has already been collected by fire departments, but there is no mechanism for compiling these data or making the results available to others.

Provincial agencies are largely unaware of the accumulated losses they could realize over the long term. Most could adopt provincial response plans and programs relating to dangerous goods incidents, but lack loss information needed to design a cost-effective system. Federal agencies, such as the TDG Directorate and Environment Canada, must evaluate the public costs and benefits of ongoing programs, but have no convenient units of measurement nor have they the data on public losses.

Because dangerous goods incidents for most companies are uncommon events, a statistically reliable sample of incidents is difficult for one organization to obtain. While there are hundreds of organizations nation-wide who share similar operations (fire departments, for example), there is no method for collecting, analyzing and sharing cost data in a meaningful way. Even industrial operators, unless they are very large, lack the means needed to accurately forecast their own losses. All of these real and current problems can be addressed by a cost database on dangerous goods incidents. A database will help industry and government find the common ground in actions that serve both profit and safety motives. At the federal level, knowing the costs of incidents could help Remedial Measures Specialists (RMSs) sell safety measures, such as emergency plans, to industry. RMSs need information on how much a typical accident will cost industry.

In short, a national cost database should be developed because there is a clear and ongoing need for data that is shared by all Canadians.

3.3 How Could the Database Be Used?

Risk managers in an increasing number of private and public organizations are called upon to evaluate and recommend risk control and financing options. To do so, they need information on their exposures, and on the frequency and severity of dangerous goods occurrences. Users could employ the database to identify exposures, the most important step in managing risk. They could calculate the expected value of future losses and forecast their expenditures to optimize attempts to reduce losses.

Currently, we foresee an interactive database under TDG operations available for use by industry, municipalities and provincial and territorial governments, as well as other federal agencies. As an analytical tool, the database could be centralized to reduce operation costs. Yet, provincial or even regional summaries could be collated and dispersed to TDG offices for other local or regional users. If designed for wide appeal the database system should lead to reductions in the cost of risk for all Canadians.

4. SCOPE OF THE RESEARCH

The scope of the study was limited to a sample of 70 road and rail transport accidents, involving flammable liquids, corrosives and compressed gasses. These products were chosen as those most likely to be involved in an accident, due to the amount transported, and those most likely to contribute significant damages. (Canadian and US data show that corrosive substances have the highest accidental release rate, primarily at loading and unloading facilities, and that gasoline/truck road accidents are the most numerous and cause, overall, the greatest dollar damage.) Table 1 lists the selection criteria used.

Table 1: Scope of research study.

PRODUCTS: Three prominent classes of products	ENTITIES: Organizations incurring loss
Class 2 - Gases, poisonous, flammable and corrosive Class 3 - Flammable Liquids Class 8 - Corrosives	- General Public - Industry - Local Government - Provincial Government - Federal Government
MODE AND: - Rail VEHICLE TYPE - Road	LOCATION: Area in which release occurs
REGIONS: Throughout Canada	- Urban Core - Residential
TIME SPAN: 1978 to date	- Commercial
NUMBER OF INCIDENTS: 70	- Industrial - Rural
COST CATEGORIES:	TIME OF DAY:
- People: evacuation, injury, death - Property: lost, damaged - Environmental damage - Liability: claims, legal costs - Net Income: cleanup, emergency response, business interruption - Costs attributable to dangerous goods	- Morning - Evening - Afternoon - Night TYPE OF OCCURRENCE: - Explosion - Leak - Fire - Spill - Contamination

Overall, three types of data were required:

* Incident Details - Time, date, place, events, products involved and primary contacts.

* Contacts and Information Sources - Organizations and individuals suffering a loss are needed to direct inquiries to the right source, and to track contacts made and pending.

* Dollar Losses - Dollar amounts to represent losses incurred. Losses that cannot easily be assigned a dollar amount, such as fatalities and environmental damage, are recorded separately.

5.   FINDINGS

One objective of this study was to develop and test an evaluation method that predicts losses from future dangerous goods incidents. Using data collected on 70 incidents, we ran a series of data manipulations to address possible losses to public and private organizations.

Before proceeding, it is important to qualify the statistical reliability of the data reported here. Predicting risks usually relies on statistical sampling of incident data. However, the 70 incidents included here do not represent a random sample. They were selected purposefully to test the data collection and evaluation methods. While these figures do not represent probability distributions, they do address the margins...the small, medium and large losses that can accompany dangerous goods events. Information on probability distributions for various loss types are available through the Transport Dangerous Goods Directorate office in Ottawa.

While the collected data are not statistically representative of dangerous goods events in general, much useful information can still be extracted. Consider the total, average and expected value of losses listed in Table 2 below that summarize the study's 70 incidents.

Considering the limited set of occurrences reported in Table 2, the total loss associated with all 70 incidents amounts to more than $41.6 million. These losses are reported in 1990 dollars, following an adjustment of older incident values to current prices. An average per incident loss for the study group is about $594,300. A more statistically reliable survey would probably see an average lower than this, but the list demonstrates the type of information the cost database can provide.

Table 2: Summary of loss data - 70 incidents.

Category	Total Loss All Incidents	No. of Incidents	Average Loss per Incident	Expected Value Given Accident*
All Incidents	$41,604,217	70	$594,300	$721,000
Rail Incidents	18,264,751	17	1,074,400	1,301,000
Road Incidents	23,339,466	53	440,400	568,000
General Public	18,807,080	70	268,700	697,000
Industry	21,535,870	70	307,700	486,000
Municipal	574,159	70	8,200	9,000
Provincial	626,368	70	8,900	21,000
Federal	60,740	70	900	1,100
Largest Loss	10,306,430	Road		
Smallest Loss	612	Road		

Intuitively, one might expect rail accidents to cost more than road accidents because of the larger and more expensive equipment in use among rail operators, and greater product volumes involved in rail transport. This is evident in the data collected. Yet, the type of incident has more bearing on the amount of loss than simply the mode of transport. Several incidents involving small leaks of gaseous products from rail cars caused losses totalling less than $50,000 each, whereas an explosion of a propane truck resulted in losses exceeding $10 million.

Industry suffered the greatest amount of direct loss among the five major groups of entities affected by dangerous goods incidents, according to the study data. Industrial firms involved in shipping, transporting and receiving hazardous materials reported losses averaging more than $300,000 per incident. Public losses were second with an average loss nearly equalling industry losses, at $269,000 per incident. Public losses include fatalities, injuries and interruption of commercial businesses due to dangerous goods occurrences, and far exceed reported costs of municipal, provincial or federal agencies. This distinction likely reflects the fact that public and industrial interests must pay to recover from direct injuries and property damage, whereas government agencies largely expend funds to respond to incidents.

There is a second important qualification to emphasize when considering these data. Dollar losses reported here represent all costs associated with an accident, including those not directly related to the hazards of dangerous goods. All losses were included here in anticipation of how the data will be used. Many decision-makers are interested in safety measures that prevent losses from all sources, not just those associated with hazardous materials. Trucking firms, for example, examine safety measures that reduce the frequency of vehicle accidents throughout the fleet, such as training courses in defensive driving. Data reports on all costs would aid in evaluating the benefits of these measures. To examine the role dangerous goods play in total costs, consider the comparisons in Table 3.

Table 3: Summary of losses attributable to dangerous goods (road and rail).

Category	Total Loss All Incidents	Loss Attributable to Dangerous Goods	% Attributable to Dangerous Goods
All Incidents	$ 41,604,000	$ 22,364,300	54 %
General Public	18,807,100	13,999,400	74 %
Industry	21,535,900	7,187,800	33 %
Municipal	574,200	498,100	87 %
Provincial	626,400	620,800	99 %
Federal	60,700	58,200	96 %

Of the total reported losses of $41.6 million, 54 percent were attributable to the presence of dangerous goods. The dangerous goods portion of all general public losses amounted to about 74 percent. Much of this loss accounted for fatalities, road closures, evacuations and legal fees connected with dangerous goods incidents. Industry losses attributable to dangerous goods mounted to 33 percent of all industry losses, reflecting the relatively high cost of vehicle damage that can occur even without hazardous materials on board. Municipal, provincial and federal costs can largely be assigned to the presence of dangerous goods where they provide response capability.

The assignment of losses to dangerous goods is fairly straight-forward for certain loss types, such as vehicle damage. The criterion used in assigning losses to dangerous goods considers the likelihood of loss if the product carried had not been a hazardous material. A tanker truck collision with another vehicle that results in no loss of product, for example, would indicate any tanker vehicle damage would not be due to dangerous goods. Other cost types are less obvious. Fire department response is one. While fire departments sometimes attend motor vehicle accidents, it is not the norm to have several companies standing by throughout the cleanup process, unless there were dangerous goods on board. Therefore, the majority of fire department and other municipal costs are attributable to the presence of dangerous goods.

As a further example of the possible reports generated by this system, consider the comparison of different types of dangerous goods presented in Table 4.

Table 4: Overall losses by product and class.

Product	PIN	Class	Expected Value Given an Account*	Number Incidents in Sample
Propane	1978	2.1	$ 1,675,000	8
Anhydrous Ammonia	1005	2.4	517,000	5
Gasoline	1203	3.1	509,000	15
Chlorine	1017	2.4	355,000	4
Diesel/Fuel Oil	1202	3.2	340,000	4
Ammonium Nitrate	2069	5.1	119,000	2
Sulphuric Acid	1830	8.0	31,000	2
Flammable Gases		2.1	1,399,000	14
Flammable Liquids		3.2	1,221,000	10
Corrosives		8.0	40,000	10

Dangerous products are often compared by evaluating their hazards, represented by a long list of physical and chemical characteristics, such as flash point, vapour pressure and solubility. Table 4 offers another means of comparison by noting the total expected losses associated with seven products and three classes of goods included in the research. Heading the list is propane, due in part to the inclusion of a single catastrophic propane incident in the database. Based on eight propane incidents, total losses associated with one future accident are expected to reach $1.7 million.

Again, note that these results are preliminary and based on skewed data; public decisions should not be made based solely on these results. The table is presented to represent the analyses made possible by the TDG Risk Management information System.

6. EVALUATION METHOD

One objective of this study was to test an evaluation method that predicts losses from dangerous goods incidents. Using data collected on 70 incidents, we conducted a series of data manipulations to address questions posed by public and private organizations.

There are two probability expressions to deal with in calculating risk. The first relates to the likelihood that an accident occurs. This information depends on accident rates per unit of transportation and distance travelled, and is considered beyond the scope of this research.

The second probability expression reflects the size of a loss given an accident. If a loss is realized, what is the probability that the financial consequences will fall within a certain cost range? Most dangerous goods incidents are relatively minor in terms of their impact and consequence. A rare few are quite large. To predict the outcome of a future unknown event, it can be helpful to analyze what has happened in the past. Loss severity distributions permit a reasonable prediction of future events based on historical records.

Building a loss severity distribution requires three steps.

1. Historic losses must be adjusted to represent common year (say 1990) dollars. This puts the losses from incidents several years apart on equal footing. Each incident is then ranked according to its adjusted 1990 loss amount.

2. The list of incident losses is arrayed into dollar value groups according to adjusted loss amounts. Divisions between the groups can be selected arbitrarily, and the number of groups can vary. Selecting ten groupings based on the maximum loss size to be evaluated can provide a convenient set of values for the analyses.

3. A loss severity distribution can be developed by summarizing the groups selected in the array. Size categories are selected by breaking the maximum loss record into ten equal intervals.

The result is a loss severity distribution constructed to represent the specific losses in mind.

Table 5 illustrates these steps using transport carrier vehicle damage data sampled from the 70 accidents under study. Of the 70 carriers contacted in the course of the study, for example, 57 reported to have suffered vehicle damage to some extent in their respective accidents. To generate an expected value for vehicle damage for all carriers, we constructed a loss severity distribution along the following lines. The largest vehicle damage loss reported was $2,089,951. This was used to construct the size categories and midpoints, as shown in Table 5.

Table 5: Expected value of carrier vehicle damage given a loss (EVL).

Size Category	Midpoint	No. of Losses in Range	Probability of Loss	Expected Value
$ 1 - 208,995	104,498	49	0.86	$ 89,831
208,996 - 417,990	313,493	0	0.00	0
417,991 - 626,985	522,488	1	0.02	9,166
626,986 - 835,980	731,483	1	0.02	12,833
835,981 - 1,044,976	940,478	2	0.04	32,999
1,044,977 - 1,253,971	1,149,473	0	0.00	0
1,253,972 - 1,462,966	1,358,468	1	0.02	23,833
1,462,967 - 1,671,961	1,567,463	2	0.04	54,999
1,671,962 - 1,880,956	1,776,458	0	0.00	0
1,880,957 - 2,089,951	1,985,453	1	0.02	34,833
	TOTALS	57	1.00	$258,494

The column denoting "Probability of Loss" indicates the probability that any given loss will fall into the range in the column "Size Category." The "Expected Value" is the total amount of loss expected in each size category given 57 incidents, represented by the product of the midpoint and the probability of loss in each range. Said another way, the expected value equals the average, weighted by probability. The weighted mean value of any single accident is predicted to be about $258,000.

Table 5 represents the expected value of a loss (EVL) given the fact that vehicle damage actually occurs. As noted, only 57 of 70 incidents involved carrier vehicle damage. To factor this probability expression into the loss severity distribution, one must account for the "zero loss" incidents, as well as greater than zero loss events. This means changing the first size category to include $0 loss, and recounting the number of accidents in this range. Table 6 illustrates this procedure.

Table 6: Expected value of carrier vehicle damage given an accident (EVA).

Size Category	Midpoint	No. of Losses in Range	Probability of Loss	Expected Value
$ 0 - 208,995	104,498	62	0.89	$ 92,555
208,996 - 417,990	313,493	0	0.00	0
417,991 - 626,985	522,488	1	0.01	7,464
626,986 - 835,980	731,483	1	0.01	10,450
835,981 - 1,044,976	940,478	2	0.03	26,871
1,044,977 - 1,253,971	1,149,473	0	0.00	0
1,253,972 - 1,462,966	1,358,468	1	0.01	19,407
1,462,967 - 1,671,961	1,567,463	2	0.03	44,785
1,671,962 - 1,880,956	1,776,458	0	0.00	0
1,880,957 - 2,089,951	1,985,453	1	0.01	28,364
	TOTALS	70	1.00	$229,895

This gives the expected value of a loss given an accident, or EVA. Note the top size category accounts for zero loss incidents, 13 zero-loss incidents have been added to this range and the total number of losses equals our sample size, 70 accidents. The expected value given an accident is somewhat lower than the value forecast given a loss. This accounts for the fact that not every accident causes each and every loss type. This expression ties more directly with accident probability data to permit calculation of overall risk.

Both expressions of expected value have their uses. Expected values given a loss (EVL) can be used to more accurately predict cumulative dollar losses of a given exposure. Expected values given an accident (EVA), on the other hand, offer a more comprehensive picture of an unknown future accident.

7.   CONCLUSIONS

Data collected through this research indicate that all sectors of Canadian society can incur financial losses due to exposure to dangerous goods accidents. Average dollar losses total nearly $600,000 per incident, although most incidents result in losses of less than $100,000. Entities at significant risk include members of the general public and industry.

The research undertaken by Transport Canada has established an effective method for forecasting future losses from dangerous goods transportation incidents. Organizations and individuals throughout society have expressed a need for information on the cost of dangerous goods occurrences. The TDG Directorate has demonstrated that a national research and database program is both realistic and feasible. Costs associated with dangerous goods accidents include a number of possible loss categories, organizations and dollar amounts. Yet, even with this diversity, this study demonstrates that realistic cost data can be collected and analyzed.

The Perceived Risks of Transporting Hazardous Materials and Nuclear Waste: A Case Study[1]

Alvin H. Mushkatel
K. David Pijawka
Arizona State University
U.S.A.

Theodore S. Glickman
Resources for the Future
Washington, D.C.
U.S.A.

1. INTRODUCTION

Public concern about the transportation of hazardous materials has emerged as a major policy issue, although little has been done to assess longitudinal trends in risk perceptions or levels of concern. The extent of public concern about hazardous materials transportation, however, is reflected in the increasing number of hearings on routing decisions and community transportation plans, the number of recent studies on public perceptions of risks, especially as it relates to shipments of nuclear materials, and legislative initiatives to promulgate transportation regulations at the federal and state levels. As with other technological hazards, the public's concern about the risks of hazardous materials transportation may be the result of its awareness of accidents, combined with media amplification of such events.

Recent studies suggest that the characteristics of hazardous materials transportation represent a technological hazard that is particularly vulnerable to high risk perceptions and amplification (Kraus and Slovic, 1988; Kasperson et al., 1988). Research by

[1] This study was funded by the Nuclear Waste Project Office, Nevada Agency for Nuclear Projects, DOE Grant Number DE-F608-85-NV10461.

Mushkatel et al. (1990) on public perceptions of the transportation of nuclear waste to the proposed repository found that current images and perceptions of transportation risks in Nevada involve more severe consequences and greater frequencies of accidents than demonstrated by the federal government's risk assessments. Based on responses to three transportation risk scenarios, it concluded that nuclear waste transportation is predisposed to amplification effects; i.e., small, inconsequential mishaps can result in higher-than-expected public perceptions of risk. Thus, even seemingly unimportant transportation accidents can have large impacts related to increased out-migration of populations from areas near routes, decreased investment, lowered perceptions of quality-of-life factors, and heightened concerns over risk. Potential adverse socioeconomic effects stemming from perceived risks of possible nuclear waste transportation accidents were also reported in a study by Easterling et al. (1990).

The paper at hand first reviews the available research on transportation risk perceptions, identifying the key findings and implications that may help to explain public reactions. Then it analyzes the results of our own survey, which focuses primarily on perceptions of nuclear waste transportation but also addresses broader issues related to hazardous materials transportation in general.

Five key aspects of transportation risk perception are addressed. First, underlying public attitudes toward the safety of transporting hazardous materials are analyzed and the question of whether these attitudes are consistent with the results of other studies that used similar measures is examined. Attitudes concerning the inevitability of accidents and the importance of personal control in transportation decisions are found to be two important variables influencing public concern about safety. Second, the relationship between distance from transportation routes and variation in risk perceptions is considered. Previous studies have analyzed people's responses as to what are acceptable distances from hazardous facilities (e.g., Lindell and Earle, 1983), but until now there have been no specific data relating distance from transportation routes and perceived risk. Third, the key risk perception dimensions that have been identified in the literature are examined. These include the ranking of transportation risk relative to other risks, the nature of the perceived consequences of hazardous materials transportation accidents, and the perceived efficacy of government in managing the risks. Fourth is the degree to which trust in government and perceptions of governmental efficacy are related to risk perceptions. Fifth is the difference in the perceptions of the risks of transporting non-nuclear hazardous materials and nuclear waste.

2. LITERATURE REVIEW

In a study examining risk perceptions under various rail transportation conditions, Kraus and Slovic (1988) found that the transportation of hazardous materials resulted in the highest level of perceived risk among respondents compared to other risk situations involving rail transportation. High levels of perceived risk were attributed to these types of transport because of the interrelationships of high scores found among several risk dimensions, including involuntariness, uncontrollability and catastrophic potential. The risk of shipping nuclear waste, for example, was viewed as potentially catastrophic in case of an accident, despite low accident probabilities associated with these shipments.

Another study of public attitudes towards hazardous materials transportation found that, even though accidents may be perceived having a low probability of occurrence, such events are believed to be inevitable; furthermore, the perception is that intervention and mitigation will not necessarily lessen the risk (Ekos Research, 1987). When these two factors combine with the perception fact that the consequences of these events are perceived as serious, irreversible and catastrophic, it should come as no surprise when serious concerns are expressed by the public along existing or proposed routes for nuclear waste transportation.

In a more recent analysis of media coverage of nuclear shipments to the proposed transuranic waste repository near Carlsbad, New Mexico (the Waste Isolation Pilot Plant, or WIPP), five major areas of concern were addressed: (1) the public's fear of transportation accidents; (2) the lack of preparedness to deal with transportation mishaps; (3) adverse impacts on economic development; (4) public mistrust of governmental capabilities; and (5) the public's fear of the consequences of a transportation accident (Latir Energy, 1992). Of particular interest is the finding that media attention not only serves as an indicator of public concern and activism, but can also stimulate concern and heighten risk perceptions.

Few studies have systematically addressed the full array of risk perception issues regarding the transportation of hazardous materials and the associated explanatory factors for these perceptions. One exception is the previously referenced study done for Transport Canada by Ekos Research (1987), which reported on a national survey of perceptions and attitudes on hazardous materials transportation, risks and options for a communications strategy.

While the body of research pertaining to WIPP transportation perceptions is preliminary and limited in its development of theoretical constructs, it does, nevertheless, provide a consistent set of findings that constitute a useful benchmark for further analytical investigations. To measure public reactions to transuranic waste shipments through Oregon, a survey was undertaken in four corridor counties and other areas to serve as the control (MacGregor and Slovic, 1991). The findings show that the concern for safety was greatest in the corridor counties and that the risks of transportation were perceived to be higher than the risks of storage. Moreover, the findings suggest that the severe, adverse images which are likely to be associated with nuclear waste shipments could result in actions intended to avoid or reduce the risks. Based on a contingent valuation methodology, Ganderton et al. (1991) concluded from their survey of New Mexico residents that over 75 percent of the population would require extremely high levels of compensation in exchange for accepting a nuclear waste transportation route in the state. In fact, the levels were so high as to indicate that nuclear waste transport was essentially not an acceptable option. The results of the survey also indicate that persons closer to transport routes place a higher value on avoiding nuclear waste shipments.

Several studies, including the one by MacGregor and Slovic (1991), have shown that nuclear waste transportation risks are perceived to be as large or even larger than the risks of storage/disposal facilities for nuclear waste. Compared to the risks of nuclear production and storage, Jenkins-Smith et al. (1991) found that the general public was more concerned about the risks related to nuclear waste transportation. Approximately 65 percent of the public perceived "substantial risk" associated with such transport, compared to 58 percent for temporary storage, 52 percent for production, and 56 percent for permanent storage stages. In another study, the analysis of public reactions to a proposed nuclear waste bypass around Santa Fe, New Mexico revealed strong perceptions of declines in property values, stigmatization of the area, and concerns over safety (ZIA Research, 1990). In this case, approximately 60 percent of the population sampled were not willing to accept any form of compensation or discount to move near the designated route.

3.   THE SURVEY AND METHODOLOGY

The survey from which the data we analyzed were drawn was part of a larger study designed to assess the socioeconomic impacts of the proposed high-level nuclear waste facility at Yucca Mountain on Las Vegas metropolitan residents. The households included in the study were selected by a random-digit dialing technique. The sample was based on the proportionate distribution of residential households within the 62 telephone prefixes that served the geographic sampling areas in 1988 (the cities of Henderson, Las Vegas, North Las Vegas, and the urbanized section of Clark County). A total sample size of 750 telephone numbers was computer-generated. Any telephone numbers that were not residential or not within the geographic boundaries were immediately replaced in the sample. All in-sample numbers were not replaced until after at least 10 callbacks had been attempted at different times, on different days, over a 4-week period.

Once a household was determined to be in the sample, a short telephone interview was conducted. A modified selection procedure was then used to identify the appropriate adult respondent in the household and a face-to-face interview with that person was scheduled. The telephone portion of the interview took approximately 10 minutes to conduct. The face-to-face interview lasted between 50 minutes to 2 hours. A final response rate of 73.5 percent was achieved with a sample size of 549 households.

4.   ATTITUDES TOWARD THE SAFETY OF TRANSPORTING HAZARDOUS MATERIALS

The social/psychological research on societal adjustments to hazards have found that broad, underlying beliefs about safety and risk influence how individuals or communities perceive the probabilities and consequences of accidents. This may help to explain the observed patterns of response in our survey. Several questions were used to discern the strength of key attitudes or beliefs associated with hazardous materials transportation safety. Four key topics addressed in this survey have also received attention in other studies. These include: (1) the extent to which hazardous materials accidents are viewed to be inevitable; (2) the attitude that hazardous materials should not be transported through populated areas; (3) the degree to which people feel that they have control over decisions to transport hazardous materials through their communities; and (4) whether it is generally safe to ship hazardous materials. Respondents were asked to indicate their level of agreement to a set of statements on a 1 to 7 Likert-type scale, where 1 represents "strong agreement" to the statement, and 7 represents "strong disagreement" with the statement.

Table 1 shows that the majority of residents of the metropolitan area expressed strong levels of agreement with attitude statements related to the inevitability of accidents involving hazardous materials, the need to avoid transporting hazardous materials through populated areas, and the lack of personal control over routing decisions. When response frequencies in levels 1 and 2 are combined as an indicator of generally strong agreement with the attitude statement, the data show that 54.2 percent of the population strongly agreed that transportation accidents involving hazardous materials are inevitable. More substantial public support (75.4 percent) was shown for the view that hazardous materials shipments should not pass through populated areas, and 5.1 percent of the metropolitan population also expressed strong agreement with the statement that they have no control over transportation decisions. Lastly, only 17.5 percent strongly agreed that it is safe to transport hazardous materials. In contrast, 39.9 percent of the sample population disagreed (levels 6 and 7) with the statement that hazardous materials transportation is safe.

Table 1: Basic attitudes toward hazardous materials safety.

| | Attitude Agreement Scale (Percent Responding) | | | | | | | |
| | Strongly agree | | | | | Strongly disagree | | |
	1	2	3	4	5	6	7	Median
Hazardous materials accidents are inevitable	40.5	13.7	14.7	10.6	8.8	6.4	5.3	2.0
Never transport through populated areas	61.7	12.8	7.7	7.7	3.9	2.4	3.9	1.0
No control over decision to transport hazardous materials through community	34.7	15.4	9.0	13.6	7.7	10.5	9.2	3.1
It is safe to transport hazardous materials	6.8	10.7	15.6	15.4	11.6	11.0	28.9	4.6

To what degree are these attitudes site-specific, reflecting sentiments that may have been influenced by the controversy over nuclear waste shipments to the proposed national repository in Nevada? To investigate this, we compared the survey data to the results of the studies by Ekos Research (1987) and MacGregor and Slovic (1991). The comparisons show that there are some differences in the strength of agreement among similar attitude measures, but that large segments of the population share these attitudes about hazardous materials transportation.

5.   PERCEIVED RISK AND DISTANCE FROM TRANSPORTATION ROUTES

The relationship between risk perception and distance from hazardous materials routes has been the subject of previous research. Risk perceptions and public concerns were found to be larger in areas where nuclear waste routes were proposed than in areas that would not be directly affected by the routes. However, these conclusions were all based on with-and-without scenarios in broad geographical areas and not on differences in response by specific distances from the route itself. The study on Canadian public attitudes toward hazardous materials transportation also suggested that familiarity with hazardous materials shipments (operationalized by persons living adjacent to transportation routes) tended to foster acceptance of the risks (Ekos Research, 1987).

We tested bivariate distance relationships from residences in order to observe whether distance from a hazardous materials route influences transportation safety attitudes or risk perceptions. The data showed no significant relationship between risk perception measures and distance from routes within the first 5 miles, the range within which the entire sample population resided. In previous research (Lindell and Earle, 1983; Mushkatel et al., 1990), preferred safe distances between residences and hazardous facilities have been found to be strongly associated with various dimensions of the perceived risk of those facilities. Table 2 shows the cumulative percent of persons preferring to live within specific distances from a highway where non-nuclear hazardous materials are transported, according to the survey we conducted. About 19 percent of the metropolitan population indicated they would feel safe living within 5 miles of such highways and about 32 percent within 10 miles. The 19 percent figure is consistent with the attitudinal measure reported earlier in Table 1, where almost 18 percent strongly agreed with the statement that "it is safe to transport hazardous materials."

Table 2: Perception of safe distance of residence from a highway where hazardous materials are transported*.

Safe Distance (miles)	Cumulative Percent
Within 5	19.1
10	31.6
20	38.3
50	57.4
75	58.9
100	78.5
200	82.6
200+	100.0
N = 545	

*It is important to note that all respondents resided within 5 miles of a hazardous materials shipment route.

6. RISK PERCEPTION OF NUCLEAR WASTE TRANSPORTATION

In their study of relative risk perceptions of different transported materials and hazardous conditions, Kraus and Slovic (1988) observed that the transportation of hazardous and nuclear materials would result in the highest perceived risks relative to all other types of transportation due to the interrelationships of high scores among several dimensions of risk, which include: involuntariness, uncontrollability, and potential catastrophic consequences. The risk of shipping chemical products was viewed as potentially catastrophic (i.e., a high number of fatalities) in case of an accident, but was seen as being controllable.

Table 3 shows the distribution of responses of urban Las Vegas residents to the question concerning the seriousness of the risks to health and safety from nuclear waste transportation. Almost 70 percent of the population perceived the risk as high (1, 2, and 3 on a 7-point scale) and 53 percent as very high (1 and 2 on the scale). The table also shows that approximately 36 percent of the population perceived the risks related to the repository to be very serious (categories 1 and 2 on the 7-point scale). This contrasts with 53 percent of the population who perceived as very serious (categories 1 and 2) the risks associated with nuclear waste transport. Within the context of the repository program, the perceived risks of transporting nuclear waste exceed that of the repository itself, which is proposed to be sited 120 miles away. This finding is similar to the one found in the New Mexico survey by Jenkins-Smith et al. (1991) mentioned above.

Table 3: Las Vegas metropolitan area residents, perceptions of the risks of nuclear waste transportation and the repository.

Seriousness of Risk		Transportation		Repository	
		Frequency	Percent	Frequency	Percent
High perceived risk	1	166	34.2	111	22.9
	2	91	18.8	64	13.2
	3	82	16.9	85	17.6
	4	51	10.5	73	15.1
	5	44	9.1	59	12.2
	6	34	7.0	54	11.2
Low perceived risk	7	17	3.5	38	7.9
Mean			2.77		3.45
N		484		485	

In comparing transportation to repository facility risks, the distance to the repository for Las Vegas residents may serve to attenuate, to some degree, the perceived effects of an unintentional release of radioactive material from the facility, where the perceived seriousness of the risks of the repository itself is very high. The data also indicate that public perceptions of transportation risks are strongly associated with perceptions of the repository. For example, of those individuals who perceive the highest level of risk in transporting nuclear waste, approximately 68 percent also perceive very serious accident risks from repository facility activities. An association of 0.62 (Tau B) was found between the two items, showing a strongly significant association between the two risk perceptions.

The research literature indicates that the concern about nuclear technologies results from the perceived consequences of accidents, despite the public's perceptions of the low probabilities of these events. To test the importance of the perceived consequences, we asked the respondents in the survey to rate their level of agreement with the statement that a transportation accident involving nuclear waste would cause "widespread damage to health and property, whatever the level of preparedness." On a 7-point scale, strong agreement with this statement (categories 1 and 2) was indicated by about 56 percent of the population (Table 4). Approximately 70 percent of the population agreed with this statement (categories 1, 2, and 3).

Table 4: Perceptions of consequences of transportation accidents involving nuclear waste.

Agreement with Statement that Accident Would Cause Widespread Health and Damage Even if Government were Prepared		Frequency	Percent
Strongly agree	1	188	38.8
	2	84	17.4
	3	63	13.0
	4	46	9.5
	5	39	8.1
	6	35	7.2
Strongly disagree	7	29	6.0

In addition to the perception of serious health risks resulting from transportation accidents involving nuclear waste, the urban residents also expressed a sense that they could not protect themselves in the case of an accident. Respondents were asked their level of agreement with this statement: "There is nothing I can do to protect myself if an accident takes place." A strong expression of fatalism was expressed by a relatively large segment of the population, with over 50 percent in strong agreement (rating of 1 and 2 on the 7-point scale) with the statement.

7. PERCEPTIONS OF MANAGEMENT CAPABILITIES

The survey data on nuclear waste transportation perceptions also reinforce the notion that perceived inevitability of accidents is a general attitude associated with hazardous materials transportation. On a scale measuring strength of agreement with the statement "If the government takes precautions against accidents I am almost certain they will work," only 22 percent agreed with this statement (categories 1 and 2, where 1 represents "strong agreement"). However, 28 percent of the sample disagreed with the statement, and the 40 percent in the remainder were ambivalent (categories 3, 4, and 5). Although there is low confidence that governmental precautions can prevent transportation accidents involving nuclear waste, the fact that almost half the population expressed uncertainty shows that there is potential for public perceptions to shift. Other data from the survey reveal that there is some confidence that nuclear waste can be transported safely and that the problems can be solved. In fact, approximately 54 percent of the respondents indicated that they have confidence that existing problems can be overcome.

Perceived management efficacy refers to the public's views of the ability of government to manage the risks of transporting nuclear materials and to reduce the risks to acceptable levels. Management activities would include accident prevention measures, development of response capabilities, and the necessary mitigation plans and policies to reduce risk and avoid serious consequences. A set of questions addressed these trust and confidence variables and their relationship to transportation risk perceptions. The Canadian study on transportation hazards (Ekos Research, 1987) found that, although most people believe that safety should be the responsibility of transportation experts, the public nevertheless doubts the credibility of industry and also wants greater direct public input into decisions about safety. In fact, the report found that 36 percent of the Canadian population agreed that they had difficulty in believing safety reports.

The data from our survey indicate that a sizeable percentage of the urban population perceive governmental management efficacy in the transportation of hazardous materials to be low. On a 4-point scale measuring the degree of confidence that government will have the technical know-how to respond to accidents, 28.9 percent of the sample was "not very confident" and 11.1 percent was "not confident at all." In contrast, 41.4 percent were "very confident." We noted earlier that a sizeable portion of the population disagreed with the idea that government precautions against accidents would work; although not in a majority, those who question the ability of government to effectively manage the risks of transporting hazardous materials are also likely to be individuals with high risk perceptions.

Two relationships were analyzed that would assist in explaining variation in transportation risk perceptions regarding nuclear waste. The first attempted to test the relationships between measures of perceived management efficacy and levels of risk in nuclear waste transport. Table 5 displays statistically significant inverse relationships between five efficacy measures and the independent risk perception variable. That is, higher risk perceptions are associated with lower management efficacy ratings.

Table 5: Relationships of perceived governmental capabilities and transportation risk perception.*

Perceived Capabilities	Kendall's Tau C Measure of Association **
Trust in federal government's safety programs	-0.338
Trust technicians and operators in risk management	-0.329
Government's ability to manage repository program	-0.334
Confidence that nuclear waste can be safely transported	-0.604
Government knows how to respond to transportation accident	-0.419

*   The dependent risk perception variable was the question that asked respondents to rate the seriousness of the risk to health and safety of transporting nuclear waste to the proposed repository.
**   All associations were significant at P<0.001.

Table 6 displays the relationship between political trust indices and repository risk perception indices. In each case, the greater the political trust, the lower the perceived risk, whether from the facility or the transportation elements. Conversely, less political trust in agencies is related to greater risk perceptions, except at the state level, where greater trust in the Nevada state government is correlated with greater public concerns. This can be explained by the state's opposition to the repository siting program and its characterization of the risks as larger than those suggested by the federal agencies.

Table 6: The relationship of trust and repository risk perceptions.*

Risk Perception Indices	Federal trust	Agency trust	State trust
Mitigation	-0.32	-0.44	0.26
Repository safety and accident probabilities	-0.33	-0.38	0.18
Risk of transportation accidents	0.24	0.31	-0.12
Transportation of hazardous materials	0.27	0.33	-0.14
Safety and health threat	-0.27	-0.34	0.17

*   All measures are significant at the 0.001 level of significance and the
    association measures are based on Tau C statistics.

8.   COMPARING TRANSPORTATION RISK ATTITUDES: HAZARDOUS MATERIALS VERSUS NUCLEAR WASTE

The survey instrument was also designed to gauge the extent to which perceptions differ between non-nuclear hazardous materials and nuclear waste transportation. Responses to three similar perception measures for both nuclear and non-nuclear hazardous materials are shown in Table 7. The risk perception measures were based on the level of agreement with three statements: (1) "transportation accidents are inevitable;" (2) "never transport these materials through populated areas;" and (3) "current transportation methods are safe." In the first case, the mean level of agreement associated with accident inevitability was 2.74 for hazardous materials and 2.38 for nuclear-related shipments. A similar pattern of differences was observed with respect to the second measure. The respondents were in stronger agreement (mean of 3.77) that current methods for transporting hazardous materials were safer than for nuclear waste (mean of 4.44). There were no significant differences in these three attitudes between nuclear material and non-nuclear hazardous materials.

9.   THE AVAILABILITY HEURISTIC AND TRANSPORTATION RISK PERCEPTIONS

Several researchers have suggested that some hazards and events are easy to imagine and that the public may perceive much higher risk levels associated with these hazards or higher probabilities of

Table 7: Perceptions of hazardous materials transportation risks versus nuclear waste transportation risks.

Attitude Measures	Accidents are Inevitable		Never Transport Through Populated Areas		Current Transport Methods are Reasonably Safe	
	Hazardous Materials	Nuclear Waste	Hazardous Materials	Nuclear Waste	Hazardous Materials	Nuclear Waste
Level of Agreement* (Percent)	68.9	77.6	82.2	83.7	36.4	34.8
Mean	2.74	2.38	2.02	1.95	3.77	4.44
N	546	540	545	540	541	537

* Respondents were asked to rate their level of agreement with three attitudinal statements asked for both hazardous materials and nuclear waste transportation, using a 7-point scale, where 1 represented strong agreement and 7 strong disagreement with the statement. A mean closer to 1 infers stronger agreement.

accident occurrences (Lichtenstein et al., 1978; Tversky and Kahneman, 1973). Based on the results of measuring responses to transportation risk scenarios, the survey data suggest that risk perceptions associated with nuclear waste transportation may be especially sensitive to images, which is a manifestation of the "availability heuristic."

The respondents were given three transportation risk scenarios and asked to respond to questions concerning level of perceived risk, social impacts, and anticipated behavior to reduce or avoid the risks. The survey was designed to assess the degree to which the risk scenario could alter perceptions and public concern. The "benign" or minimal risk scenario envisioned a 30-year repository program with several small transportation incidents resulting in no release of radioactive substances. Interestingly, many respondents indicated that their images of transportation accidents involving nuclear waste were much more serious and pronounced than the scenarios that were presented.

Responses to the scenario conditions were as high or higher than the pre-scenario responses for most perception measures. Table 8, for example, shows that introducing the "benign" scenario or lowest risk scenario had the effect of increasing the minimum acceptable distance people were willing to live from routes carrying nuclear waste shipments. The first column represents the cumulative percent of persons currently preferring to live at certain distances from a nuclear corridor. The second column represents the cumulative percent after the information scenario was introduced. The baseline data indicate that 18.9 percent of the population would feel safe living within 5 miles from a route. This percentage increases to 31.3 percent within 10 miles from such routes. The scenario-driven data indicate pronounced shifts in the percentage willing to live within these distances. After the scenario was introduced, only 1.5 percent of the population was willing to live within 5 miles of the route and 15.8 percent within 10 miles.

Table 8: Cumulative acceptable distances residents are willing to live from a nuclear waste route.

Miles	Current Perceptions (Cumulative Percent)	Post-Scenario Perceptions (Cumulative Percent)
5 miles or less	18.9	1.5
10	31.3	15.8
20	38.1	21.8
50	57.0	36.8
75	58.5	38.3
100	78.0	67.7

10. FINDINGS

The consistency between our findings with the existing research on transportation risk perceptions provides strong support for arguing that the public is very concerned about the potential harmful effects from the transportation of hazardous materials. These concerns exist despite the low probability and high geographic randomness of accident events and a historical record of few fatalities. Although this particular study did not find a significant relationship between residential distance and risk perceptions within 5 miles of existing routes, the data demonstrate that most people would prefer to live further from hazardous materials transportation routes. These findings are consistent with several other studies which show that the closer the routes, the stronger is the likelihood that people will oppose routes or have greater levels of concern. The study does not support the finding of the Transport Canada survey (Ekos Research, 1987) that knowledge or familiarity with transportation risks on the basis of living closer to the routes will reduce concern. Changes in values and information on risk during the last 5 years, however, may have altered these public perceptions. The data from this study suggest instead that awareness of risks and information on possible mishaps may in fact heighten levels of perceived risk. There is extremely low acceptance of having to reside close to shipment routes involving hazardous materials. However, it is important to note that the responses were to hazardous materials in general. The acceptability of living closer to routes might be higher for specific products such as gasoline.

Public attitudes about hazardous materials transportation are consistent, strong and not easily changed. Substantial segments of the public feel that accidents involving hazardous materials are inevitable, that there is little personal control in decisions, and that hazardous materials shipments, in general, should not go through populated areas.

High public risk perceptions are associated with event inevitability and with the perceived severity and catastrophic potential of accident consequences. The data in this case study data suggest that the "imageability" of transportation accidents involving hazardous materials (and especially nuclear materials) is large; transportation risk perceptions are particularly sensitive to amplification effects.

Despite high risk perceptions, the respondents expressed confidence that problems in transportation can be overcome and risks can be made acceptable. A large segment of the sample indicated that nuclear waste can be transported in a way that is "acceptably safe." However, large segments of the sampled public are not confident that the government has the technical know-how to respond to accidents involving nuclear waste or that government precautions against accidents would work. Low levels of trust in government and its management of risk are strongly associated with higher levels of perceived risk. The findings imply that changes in trust perceptions and greater confidence in risk management programs may influence risk perceptions in the transportation area. However, the findings also suggest that such changes will require much more than a risk communication program to inform the public about the risks: it may require fundamental structural changes in risk management programs and processes due to low levels of confidence in governmental capabilities.

Differences between perceptions of nuclear waste transport and non-nuclear hazardous materials according to three attitudinal measures were found to be insignificant. This might be a function of the specific measures used, but it certainly underlines the need for more comparative analyses between these two types of products and the development of theoretically based units of comparison. Moreover, greater understanding is needed regarding perception differences by mode of transport, geographic region, and sociodemographics. The public's perceptions and preferences for mitigation strategies and management programs also need to be assessed.

11. REFERENCES

EASTERLING, D., MORWITZ, V., & KUNREUTHER, H. (1990, Feb.). Estimating the Economic Impact of a Repository from Scenario-Based Surveys: Models of the Relation of Stated Intent and Actual Behavior.University of Pennsylvania.

EKOS RESEARCH ASSOCIATES, INC. (1987). Risky Business: Rethinking the Federal Communications Role in Light of Public Perceptions of the Transportation of Dangerous Goods. Submitted to Transport Canada.

GANDERTON, P., MCGUCKIN, T., CUMMINGS, R., & HARRISON, G. (1991, Feb.). Assessing Risk Costs for Nuclear Waste Transportation. Project No. WERC-90-063. New Mexico Waste Management Education and Research Consortium.

JENKINS-SMITH, H.C., et al. (1991, June). Perceptions of Risk in the Management of Nuclear Waste. Sandia National Laboratories, New Mexico.

KASPERSON, R., et al. (1988). "The Social Amplification of Risk: A Conceptual Framework." Risk Analysis, 8, 2.

KRAUS, N.N. & SLOVIC, P. (1988). "Taxonomic Analysis of Perceived Risk: Modeling Individual and Group Perceptions Within Homogeneous Hazard Domains." Risk Analysis, 8, 3.

LATIR ENERGY CONSULTANTS (1992, Jan.). Transportation of radioactive Waste to the WIPP: New Mexicans' Perceptions 1988-1992.

LICHTENSTEIN, S., SLOVIC, P., FISCHHOFF, B., LAYMAN, H., & COMBS, B. (1978). "Judged Frequency of Lethal Events." Journal of Experimental Psychology.

LINDELL, M. & EARLE, T.C. (1983). "How Close is Close Enough: Public Perceptions of the Risks of Industrial Facilities." Risk Analysis, 3, 4.

MACGREGOR, D., & SLOVIC, P. (1991). Radioactive Waste Transport Through Oregon: Oregon Survey Results. Oregon Department of Energy.

MUSHKATEL, A.H., PIJAWKA, K.D., & DANTICO, M. (1990). Risk-induced Social Impacts: The Effects of the Proposed Nuclear Waste Repository on Residents of the Las Vegas Metropolitan Area. State of Nevada, Agency for Nuclear Projects.

TVERSKY, A., & KAHNEMAN (1973). "Availability: A Heuristic for Judging Frequency and Probability." Cognitive Psychology 5, 207-32.

ZIA RESEARCH ASSOCIATES (1990). Santa Fe Property Value Opinion Research Survey Regarding the WIPP Bypass.

Concluding Remarks

F.F. Saccomanno
Institute for Risk Research
University of Waterloo
Waterloo, Ontario
Canada

K. Cassidy
Major Hazards Analysis Unit
Health and Safety Executive
United Kingdom

1. INTRODUCTION

This Conference provided an open forum to review issues concerning QRA and the need to make the process more understandable to decision-makers, who are not always familiar with its technical complexity and ramifications. The Conference was successful in bringing together two groups (often referred to as "two solitudes"): the technical analyst and the decision-maker (who may be non technical) but who has to reflect non-technical users in his judgements. These papers have addressed several of the issues that were raised by these groups on the role and relevance of QRA in decision-making. The lessons that were learned from this exercise can provide important guidance to future initiatives on the assessment of risks from the transport of hazardous materials.

2. ROLE OF QRA

Decision-making in a risk environment is a four stage process: identifying hazards, quantifying these hazards objectively, assessing their risk tolerance in terms of community standards and developing a policy for their control. QRA is mainly used to identify and measure the hazards and risks, and to help evaluate alternative control strategies. The process consists of five components: accident/incident occurrence, probability of a breach of containment, volume and rate of material released given an accident/incident, the hazard areas associated with different levels of threat, and finally the population and property placed at risk normally at a given location.

QRA is a policy tool providing risk estimates to aid and inform decision-makers and policy planners. It has, however, a multi-dimensional role that extends beyond the initial estimation exercise. Risk estimates must be reliable and they must convey effectively the actual threat posed. It is precisely these issues of reliability and reporting that have been most problematic in making the process more acceptable for decision-making.

3.   CONCERNS REGARDING THE PROCESS

QRA risk estimates can be reported from two perspectives: individual and societal. Individual risks are expressed in terms of the probability of one person incurring specified damage at a given location within a specified time interval of exposure. Societal risks, on the other hand, reflect the cumulative threat to all individuals residing within a defined hazard area for a given level of transport activity. For societal risks, all locations that are subject to a potential threat are specifically considered, although it is accepted that societal risk can also have a wider dimension. Since individual risks in the transport of dangerous goods are generally found to be **de minimis**, risks have tended to be expressed from a societal perspective.

Societal risks can be reported in terms of F - N curves (graphs of the probability of exceeding a certain number of fatalities versus the number of fatalities) or as an expectation of damage for a given level of transport activity (or the mean of the F - N curve).

Decision-makers, however, have found these methods of reporting risk to be deficient. What is of concern to them is not the risk numbers per se, but rather technical guidelines for possible courses of action. To avoid misuse of the raw estimates, it is becoming imperative that numbers are not released without a detailed explanation of their meaning, their sensitivity to underlying assumptions, their uncertainties and an adequate understanding of the consequences of taking alternative courses of action.

One of the driving forces behind this International Consensus Conference has been a concern about inconsistencies in the reported estimates from various model sources. It was felt that these inconsistencies in the estimates could create problems of credibility from the perspective of the users of the models. Possible sources of inconsistency in estimation can be taken into account by documenting the underlying assumptions and technical underpinnings of the various stages of QRA. Unexplainable, random uncertainty in the estimates must be treated statistically and reported in the results. Currently, the outputs of many QRA models fail to reflect this uncertainty in estimation, such that risk estimates have been reported without their corresponding confidence intervals.

QRA estimates can be reported either in absolute or in relative terms. The most appropriate form of reporting depends on the nature of the problem being investigated. While for some studies the reporting of relative risks is acceptable, for other studies only absolute risks have any real relevance. QRA models need to recognize that in reporting absolute measures of risk, the complexity of the model (or lack of it) will influence the types of questions that the decision-maker can address. Simple models can be very useful, but they are also very limited in providing answers to specific, complex questions about possible courses of action.

4.   GUIDELINES FOR QRA

These Proceedings should provide useful guidelines for future development of the QRA models. The critical issue remains how to make the process more practicable and accessible to decision-makers and the public at-large.

Other Publications by the Institute for Risk Research [*]

Development of Environmental Health Status Indicators,
R.S. McColl, ed., 1992.

Energy for 300 Years: Benefits and Risks, J. Nathwani, E. Siddall,
N.C. Lind, 1992.

Managing Risks in the Public Interest, N.C. Lind, J. Nathwani,
E. Siddall, 1991.

*Municipal Solid Waste Management: Making Decisions in the
Face of Uncertainty,* M.E. Haight, ed., 1991, 2nd printing 1992.

*Municipal Solid Waste Management: Making Decisions in the
Face of Uncertainty. Workshop Summary Report,* 1991.

Energy Alternatives: Benefits and Risks, H.D. Sharma, ed., 1990.

Risk Management for Dangerous Goods, J.H. Shortreed, ed., 1989.

Prospects and Problems in Risk Communication, W. Leiss, ed.,
1989, 2nd printing 1992.

*Risk Assessment and Management: Emergency Planning
Perspectives,* L.R.G. Martin and G. Lafond, eds., 1988.

Processing Doubtful Information, E. Rosenblueth, C. Ferregut,
M. Ordaz and X. Chen, 1987.

Environmental Health Risks: Assessment and Management,
R.S. McColl, ed., 1987.

Reliability and Risk Analysis in Civil Engineering, Vols. I and II,
N.C. Lind, ed., 1987.

[*] Niels C. Lind & Jatin Nathwani, members of the Institute for Risk Research, are editors of *Risk Abstracts*, a quarterly journal of abstracts, reviews and references published by Cambridge Scientific Abstracts.

Institute for Risk Research   Tel.: (519) 885-1211, ext. 3355
University of Waterloo   Fax: (519) 888-6197
Waterloo, Ontario
Canada N2L 3G1